Basic Microwave Engineering

Basic Microwave Engineering

As per the latest Syllabus
JNTU, Hyderabad, Kakinada and Anantpur (A.P.)
(Electronic and Communicative Engineering)

M.L. Sisodia
Advisor and Professor, ECE
Global Institute of Technology
Jaipur, Rajasthan

NEW AGE
JNTU

An Imprint of

NEW AGE INTERNATIONAL (P) LIMITED, PUBLISHERS
New Delhi • Bangalore • Chennai • Cochin • Guwahati • Hyderabad
Jalandhar • Kolkata • Lucknow • Mumbai • Ranchi
Visit us at **www.newagepublishers.com**

Published by New Age International (P) Ltd., Publishers
First Edition: 2011

Branches:

- 37/10, 8th Cross (Near Hanuman Temple), Azad Nagar, Chamrajpet, **Bangalore**-560 018. Tel.: (080) 2675 6823, Telefax: 2675 6820, E-mail: bangalore@newagepublishers.com
- 26, Damodaran Street, T. Nagar, **Chennai**-600 017. Tel.: (044) 24353401, Telefax: 24351463 E-mail: chennai@newagepublishers.com
- CC-39/1016, Carrier Station Road, Ernakulam South, **Cochin**-682 016. Tel.: (0484) 2377004, Telefax: 4051303 E-mail: cochin@newagepublishers.com
- Hemsen Complex, Mohd. Shah Road, Paltan Bazar, Near Starline Hotel, **Guwahati**-781 008. Tel.: (0361) 2513881 Telefax: 2543669, E-mail: guwahati@newagepublishers.com
- 105, 1st Floor, Madhiray Kaveri Tower, 3-2-19, Azam Jahi Road, Nimboliadda, **Hyderabad**-500 027. Tel.: (040) 24652456, Telefax: 24652457, E-mail: hyderabad@newagepublishers.com
- RDB Chambers (Formerly Lotus Cinema)106A, 1st Floor, S.N. Banerjee Road, **Kolkata**-700 014. Tel.: (033) 22273773, Telefax: 22275247, E-mail: kolkata@newagepublishers.com
- 16-A, Jopling Road, **Lucknow**-226 001. Tel.: (0522) 2209578, 4045297, Telefax: 2204098 E-mail: lucknow@newagepublishers.com
- 142C, Victor House, Ground Floor, N.M. Joshi Marg, Lower Parel, **Mumbai**-400 013. Tel.: (022) 24927869 Telefax: 24915415, E-mail: mumbai@newagepublishers.com
- 22, Golden House, Daryaganj, **New Delhi**-110 002. Tel.: (011) 23262370, 23262368, Telefax: 43551305 E-mail: sales@newagepublishers.com

---

ISBN : 978-81-224-3241-1

₹ 190.00

C-11-02-5338

Printed in India at Ajit Printers, Delhi.
Typeset at Pagitek Graphics, Delhi.

PUBLISHING FOR ONE WORLD
NEW AGE INTERNATIONAL (P) LIMITED, PUBLISHERS
4835/24, Ansari Road, Daryaganj, New Delhi-110002
Visit us at **www.newagepublishers.com**

Preface

The central theme of this book concerns the basic principles and applications of Microwave Transmission Lines, Waveguide Components, Microwave Tubes, Microwave Solid State Devices and Microwave Measurements. The scope is governed by the revised syllabus prescribed for the subject by Jawahar Lal Technological University (JNTU) Hyderabad for III Year B. Tech. ECE-II Semester students. The book is organized into 11 chapters. The topic included in Unit 1 of the syllabus are discussed in chapters 1 and 2. Chapter 3 describes Circular Waveguides and Microstrip Lines. Chapter 4 treats Cavity Resonators and Reentrant Cavities. Chapter 5 describes Waveguide Components and Application 1 as per Unit III and topics, included in Unit IV have been discussed in chapter 6. Chapters 7, 8 and 9 describe Microwave Tubes—Klystrons, Helix TWT and M-type Tubes (Mangerons) respectively. The topics included in Unit VII of the syllabus—Microwave Solid State Devices are discussed in chapter 10. Basic Microwave Measurements are explained in chapter 11, which is Unit VIII of the syllabus.

The book has been written in a simple language that can be understood by undergraduate students in Engineering, coming from different background. Any suggestions towards improvement of this edition of the book shall be highly appreciated.

AUTHORS

Preface

The central theme of this book concerns the basic principles and applications of Microwave Transmission Lines, Waveguide Components, Microwave Tubes, Microwave Solid State Devices, and Microwave Measurements. The scope is governed by the revised syllabus prescribed for the subject by Jawaharlal Nehru Technological University (JNTU) Hyderabad for III Year B. Tech. ECE-II Semester students. The book is organized into 11 chapters. The topics included in Unit I of the syllabus are discussed in chapters 1 and 2. Chapter 3 describes Circular Waveguides and Microstrip Lines. Chapter 4 treats Cavity Resonators and Reentrant Cavities. Chapter 5 describes Waveguide Components and Applications I as per Unit III and Topics included in Unit IV have been discussed in chapter 6. Chapters 7, 8 and 9 describe Microwave Tubes—Klystrons, Helix TWT and M-type Tubes (Magnetrons) respectively. The topics included in Unit VII of the syllabus—Microwave Solid State Devices are discussed in chapter 10. Basic Microwave Measurements are explained in chapter 11, which is Unit VIII of the syllabus.

The book has been written in a simple language that can be understood by undergraduate students in Engineering, coming from different background. Any suggestions towards improvement of this edition of the book shall be highly appreciated.

AUTHORS

Syllabus

JAWAHAR LAL NEHRU TECHNOLOGICAL UNIVERSITY HYDERABAD

III Year B.Tech. ECE. II-Sem **Chapters**

MICROWAVE ENGINEERING

UNIT I

MICROWAVE TRANSMISSION LINES: Introduction, Microwave Spectrum and Bands, Applications of Microwaves. Rectangular Waveguides — TE/TM mode analysis, Expressions for Fields, Characteristic Equation and Cut-off Frequencies, Filter Characteristics, Dominant and Degenerate Modes, Sketches of TE and TM mode fields in the cross-section, Mode Characteristics — Phase and Group Velocities, Wavelengths and Impedance Relations; Power Transmission and Power Losses in Rectangular Guide. Related Problems. 1 & 2

UNIT II

CIRCULAR WAVEGUIDES: Introduction, Nature of Fields, Characteristic Equation, Dominant and Degenerate Modes. Impossibility of TEM mode. Microstrip Lines— Introduction, Zo Relations, Effective Dielectric Constant, Losses, Q factor. Cavity Resonators— Introduction, Rectangular and Cylindrical Cavities, Dominant Modes and Resonant Frequencies, Q factor and Coupling Coefficients. Related Problems. 3 & 4

UNIT III

WAVEGUIDE COMPONENTS AND APPLICATIONS - I : Coupling Mechanisms — Probe, Loop, Aperture types. Waveguide Discontinuities — Waveguide irises, Tuning Screws and Posts, Matched Loads. Waveguide Attenuators — Resistive Card, Rotary Vane types; Waveguide Phase Shifters — Dielectric, Rotary Vane types. Waveguide Multiport Junctions — E plane and H plane Tees, Magic Tee, Hybrid Ring; Directional Couplers — 2 Hole, Bethe Hole types. 5

UNIT IV

WAVEGUIDE COMPONENTS AND APPLICATIONS - II : Ferrites— Composition and Characteristics, Faraday Rotation; Ferrite Components — Gyrator, Isolator, Circulator. Scattering Matrix— Significance, Formulation and Properties. *S* Matrix Calculations for — 2 port Junction, *E* plane and *H* plane Tees, Magic Tee, Directional Coupler, Circulator and Isolator. Related Problems. 6

UNIT V

MICROWAVE TUBES — I: Limitations and Losses of conventional tubes at microwave frequencies. Microwave tubes — O type and M type classifications. O-type tubes : 2 Cavity Klystrons — Structure, Reentrant Cavities, Velocity Modulation Process and Applegate Diagram, Bunching Process and Small Signal Theory — Expressions for o/p Power and Efficiency. Reflex Klystrons — Structure, Applegate 7

Diagram and Principle of Working, Mathematical Theory of Bunching, Power Output, Efficiency, Electronic Admittance; Oscillating Modes and o/p Characteristics, Electronic and Mechanical Tuning. Related Problems.

UNIT VI

HELIX TWTS: Significance, Types and Characteristics of Slow Wave Structures; Structure of TWT and Amplification Process (qualitative treatment), Suppression of Oscillations, Nature of the tour Propagation Constants, Gain Considerations. 8

M-type Tubes

Introduction, Cross-field effects, Magnetrons — Different Types, 8-Cavity Cylindrical Travelling Wave Magnetron — Hull Cut-off and Hartree Conditions, Modes of Resonance and PI-Mode Operation, Separation of PI-Mode, o/p characteristics. 9

UNIT VII

MICROWAVE SOLID STATE DEVICES: Introduction, Classification, Applications. TEDs — Introduction, Gunn Diode — Principle, RWH Theory, Characteristics, Basic Modes of Operation, Oscillation Modes. Avalanche Transit Time Devices — Introduction, IMPATT and TRAPATT Diodes — Principle of Operation and Characteristics. 10

UNIT VIII

MICROWAVE MEASUREMENTS: Description of Microwave Bench — Different Blocks and their Features, Precautions; Microwave Power Measurement — Bolometer Method. Measurement of Attenuation, Frequency, VSWR, Cavity Q. Impedance Measurements. 11

Contents

CHAPTER 1

Introduction

This chapter is intended to provide basic principles of microwave engineering which is treated in greater detail in the text.

1.1 MICROWAVE SPECTRUM AND BANDS

The whole electromagnetic-spectrum (Table 1.1) that extends from d.c. to γ-rays and beyond is broadly divided into two regions, namely, the radio-spectrum from d.c. to 300 GHz and the optical spectrum extending from 300 GHz to infinity. The term microwave* is commonly used to designate frequencies ranging from 300 MHz to 300 GHz or having wavelengths ranging from 100 cm to 1 mm in radio-spectrum. The microwave region has been further divided into three regions, namely, Ultrashort wave (Ultra high frequency-UHF), Supershort wave (Super high frequency-SHF) and Extreme shortwave (Extreme high

Table 1.1 The electromagnetic spectrum

Region	*Frequency limits (GHz)*		*Wavelength limits (cm)*		Remarks
	Minimum	*Maximum*	*Maximum*	*Minimum*	
(*a*) Radio-spectrum	d.c.	300	Infinity	0.1	VLF/LF/MF/HF/ VHF/UHF/SHF etc.
(*b*) Optical spectrum					
(*i*) Infrared	300	375×10^3	0.1	8×10^{-5}	Heat and invisible light
(*ii*) Visible	375×10^3	790×10^3	8×10^{-5}	38×10^{-6}	Light-red to violet
(*iii*) Ultraviolet	790×10^3	225×10^5	38×10^{-6}	12×10^{-7}	Chemically invisible
(*iv*) X-rays	225×10^5	450×10^8	12×10^{-7}	6×10^{-10}	–
(*v*) γ-rays	450×10^8	270×10^9	6×10^{-10}	1×10^{-10}	–
(*vi*) Cosmic rays	270×10^9	Indefinite	1×10^{-10}	Indefinite (smaller than 10^{-12} cm)	–

* Corrara in 1932 first used the term 'microwaves' to designate electromagnetic waves of 30 cm wavelength, which is also the term appeared for the first time in the proceedings of the IRE (Now it is IEEE).

frequency-EHF) according to the recommendations of the International Radio Consultative Committee. The position of microwave bands in entire radio-spectrum is shown in Table 1.2. The region of microwaves has been pushed into wavelengths lower than 1 mm and are known as submillimetre waves (> 300 GHz).

Table 1.2 Position of microwave bands in the entire radio-spectrum

S. No.	*Frequency band*	*Wavelength band*	*Designation (I.R.C.C. band)*	*Remarks*
1.	0 to 30 kHz	Infinity to 10^4 m	Very low frequency (VLF)/Very long waves	–
2.	30 to 300 kHz	10^4 to 10^3 m	Low frequency (LF)/Long waves	–
3.	300 to 3000 kHz	10^3 to 10^2 m	Medium frequency (MF)/Medium waves	–
4.	3 to 30 MHz	100 to 10 m	High frequency (HF)/Short waves	–
5.	30 to 300 MHz	10 to 1 m	Very high frequency (VHF)/ Very short waves	–
6.	0.3 to 3 GHz	1 to 0.1 m	Ultra high frequency (UHF)/ Ultrashort waves	Microwaves
7.	3 to 30 GHz	10 to 1 cm	Super high frequency (SHF)/ Supershort waves	Microwaves
8.	30 to 300 GHz	10 to 1 mm	Extreme high frequency (EHF)/ Extreme short-waves	Microwaves
9.	300 to 3000 GHz	1 to 0.1 mm	Submillimetre	–

The Institute of Electrical and Electronics Engineers (IEEE) recommended microwave bands designations as listed in Table 1.3.

Table 1.3 IEEE microwave frequency bands

Frequency range (GHz)	*Approximate band designation*
0.003 to 0.030	HF
0.030 to 0.300	VHF
0.300 to 1.000	UHF
1.000 to 2.000	L band
2.000 to 4.000	S band
4.000 to 8.000	C band
8.000 to 12.000	X band
12.000 to 18.000	Ku band
18.000 to 26.000	K band
26.000 to 40.000	Ka band
40.000 to 300.000	Millimetre
>300.00	Submillimetre

The radar band classification, as listed in Table 1.4, came into use during World War II and is still in use today even though the new military band designation, as listed in Table 1.5 has been adopted by the U.S. Department of Defence since 1970.

Table 1.4 U.S. Military microwave bands (Radar bands)

Frequency range (GHz)	*Designation*
0.225 to 0.390	P band
0.390 to 1.550	L band
1.550 to 3.900	S band
3.900 to 6.200	C band
6.200 to 10.900	X band
10.900 to 36.000	K band
36.000 to 46.000	Q band
46.000 to 56.000	V band
56.000 to 100.000	W band

Table 1.5 U.S. New military microwave bands (New radar bands)

Frequency range in (GHz)	*Designation*	*Frequency range in (GHz)*	*Designation*
0.100 to 0.250	A band	6.000 to 8.000	H band
0.250 to 0.500	B band	8.000 to 10.000	I band
0.500 to 1.000	C band	10.000 to 20.000	J band
1.000 to 2.000	D band	20.000 to 40.000	K band
2.000 to 3.000	E band	40.000 to 60.000	L band
3.000 to 4.000	F band	60.000 to 100.000	M band
4.000 to 6.000	G band		

1.2 UNIQUE CHARACTERISTICS OF MICROWAVES

The microwaves display certain unique characteristics that make them distinct from the waves of other adjacent bands, namely, (**1**) Short wavelengths, (**2**) More bandwidth (information carrying capacity). For instance, a frequency band 10^9 to 10^{12} Hz (microwave region) contains thousand sections such as the entire frequency band from 0 to 10^9 Hz. Then any one of these thousand sections may be used to transmit all radio, television or other communication that is presently transmitted by 0 to 10^9 Hz band. A 10% bandwidth at 4 GHz carrier is 400 MHz which can carry 1,00,000 voice channels (bandwidth 4 kHz) or 66 television channels (bandwidth 6 MHz), or 8000 – 266 digital data channels (50 kHz – 1.5 MHz bandwidth). (**3**) Small antenna size with large antenna gain (narrow beam antenna), (**4**) Travel by line of sight propagation through ionosphere with negligible absorption and reflection. For instance, a 140 cm diameter parabolic reflector type antenna produces a beam of 1° beam width (angular beam width = $140°/(D/\lambda)$; where D is diameter of the parabola and λ is the wavelength) at 30 GHz ($\lambda = 1$ cm). At 300 MHz ($\lambda = 100$ cm) the diameter of the parabolic antenna becomes 140 m to provide a beam of 1° beam width. The size is too large to be carried aboard in an aeroplane. (**5**) Reflection from metallic surfaces, (**6**) Microwave heating, (**7**) Molecular, atomic and nuclear resonance, etc. These features have provided unique opportunities for several useful applications. Some important and typical applications of microwaves are briefly described here under.

1.3 APPLICATIONS OF MICROWAVES

1.3.1 Microwave Communication Systems

Because of the increase in bandwidth microwaves are extensively used to carry voice (4 kHz bandwidth), digital data (50 kHz – 1.5 MHz bandwidth), television signals (6 MHz bandwidth) or telephonic traffic etc., over long distances links on the ground (ground communication) to deep-space spacecraft (space communication). Broadly, microwave communication systems are of two types, namely, (*i*) **ground-wave system** where signal is transmitted over a low loss cable or waveguide. With the commercial availability of high power lasers, fiber optical cables are increasingly used for long-distance telephonic traffic, as they are characterised by low-loss, very high bandwidth and a high degree of ruggedness, and (*ii*) **Radio-links** where the signal propagates through space. This is because of the fact that microwave signals travel by line of sight and are not bent/reflected by the ionosphere. However, for long distance links on the ground, repeater stations are required at frequent intervals, to receive and retransmit the signal. The newer systems, like (*i*) direct broadcast satellite television (DBST), (*ii*) personal communication systems (PCSs), (*iii*) wireless local area computer networks (WLANs), (*iv*) cellular video (CV) systems and (*v*) global positioning satellite (GPS) systems operate in microwave band.

1.3.2 Radar Systems and Countermeasures

Microwaves are widely used for directive signal transmission (navigation) and locating and ranging objects in space (radar or radio-detection and ranging). It is because extremely narrow microwave beam could be produced by the microwave antennas. Many varied forms of radar systems have been developed and are in use. Besides, military applications of radar systems such as (*i*) air and marine navigation, (*ii*) detection and tracking of aircraft, missiles, spacecraft, (*iii*) missile guidance, (*iv*) fire control for missiles and artillery, (*v*) weapon fuses, and (*vi*) reconnaissance, there are civilian and scientific applications of radar systems some of which are listed below:

Civilian applications—(*i*) Airport surveillance, (*ii*) marine navigation, (*iii*) weather detecting radar, (*iv*) altimetry, (*v*) aircraft landing radar, (*vi*) automobile speed measuring radar (Police radar) etc.

Scientific applications—(*i*) Remote sensing of natural resources, (*ii*) mapping and imaging radar, (*iii*) astronomy etc.

Countermeasures—Countermeasures involve the radiation of noise power in the operating band of the radar to confuse or deceive the radar as communication system. Electronic support measures (ESM), Electronic countermeasures (ECM) and Electronic counter-counter-measures (ECCM) have been developed using microwave technology.

1.3.3 Radiometry/Radio-astronomical Research

Very sensitive microwave receivers are being used to develop information about a target solely from the microwave portion of the black body radiations (noise) that it either emits directly or reflects from surroundings. It is known as radiometry and the device is called radiometer. Radiometry is being used by radio astronomers in (*a*) planetary mapping, (*b*) solar emission mapping, (*c*) mapping of galactic objects, and (*d*) measurement of cosmological background radiation. The technique has been extended to environment studies (remote sensing) such as (*a*) measurements of soil moisture, (*b*) flood mapping, (*c*) snow cover/ice cover mapping, (*c*) ocean surface wind speed, and (*d*) developing atmospheric temperature and humidity profiles. Radiometry is extensively used in military operations such as (*a*) target detection and recognition, (*b*) surveillance, and (*c*) mapping. Radio-astronomical applications of microwaves are due to the fact that ionosphere is a transparent media for propagation of microwaves.

1.3.4 Industrial Applications

During the last one and half decades, the industrial applications of microwaves have expanded greatly. Microwave heating has entered into commercial use in the form of microwave ovens, in which cooking is done quickly and uniformly by waves, inside and outside the food-material simultaneously. To the average consumer, the "microwave" connotes a microwave oven, which is getting increasingly popular in households for heating food. Microwave drying machines are used in textile and paper industries. Industrial applications are particularly evident in the food processing industry, the rubber industry and foundries. Other industrial applications include the central and on-line measurement of expensive materials during processing procedures in environments hostile to opto-electronic techniques. Recently, microwaves have also been used for non-destructive testing of metals such as thickness measurements, and measurement of moisture content in paper and textile industry and in liquids etc.

1.3.5 Microwaves in Basic and Applied Research

Microwaves have provided a very powerful experimental probe for the study of basic properties of matter as various molecular atomic and nuclear resonance occurs at microwave frequencies. Consequently, microwaves are capable of energetically interacting with matter. This feature is widely used in microwave and radio frequency spectroscopy for structural analysis. Microwave absorption spectra provide information about molecular structure and energy levels. Useful molecular resonances exist at microwave frequencies in the diodes of certain crystal materials. The resonant interaction of microwaves with crystals has been used for the generation of microwave power. Impact conjugation avalanche transist time (IMPATT) device, the non-reciprocal ferrite devices and masers are yet another examples of microwave resonances in molecules.

The interaction of an electron beam with periodic slow-wave microwave structures has been used to design high power linear accelerators that are indispensable instruments in nuclear research.

1.3.6 Biomedical Applications

The potential of microwaves for used in medicine is immense. The exact location of deep cancerous tissue, in particular, can be known by means of microwave radiometers. Microwave diathermy machines are used to remove rheumatic pains by producing heat inside the muscle without affecting the skin. Patients afflicted with uncontrollable pain or random muscle movements can be treated using microwave irradiation which creates thermal blocks in the nerve network.

Microwave radiations are being used for cancer therapy—hyperthermia—local, regional and whole body. Electromagnetic transmission through a human body has been used for monitoring heartbeat, lung-water detection, etc. Biomedical applications of microwaves are ever increasing.

1.3.7 Potential Application in the Field of Energy Transfer

It has a potential application in the field of energy transfer, *i.e.,* electrical energy can be transferred, without use of transmission lines, by converting it into microwave power and radiating through antenna in a narrow beam. At the receiving station it could be converted back to electric energy. Such a system has been proposed in the form of satellite power stations (SPS) as a method of tapping solar energy on a 24 hour-a-day, 365-days-a-year basis for the present century and beyond.

Although the field of microwave engineering is already well developed, the scope of its applications in communication, industry and basic research is ever increasing. The extension of microwave techniques into the field of optics is one such example.

However, varied uses of microwaves have increased microwave pollution and the consequent risk to health. This has stimulated the study of the biological effects and safety in several large research projects.

Table 1.6 summarises the typical and unique applications of microwaves. The scope of microwave applications in communication industry and basic and applied research is ever increasing. The extension of microwave techniques into the field of development of optical communication system is one such example.

Table 1.6

S. No.	*Applications*	*Frequency band*
1.	Television, satellite communication, surveillance radar, navigational aids, food industry (microwave-ovens) point to point communication.	0.3 to 3 GHz
2.	Altimeter, air and ship borne radar, microwave links, common-carrier land mobile communication, satellite communication, navigation, basic research-microwave spectroscopy.	3 to 30 GHz
3.	Radar, radio-astronomy, radio-meteorology, space research, nuclear physics, nucleonics.	30 to 300 GHz

1.4 TYPICAL MICROWAVE SYSTEM

Figure 1.1 shows a typical microwave system. It has a transmitter subsystem, which normally consists of microwave generator, waveguides, wavemeter, attenuator and transmitting antenna, and a receiver subsystem, which normally consists of receiving antenna, waveguides, a microwave amplifier, detector. Another typical radar system is shown in Fig. 1.2 which uses single antenna for transmission and reception.

Therefore, a first course on Microwave Engineering should include four major areas of study, namely (*i*) Microwave Transmission Lines – Waveguides, (*ii*) Waveguide Component and Applications, (*iii*) Microwave Sources – Tubes and Solid State Devices, and (*iv*) Microwave Measurements.

These topics are discussed in the text keeping a balance between mathematical and physical approach.

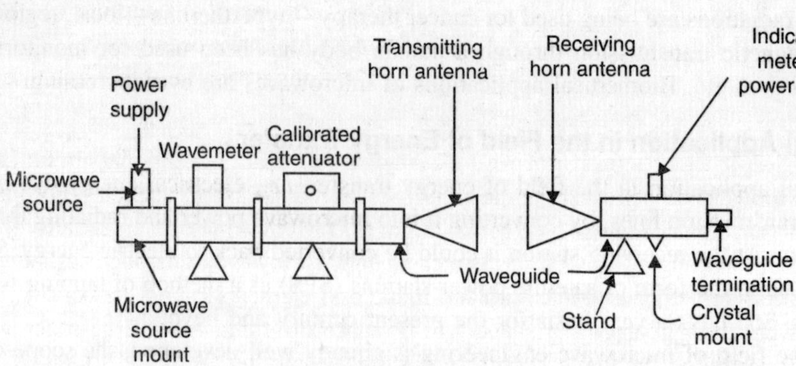

Fig. 1.1 Typical microwave system

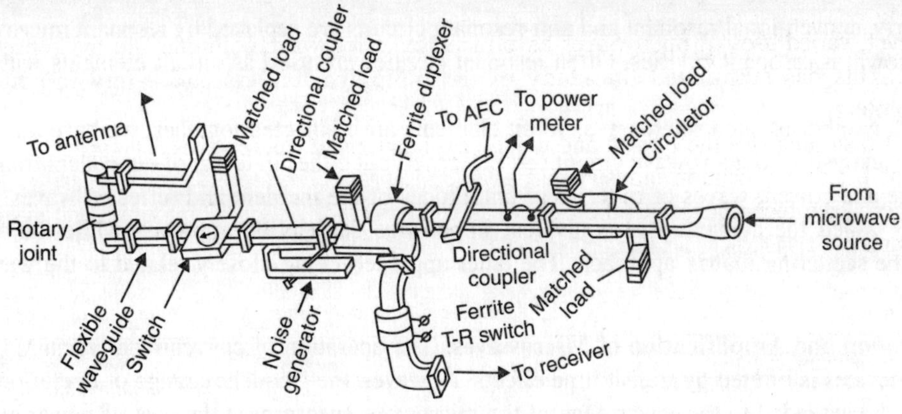

Fig. 1.2 Typical radar system

1.5 BASIC MICROWAVE CONCEPTS

(i) Microwave Transmission: The principles of microwave transmission cannot be derived by mere extensions of either low frequency radio or high frequency optical concepts, although they are all based upon the fundamental laws of electromagnetism. For instance, if microwave power is fed in a conventional two conductor lines where the longitudinal and transverse dimensions of the line are comparable to the wavelength of the propagating signal, it leads to a series of interesting effects that fall outside the scope of problems examined by classical theory of long transmission lines. It turns out that such a line cannot be used for microwave transmission. One has, therefore, to use hollow metal tubes called waveguides. The energy propagation in these structures is basically a reflection phenomena. Conversely, a hollow-pipe waveguide cannot be treated by the rules of low-frequency electricity, for by these rules opposite currents cannot flow in the same metallic conductor without coalescing into one net current, yet we see that opposite currents can flow in the same conductor in waveguides. Another class of waveguides of most recent origin is called surface waveguides which are absolutely uncommon to low frequency transmission.

The waveguide transmission of microwaves is associated with a number of interesting problems such as coupling of power to another system say from generator to line, exciting of waves in a waveguide, etc. To overcome these, three basic coupling methods, viz. electrical coupling (probe), magnetic coupling (loop), and aperture coupling (waveguide to waveguide) have been evolved. The basic feature of these methods is that one can control the amount of coupling because these structures have small antennas that radiate into the waveguide to be coupled.

(ii) Microwave Circuit Elements: It is a well-established fact that conventional circuit elements such as resistors, inductors and capacitors do not respond well at microwave frequencies. For instance, a coil of wire may be an excellent inductor at 1 MHz, but at 500 MHz it may be an equally good capacitor because of the predominating effect of inter-turn capacitance. However, this does not mean that energy dissipating (resistors) and storing (capacitors and inductors) elements cannot be constructed at microwave frequencies but their geometrical shape will be quite different. As will be seen, a section of a microwave line (distributed parameters), offers reactances varying from $-\infty$ to $+\infty$ if its length is suitably chosen.

Similarly, conventional resonant and anti-resonant circuits are replaced by resonant microwave line sections known as resonant cavities. Often resonant cavities are used as circuit elements with varying properties.

When a number of such microwave circuit elements are connected together, we have a microwave circuit. The analysis of a microwave circuit can be carried out either in terms of equivalent transmission line, voltage and current waves or in terms of amplitudes of the incident and reflected waves. The first approach is called the conventional equivalent impedance description approach while the second is known as the scattering matrix approach. The latter approach being closely related to the wave nature of fields.

(iii) Generation and Amplification of Microwaves: The operation of conventional vacuum tubes and solid state devices is limited by transit time effects. However, the frequency range of operation of these devices can be extended to the lower edge of the microwave spectrum at the cost of power output and noise characteristics. Therefore, the development of new devices was essential to exploit this frequency region. Fortunately, number of new principles of operation such as velocity modulation, interaction of space charge waves with electromagnetic fields were proposed. It involves transfer of power from a source of direct voltage to a source of alternating voltage by means of a density-modulated stream of electrons resulting in the development of klystron (in 1939 by R.H. Varian and S.F. Varian [1]), magnetron (in 1921 by Hull and an improved model in 1940 by Boot and Randal [2, 3]) and travelling-wave tube (TWT) (in 1944 by R. Komfner [4].

The interaction of the impact ionization avalanche and the transit time of charge carriers was used to develop Reed diode in 1958 [5], IMPATT by Johnston et al. in 1965 [6], TRAPATT by Prager et al. in 1967 [7]. Quantum mechanical tunneling was used to develop Tunnel Diode by Esaki in 1958 [8]. Transferred electron techniques were used to develop transfer electron devices by Ridley et al. in 1961 [9] and Hilsum in 1962 [10]. Gunn oscillator was developed in 1963 [11] which operate simply by the application of a dc voltage to a bulk. In all solid-state devices the negative resistance characteristics is exploited for microwave generation and amplification.

REVIEW QUESTIONS

1. Explain how microwave engineering is different from low frequency electronic engineering. (JNTU, 2008)
2. Describe in brief basic microwave concepts.
3. List unique characteristics of microwaves.
4. List the typical application of microwave.
5. Explain with an example how microwave has large information carrying capacity.
6. Explain "Microwaves are widely used for directive signal transmission and locating and ranging objects in space".

SELECTED TITLES

1. Collin, R.E., **Foundations for Microwave Engineering,** McGraw-Hill Book Company, New York, 1966.
2. Gandhi, O.P., **Microwave Engineering and Applications,** Pergamon Press, New York, 1981.
3. Liao, S.Y., **Microwave Devices and Circuits,** Prentice-Hall Inc, Englewood Cliffs, N.J., U.S.A., 1995 [Indian reprint].
4. Soohoo, R.F., **Microwave Electronics,** Addison-Wesley Publishing Company, Reading, Mass., 1971.

5. Reich, H.J., et al., **Microwave Principles,** D. Van Nostrand Reinhold Company, New York, 1957 (an East-West Edition).
6. Reich, H.J., et al., **Microwave Theory and Techniques,** D. Van Nostrand Reinhold Company, New York, 1953.
7. Brownwell, A.B., and R.E. Beam, **Theory and Application of Microwaves,** McGraw-Hill Book Co., New York, 1947.
8. Chatterjee, R., **Elements of Microwave Engineering,** East-West Press, New Delhi, 1984.
9. Rizzi, P.A., **Microwave Engineering: Passive Circuits,** Prentice-Hall, Inc., Englewood Cliffs, N.J., 1988.
10. Sisodia, M.L. and G.S. Raghuvanshi, **Microwave Circuits and Passive Devices,** John Wiley & Sons, New York, 1987 (Wiley Eastern Limited, New Delhi, 1987, now New Age, 2003).
11. Sisodia, M.L. **Microwave Active Devices,** Vacuum and Solid State, New Age International Publishers, New Delhi, 2003.
12. Sisodia, M.L. and G.S. Raghuvanshi, **Basic Microwave Techniques and Laboratory Manual,** John Wiley & Sons, 1987. (Wiley Eastern Limited, 2000, New Delhi, now New Age).
13. Sisodia, M.L. and Vijay Laxmi Gupta, **Microwaves: Introduction to Circuits, Devices and Antennas**, New Age International Publishers, New Delhi, 2003.

REFERENCES

1. Warnecke, R.R., et al., *Velocity Modulated Tubes: The Advances in Electronics*, Vol. 3, Academic Press, New York, 1951.
2. Hull, A.W., Phys. Rev. **18**, 31 (1921).
3. Okress, E., Editor, *Crossed Field Microwave Devices*, Academic Press, New York, Vols. I and II, 1961.
4. Kompfner, R., *The Travelling Wave Tube as Amplifier at Microwaves*, Pros. IRE **35**, 124-127, February 1947.
5. Read, W.T., *A Proposed High Frequency Negative-Resistance Diode*, Bell System tech. J., **37**, 401-446, 1958.
6. Johnston, R.L., B.C. Deloach, and G.B. Cohen, *A Silicon Diode Microwave Oscillator*, Bell System Tech., J., **44**, 369-372, February 1965.
7. Prager, H.J., et al., *High-power, High-efficiency Silicon Avalanche Diodes at Ultra High Frequencies*, Proc. IEEE (letters), **55**, 586-587, April 1967.
8. Esaki, L., *New Phenomenon in Narrow Ge p-n Junctions*, Phys. Rev., **109**, 603, 1958.
9. Ridley, B.K. and T.B. Watkins, *The Possibility of Negative Resistance Effect in Semiconductors*, Proc. Phys. Soc., **78**, 293-304, August 1961.
10. Hilsum, C., *Transferred Electron Amplifiers and Oscillators*, Proc. IEEE., **50**, 185-189 February 1962.
11. Gunn, J.B., *Instabilities of Current in III-V Semiconductors*, IBM J. Res. Develop., **8**, 141-159, April 1964.

CHAPTER 2

Microwave Transmission Lines I: Rectangular Waveguides

2.1 INTRODUCTION

Hollow waveguides are generally used as transmission lines at frequencies around 1 GHz and above. Waveguides have certain advantages in comparison to coaxial lines. These are

1. Higher power handling capability.
2. Lower loss per unit length.
3. A simpler, lower cost structure.

Moreover, the reflections caused by the flanges used in connecting waveguide sections is much less than that associated with coaxial connectors. It is precisely because of the low loss factor, that waveguides have edge over other kinds of transmission lines at higher frequencies. By definition, a waveguide is simply a pipe of virtually any consistent cross-sectional shape through which an electromagnetic (em) wave travels by reflection, and not by conduction. It is because of this reason that we always speak of electric and magnetic fields rather than voltage and current as was the case with transmission lines. Since the method of propagation in a waveguide is by means of reflection, hence the interior surfaces should be smooth, often silvered, free of moisture and abrupt changes in shape or direction otherwise

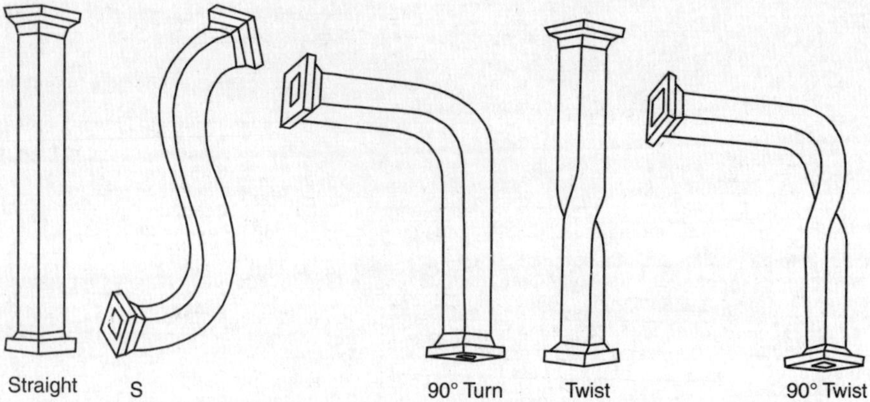

Fig. 2.1 Typical waveguide shapes

they will cause reflections to occur back towards the source. A few typical waveguide shapes are shown in Fig. 2.1.

2.2 RECTANGULAR WAVEGUIDES

Rectangular waveguides were one of the earliest types of transmission lines used to carry microwave signals. The hollow rectangular waveguides can propagate in Transverse Electric (TE) or Transverse Magnetic (TM) modes, but not in Transverse Electromagnetic (TEM) mode, since only one conductor is present. As mentioned earlier the rectangular waveguides have cut-off frequencies below which propagation is not possible. The waveguide shown in Fig. 2.2 is filled with a source free lossless dielectric material ($\sigma_d = 0$) and its walls are perfectly conducting ($\sigma_c \approx \infty$).

Figure 2.2*b* shows a parallel plate transmission line. The width of the metal strip is *w* and their spacing is *b*. Power is delivered to the load via longitudinal current flow along the two plates. If now a pair of shorted stubs of lengths l_s are connected to the parallel strip as shown in the Fig. 2.2*b*, and the stubs are quarter wavelength long, they will have no effect on the transmission of power since they present an infinite impedance in shunt with the line. The current flow in the stub is reactive as no real power can be delivered to the shorts. Power can only be delivered by longitudinal currents, which is why they are called power currents. Note that the direction of power and reactive currents are perpendicular to each other. We can connect an infinite set of such quarter wavelength shorted stubs along the length of the parallel plate transmission line. The resultant configuration will be exactly similar to the rectangular waveguides shown in Fig. 2.2*a*, where $a = w + 2l_s$.

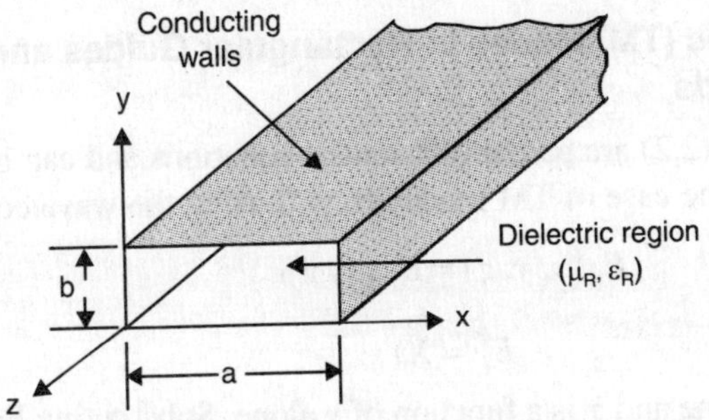

Fig. 2.2 (*a*) Rectangular waveguide

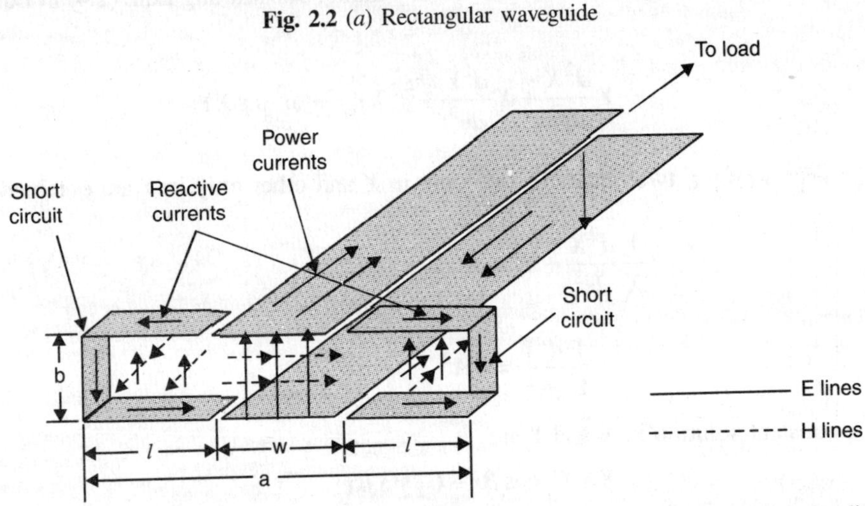

Fig. 2.2 (*b*) The development of rectangular waveguide from a parallel-plate transmission line

Since $l_s = \lambda/4$ and the width w must be finite so that power currents can flow, the condition for power transmission is

$$a > \lambda/2$$

In other words, electromagnetic wave propagation in rectangular waveguides can only occur at frequencies high enough to satisfy this inequality. As the frequency decreases, λ increases and hence $\lambda/4$ stubs take up a greater portion of the guide width a, leaving less room for w the parallel-plate line. Decreasing the frequency to the point where $a = 2l_s = \lambda/2$ results in $w = 0$, which prevents the flow of power currents. The frequency at which this occurs is called the cut-off frequency (f_c) of the guide.

2.2.1 Solutions of Wave Equation in Rectangular Coordinates

The electric and magnetic wave equations in frequency domain are given by

$$\nabla^2 \mathbf{E} + \bar{\gamma}^2 \mathbf{E} = 0 \text{ and } \nabla^2 \mathbf{H} + \bar{\gamma}^2 \mathbf{H} = 0 \qquad ...(2.1)$$

which can be written for TM and TE waves as

$$\left.\begin{aligned} \frac{\partial^2 E_z}{\partial x^2} + \frac{\partial^2 E_z}{\partial y^2} + \bar{\gamma}^2 E_z = -\omega^2 \mu \varepsilon E_z \\ \frac{\partial^2 H_z}{\partial x^2} + \frac{\partial^2 H_z}{\partial y^2} + \bar{\gamma}^2 H_z = -\omega^2 \mu \varepsilon H_z \end{aligned}\right\} \qquad ...(2.2)$$

2.2.2 Transverse Magnetic (TM) Modes in Rectangular Guides and Expression for Fields

The wave equations in Eqn. (2.2) are partial differential equations and can be solved by the method of separation of variables. For the case of TM modes $H_z = 0$ using the wave equation in E_z, we can write

$$E_z(x, y, z) = E_z^0(x, y) e^{-\bar{\gamma} z} \qquad ...(2.3)$$

Let

$$E_z^0 = XY \qquad ...(2.4)$$

where X is a function of x alone and Y is a function of y alone. Substituting Eqn. (2.4) in Eqn. (2.2), one obtains

$$Y\frac{d^2 X}{dx^2} + X\frac{d^2 Y}{dy^2} + \bar{\gamma}^2 XY = -\omega^2 \mu \varepsilon XY$$

Using $h^2 = \bar{\gamma}^2 + \omega^2 \mu \varepsilon$ two equations one only in X and other only in Y are obtained.

$$\frac{1}{X}\frac{d^2 X}{dx^2} + h^2 = A^2$$

$$\frac{1}{Y}\frac{d^2 Y}{dy^2} = -A^2$$

where A^2 is a constant solution in X and Y are

$$X = C_1 \cos Bx + C_2 \sin Bx$$

where $B^2 = h^2 - A^2$

$$Y = C_3 \cos Ay + C_4 \sin Ay$$

This gives

$$E_z^0 = XY$$

The solution for this equation on applying the following boundary conditions:

$E_z = 0$, when $x = 0$, $x = a$, $y = 0$, $y = b$ gives

$$E_z^0 = C \sin\left(\frac{m\pi}{a}\right) x \sin\left(\frac{n\pi}{b}\right) y \quad \text{...(2.5)}$$

where C is a constant. The other field components are obtained as

$$E_x^0 = -\frac{j\bar{\beta}C}{h^2}\left(\frac{m\pi}{a}\right)\cos\left(\frac{m\pi}{a}\right)x \sin\left(\frac{n\pi}{b}\right)y \quad \text{...(2.6a)}$$

$$E_y^0 = -\frac{j\bar{\beta}C}{h^2}\left(\frac{n\pi}{b}\right)\sin\left(\frac{m\pi}{a}\right)x \cos\left(\frac{n\pi}{b}\right)y \quad \text{...(2.6b)}$$

$$H_x^0 = \frac{j\omega\varepsilon C}{h^2}\left(\frac{n\pi}{b}\right)\sin\left(\frac{m\pi}{a}\right)x \cos\left(\frac{n\pi}{b}\right)y \quad \text{...(2.6c)}$$

$$H_y^0 = -\frac{j\omega\varepsilon C}{h^2}\left(\frac{m\pi}{a}\right)\cos\left(\frac{m\pi}{a}\right)x \sin\left(\frac{n\pi}{b}\right)y \quad \text{...(2.6d)}$$

The propagation constant γ in this case is given by

$$\bar{\gamma} = \bar{\alpha} + j\bar{\beta} = \sqrt{\left(\frac{m\pi}{a}\right)^2 + \left(\frac{n\pi}{b}\right)^2 - \omega^2\mu\varepsilon} \quad \text{...(2.7)}$$

When the attenuation constant $\bar{\alpha}$ is equal to zero, the phase constant $\bar{\beta}$ will have the value

$$\bar{\beta} = \sqrt{\omega^2\mu\varepsilon - \left(\frac{m\pi}{a}\right)^2 - \left(\frac{n\pi}{b}\right)^2} \quad \text{...(2.8)}$$

The cut-off frequency ω_c is obtained when $\bar{\beta} = 0$

$$\omega_c = \frac{1}{\sqrt{\mu\varepsilon}}\sqrt{\left(\frac{m\pi}{a}\right)^2 + \left(\frac{n\pi}{b}\right)^2} \quad \text{...(2.9)}$$

This is the frequency below which wave propagation will not occur.

$$f_c = \frac{1}{2\pi\sqrt{\mu\varepsilon}}\sqrt{\left(\frac{m\pi}{a}\right)^2 + \left(\frac{n\pi}{b}\right)^2} \quad \text{...(2.10)}$$

The corresponding cut-off wavelength is

$$\lambda_c = \frac{2}{\sqrt{\left(\frac{m}{a}\right)^2 + \left(\frac{n}{b}\right)^2}} \qquad \text{...(2.10}a\text{)}$$

from which it can be seen that

$$f_c \lambda_c = v_0$$

The propagation constant can be expressed as

$$\overline{\beta} = \omega\sqrt{\mu\varepsilon}\sqrt{1 - \left(\frac{f_c}{f}\right)^2} \qquad \text{...(2.10}b\text{)}$$

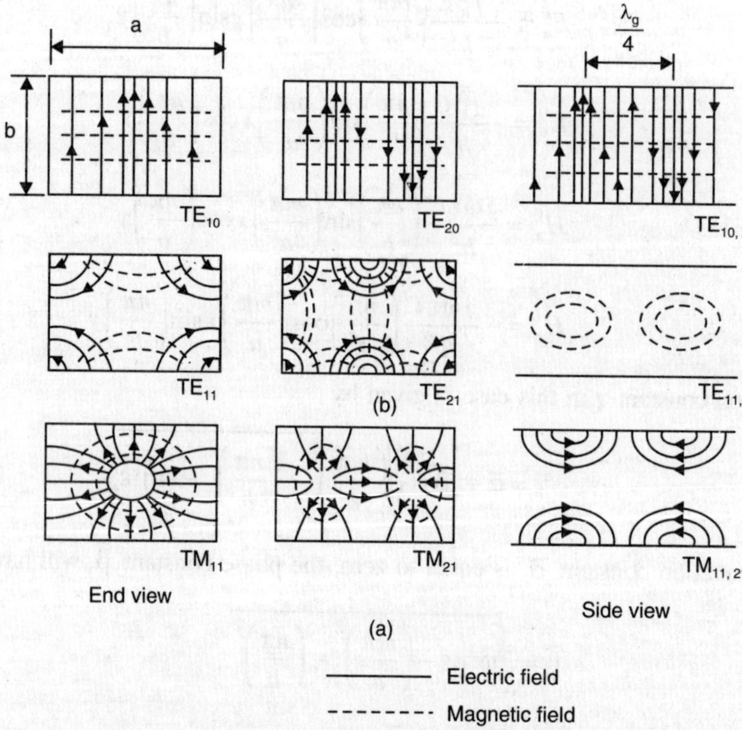

Fig. 2.3 Field patterns for the lower order modes: Rectangular waveguide (TM and TE)

The phase velocity is given by

$$\overline{v}_p = \frac{\omega}{\overline{\beta}} = \frac{\omega}{\sqrt{\omega^2\mu\varepsilon - \left(\frac{m\pi}{a}\right)^2 - \left(\frac{n\pi}{b}\right)^2}} \qquad \text{...(2.11)}$$

This expression indicates that the velocity of propagation of the wave in the guide is greater than the phase velocity in the free space. As the value of frequency is increased above cut-off, the phase velocity decreases from an infinitely large value and approaches c, the velocity of free space. Since the wavelength in the guide is given by $\bar{\lambda} = \bar{v}_p / f$, it will be longer than the corresponding free space wavelength. The lowest possible value for m and n for TM mode is 1. Substituting for $m = n = 1$ in Eqn. (2.6), we get fields for the lowest particular mode called the TM_{11}. Higher-order modes require higher frequencies in order to be propagated along the guide of given dimensions.

The wave configuration for different TM modes in a waveguide are shown in Fig. 2.3. The subscripts indicate the number of half sine wave variations of the field components in the x and y directions respectively.

2.2.3 Transverse Electric (TE) Modes

The solutions for transverse electric waves are derived in similar manner to that for transverse magnetic waves. The H_z component will be given as

$$H_z(x, y, z) = H_z^0(x, y)e^{-\bar{\gamma} z}$$

Let $$H_z^0 = XY$$

Boundary conditions cannot be applied to H_z directly. Hence it is differentiated with respect to x and y to obtain E_x^0, E_y^0, H_x^0 and H_y^0. The boundary conditions are then applied to these field components as

$$E_x^0 = 0 \text{ at } y = 0, b$$

$$E_y^0 = 0 \text{ at } x = 0, a$$

The field expressions result in the following relations:

$$H_z^0 = C\cos\left(\frac{m\pi}{a}\right)x\cos\left(\frac{n\pi}{b}\right)y \quad ...(2.12)$$

$$H_x^0 = \frac{j\bar{\beta}}{h^2}C\left(\frac{m\pi}{a}\right)\sin\left(\frac{m\pi}{a}\right)x\cos\left(\frac{n\pi}{b}\right)y \quad ...(2.12a)$$

$$H_y^0 = \frac{j\bar{\beta}}{h^2}C\left(\frac{n\pi}{b}\right)\cos\left(\frac{m\pi}{a}\right)x\sin\left(\frac{n\pi}{b}\right)y \quad ...(2.12b)$$

$$E_x^0 = \frac{j\omega\mu}{h^2}C\left(\frac{n\pi}{b}\right)\cos\left(\frac{m\pi}{a}\right)x\sin\left(\frac{n\pi}{b}\right)y \quad ...(2.12c)$$

$$E_y^0 = -\frac{j\omega\mu}{h^2}C\left(\frac{m\pi}{a}\right)\sin\left(\frac{m\pi}{a}\right)x\cos\left(\frac{n\pi}{b}\right)y \quad ...(2.12d)$$

2.2.4 Dominant and Degenerate Modes

Dominant Mode

In the above expressions $\bar{\gamma}$ has been put equal to $j\bar{\beta}$, as it is valid for frequencies above cut-off. From Eqns. (2.12 *a, b, c* and *d*) it is possible to make either m or n but not both equal to zero without causing all the fields to vanish. That is lower order is possible than in the case of TM wave. The lowest order

TE wave in rectangular guides is, therefore, the TE_{10} mode. This wave which has the lowest cut-off frequency is called the dominant mode. Because of the practical importance of the TE_{10} mode, the field components, when $m = 1$ and $n = 0$ case are given as follows

$$H_z = C\cos\frac{\pi x}{a} \qquad \text{...(2.13}a\text{)}$$

$$H_x^0 = \frac{j\beta aC}{\pi}\sin\frac{\pi x}{a} \qquad \text{...(2.13}b\text{)}$$

$$E_y^0 = -\frac{j\omega\mu aC}{\pi}\sin\frac{\pi x}{a} \qquad \text{...(2.13}c\text{)}$$

$$E_x^0 = H_y = 0 \qquad \text{...(2.14)}$$

and

$$\overline{\beta} = \sqrt{\omega^2\mu\varepsilon - \left(\frac{\pi}{a}\right)^2} \qquad \text{...(2.15)}$$

$$f_c = \frac{c}{2a}, \quad \lambda_c = 2a \text{ and } h = \frac{\pi}{a}$$

For the TE_{10} mode the cut-off frequency is independent of the dimension b. Field configurations for some of the TE modes are shown in Fig. 2.3.

Degenerate Modes

Some of the higher order modes, having the same cut-off frequency, are called degenerate modes. It can be seen that in a waveguide the possible TE_{mn} and TM_{mn} modes (both m and n cannot be zero) are always degenerate. The wave guide dimensions are always selected in such a way that only the desired mode (generally the dominant TE_{10} or TM_{11}) propagate and higher are not supported.

2.3 EXCITATION OF MODES IN RECTANGULAR WAVEGUIDES

We present certain aspects of excitation of various modes in a guide using a typical coaxial line-waveguide probe coupling. In order to launch a particular mode, probe is chosen which will produce lines of E and H that are roughly parallel to the lines of E and H of the particular mode. Figure 2.4 shows that the short-circuit position and probe depth can be adjusted to achieve maximum power transfer from the coaxial line into the waveguides. The centre conductor of the coaxial line extends into the waveguides to form as electric probe. For maximum coupling of the dominant TE_{10} mode in a rectangular guide, the probe should extend into the guide through the centre of the broad face so that to coincide with the positions of maximum electric field for the TE_{10} mode. The evanescent modes that are also excited are localized fields that store reactive energy. These give the junction its reactive properties.

All TE_{mo} modes with m even has a zero field at the centre and field polarities reverse alternately in the transverse direction. These modes can be excited by two probes fed with opposite phase currents. Figure 2.4b illustrates loop or inductive coupling. The loop is placed at a point where magnetic field (H) is maximum. There are a number of locations possible for H to be maximum. Aperture coupling is shown in Fig. 2.4c, slot X is at a position of maximum E field while slot Y is in a region of maximum H field density.

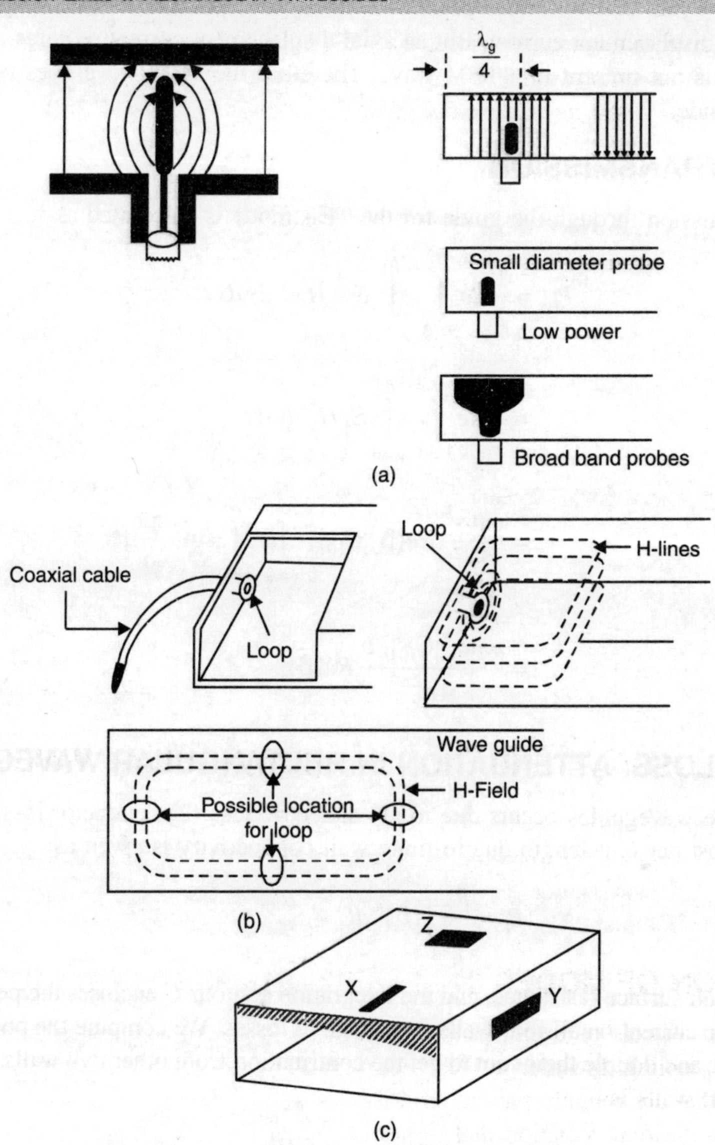

Fig. 2.4 Methods of coupling to and from a rectangular waveguide

We have seen that only TE ($E_z = 0$) and TM ($H_z = 0$) modes can propagate in hollow rectangular and cylindrical waveguides. The familiar TEM mode for which there in no axial component of either E or H does not propagate. Suppose a TEM wave is assumed to exist within a hollow guide of any shape. Then lines of H must lie entirely in the transverse plane. In a non-magnetic material,

$$\nabla \cdot H = 0$$

which requires that lines of H should be closed loops. Hence, if a TEM wave exists in a hollow guide then the lines will be closed loops in the plane perpendicular to the axis. But by Maxwell's first equation the magnetomotive force around each of these closed loops must equal to the axial current through the loop. In the case of a coaxial transmission line, this axial current through the H loops is the conduction current in the inner conductor. But as the hollow waveguides do not have an inner conductor, this axial current must be a displacement current. But an axial displacement current requires an axial component of E field, which is not present in a TEM wave. Therefore the TEM wave does not exist in a single conductor waveguide.

2.4 POWER TRANSMISSION

The power transmission through the guide for the TE_{10} mode is calculated as

$$P_{10} = \frac{1}{2} Re \int_{x=0}^{a} \int_{y=0}^{b} E \times H \cdot z \, dy\, dx \qquad ...(2.16)$$

$$= \frac{1}{2} Re \int_{x=0}^{a} \int_{y=0}^{b} E_y H_x^* \, dy\, dx$$

$$= \frac{\omega\mu a^2}{2\pi^2} Re\left(\bar{\beta}\right)\left|C_{10}\right|^2 \int_{x=0}^{a} \int_{y=0}^{b} \sin^2 \frac{\pi x}{a} dx \qquad ...(2.17)$$

$$= \frac{\omega\mu a^3 \left|C_{10}\right|^2 b}{4\pi^2} Re\left(\bar{\beta}\right)$$

2.5 POWER LOSS: ATTENUATION IN RECTANGULAR WAVEGUIDE

Attenuation in the waveguides occurs due to (1) dielectric loss (2) conductor loss.

The power lost per unit length due to finite wall conductivity is given as

$$P_l = \frac{R_s}{2} \int_C \left|J_s\right|^2 dl \qquad ...(2.18)$$

where R_s is the wall surface resistance, and the integration contour C encloses the perimeter of the guide walls. The surface current on all four walls contribute to losses. We compute the power lost in the walls at $x = 0$ and $y = 0$, and double their sum to get the contribution from other two walls. The surface current on the $x = 0$ (left) walls is

$$J_s = \hat{n} \times H|_{x=0} = -\hat{y} H|_{x=0} = -\hat{y} C e^{-\bar{\beta} z} \qquad ...(2.19)$$

The surface current on the bottom wall at $y = 0$ is

$$J_s = \hat{n} \times H|_{y=0} = \hat{y} \times \left(\hat{x} H_x|_{y=0} + \hat{z} H_z|_{y=0}\right)$$

$$= \hat{z} \frac{j\bar{\beta} a}{\pi} C \sin\frac{\pi x}{a} e^{-j\bar{\beta} z} + \hat{x} C \sin\frac{\pi x}{a} e^{-j\bar{\beta} z} \qquad ...(2.20)$$

On substituting Eqn. (2.20) in Eqn. (2.18), we obtain power loss as

$$P_l = R_s \int_{y=0}^{b} \left|J_{sy}\right|^2 dy + R_s \int_{x=0}^{a} \left[\left|J_{sx}\right|^2 + \left|J_{sz}\right|^2\right] dx$$

$$= R_s |C|^2 \left(b + \frac{a}{2} + \frac{\bar{\beta}^2 a^3}{2\pi^2} \right) \qquad ...(2.21)$$

The attenuation due to conductor loss is obtained by perturbation method. According to this method the field magnitude decreases as $e^{-\alpha z}$ and the power decreases as $P_0 e^{-2\bar{\alpha} z}$, where P_0 is the power at the sending end $z = 0$.

The rate of decrease of power for any mode (TE_{mn} or TM_{mn}) is given by

$$-\frac{dP_{mn}}{dz} = 2\alpha_c P_{mn} = P_l \qquad ...(2.22)$$

where P_l is the power loss per unit length of the guide and P_{mn} is the power transmitted through the guide. Hence the attenuation constant is given by

$$\alpha_c = \frac{P_l}{2P_{mn}} \qquad ...(2.23a)$$

We have already obtained P_l and P_{10}, substituting these in Eqn. (2.23*a*) the attenuation for TE_{10} mode is given as

$$\alpha_c = \frac{P_l}{2P_{10}} = \frac{2\pi^2 R_s \left[b + a/2 + \left(\bar{\beta}^2 a^3 / 2\pi^2 \right) \right]}{\omega\mu a^3 b \beta} \qquad ...(2.23b)$$

Attenuation versus frequency curves are drawn in Fig. 2.5.

Attenuation due to dielectric loss is computed from the propagation constant, using complex dielectric constant

$$\bar{\gamma} = \bar{\alpha}_d + j\bar{\beta} = \sqrt{h^2 - \omega^2 \mu\varepsilon}$$

$$= \sqrt{h^2 - \omega^2 \mu\varepsilon (1 - j \tan\delta)} \qquad ...(2.24)$$

For most dielectric materials tan δ < < 1, Eqn. (1.56) can be written as

$$\bar{\gamma} = \sqrt{h^2 - \omega^2 \mu\varepsilon + j\omega^2 \mu\varepsilon \tan\delta}$$

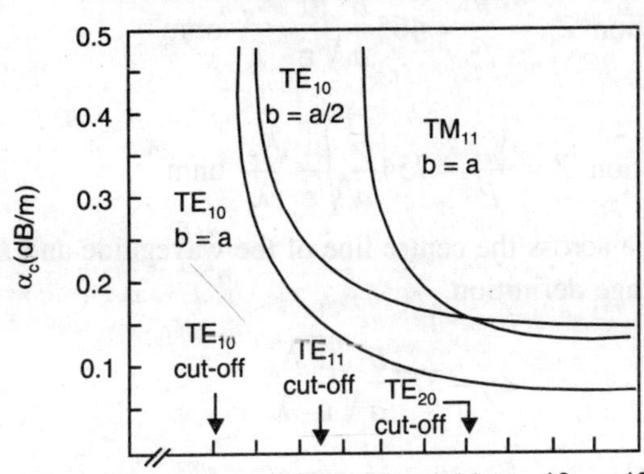

Fig. 2.5 Attenuation vs frequency curves for various modes in a typical rectangular brass guide

$$= \sqrt{h^2 - \omega^2 \mu\varepsilon}\left(1 + \frac{j\omega^2 \mu\varepsilon \tan\delta}{\sqrt{h^2 - \omega^2 \mu\varepsilon}}\right)^{1/2}$$

$$= \sqrt{h^2 - \omega^2 \mu\varepsilon} + \frac{j\omega^2 \mu\varepsilon \tan\delta}{2\sqrt{h^2 - \omega^2 \mu\varepsilon}}$$

$$= \frac{\omega^2 \mu\varepsilon \tan\delta}{2/\bar{\beta}} + j\bar{\beta} \qquad ...(2.25)$$

where $$j\bar{\beta} = \sqrt{h^2 - \omega^2 \mu\varepsilon}$$

From Eqn. (2.25) it is observed that when the loss is small the phase constant $\bar{\beta}$ is unchanged, while the attenuation constant due to dielectric loss is given by

$$\bar{\alpha}_d = \frac{\omega^2 \mu\varepsilon \tan\delta}{2\bar{\beta}} \text{ Np/m for both TE and TM modes.}$$

For a waveguide operated well below cut-off, it may be shown that the attenuation in dB for a l length of waveguide is

$$\alpha_{dB} = \frac{54.5l}{\lambda_c} \text{dB}$$

where l = length of the waveguide; and λ_c = cut-off wavelength of waveguide.

2.6 WAVEGUIDE IMPEDANCE

The characteristic impedance of a waveguide is not uniquely defined. The reason for this is that for a given field pattern voltage can be defined in a variety of ways. The commonly used definitions are based upon power flow along the waveguide. These are

1. Power current definition $Z_0 = \frac{P}{I_z^2} = 465\frac{b}{a}\sqrt{\frac{\mu}{\varepsilon}}\frac{\lambda_g}{\lambda}$ ohm

2. Power voltage definition $Z = \frac{V_{cl}^2}{P} = 754\frac{b}{a}\sqrt{\frac{\mu}{\varepsilon}}\frac{\lambda_g}{\lambda}$ ohm

   Here V_{cl} is the voltage across the centre line of the waveguide and I_z is the longitudinal current.

3. Modified power voltage definition

$$Z_0 = 377\frac{b}{a}\sqrt{\frac{\mu}{\varepsilon}}\frac{\lambda_g}{\lambda}$$

where $$\lambda_g/\lambda = \frac{1}{\sqrt{1-(f_c/f)^2}}$$

Note that all these definitions differ only by a constant. Most commonly used definition is the modified power-voltage one.

2.7 MICROWAVE FILTERS

Microstrip band pass filters: Band pass filter can be constructed in microstrip configuration using end coupled, parallel coupled or the interdigital form as shown in Figs. 2.6 and 2.7. End coupled microstrip band pass filter is shown in Fig. 2.6. The tuned circuits are realised from open or short circuited microstrip transmission line sections of one quarter or half wavelength long. The resonators are coupled by means of gap capacitances between the resonator sections. The resonator length l and the coupling space S can be calculated from the relations available in literature [1]. The gaps in the structure are made very small compared with substrate height to ensure tight coupling and low insertion loss in the pass band. The resonator lengths depend on the coupling reactance, the gap capacitance and guide wavelength. This structure is large, hence is not preferred.

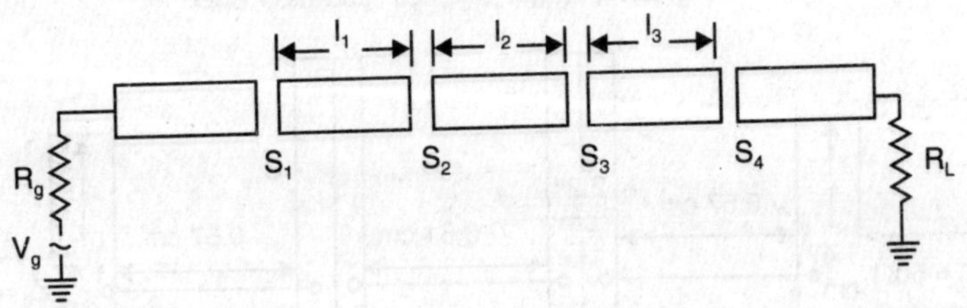

Fig. 2.6 End coupled microstrip band-pass filter

Microstrip parallel coupled filter: The parallel coupled filter shown in Fig. 2.7 is more compact than the end coupled filter described earlier. The coupling between resonators occur over a quarter-wave-long side of each resonator, hence the slot width is larger and the tolerance on the slot width is not very critical. The ends of each resonator section may be open-circuited or short-circuited. Using even and odd mode excitations, the expressions for design of a Chebyshev response n section band pass filter terminated in input and output lines of characteristic impedance Z_0 are summarised below. The maximum effective length of each resonator is calculated, the capacitive end loading effect is corrected for

$$I_e = \frac{\lambda_g}{4} - I_{eo} \qquad \text{...(2.26)}$$

where I_{eo} is extension of the line length due to fringing field at the ends, is given by

$$I_{eo} = 4.12 \ln\left(\frac{\varepsilon_{eff} + 0.3}{\varepsilon_{eff} - 0.258}\right)\left(\frac{W/h + 0.262}{W/h + 0.813}\right) \qquad \text{...(2.27)}$$

$$\left(\frac{I_{eo}}{h}\right)_{\max} = \frac{2}{\pi}\ln 2 \approx 0.441 \qquad \text{...(2.28)}$$

The band pass circuit consists of series resonant circuit in the series arm and parallel resonant circuit in the shunt arm.

Strip line filters: Strip line filter is an important class of microwave filters. There are standard techniques available by which a lumped or a distributed circuit may be transformed into its equivalent strip line version. Figure 2.8 shows a strip line version filter of a 2 GHz low pass filter.

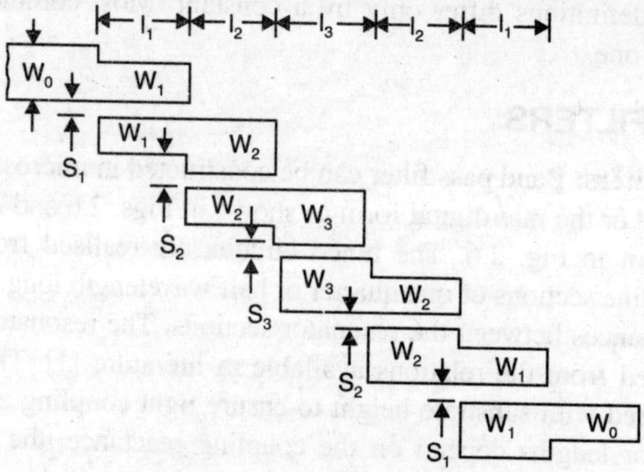

Fig. 2.7 Parallel coupled land-pass filter

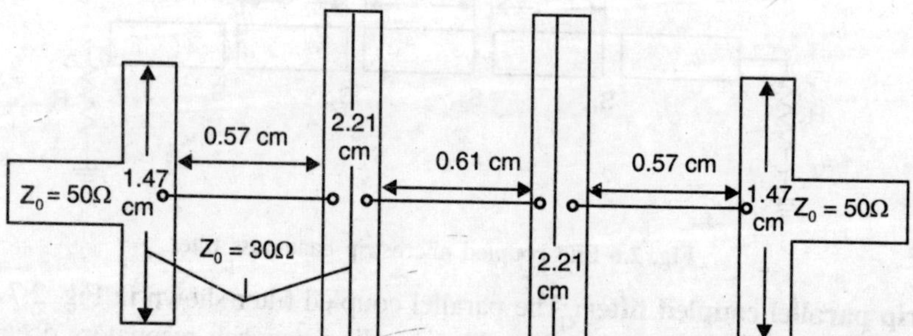

Fig. 2.8 Strip line version filter of a 2 GHz low pass filter

Example 2.1. A rectangular waveguide has $a = 4$ cm, $b = 2$ cm. Determine the minimum frequency which can be transmitted if a dielectric of $\varepsilon = 4$ be inserted inside the waveguide.

Solution:

$$\lambda_{c(\max)} = 2a = 8 \text{ cm} = 0.08 \text{ m}$$

$$\therefore \qquad f_{c(\min)} = \frac{3 \times 10^8}{0.08} = 3.75 \text{ GHz}$$

That is minimum frequency which can be passed without dielectric is 3.75 GHz. When we insert a dielectric.

$$\lambda_{(\text{dielectric})} = \frac{\lambda_{(\text{free space})}}{\sqrt{\varepsilon}} = \frac{0.08}{\sqrt{4}}$$

$$= 0.04 \text{ m}$$

and

$$f_c = \frac{3 \times 10^8}{0.04} = 7.5 \text{ GHz}$$

That is the minimum frequency that can be passed with dielectric ($\varepsilon = 4$) in the waveguide is therefore 7.5 GHz.

Example 2.2. Show that a 1 GHz signal cannot propagate in the TE_{10} mode in a rectangular waveguide having a = 50 mm.

Solution: The cut-off wavelength λ_c for TE_{mm} mode is given by

$$\lambda_c = \frac{2}{\sqrt{(m/a)^2 + (n/b)^2}}$$

$$= \frac{2a}{m} ... n = 0 \text{ for } TE_{10} \text{ mode}$$

$$= \frac{2 \times 50}{1} = 100 \text{ mm} = 10 \text{ cm}$$

Free space wavelength $\lambda_0 = \dfrac{3 \times 10^{10}}{1 \times 10^9} = 30$ cm

As $\lambda_0 > \lambda_c$ for TE_{10} mode, therefore this mode cannot propagate in the given waveguide.

Example 2.3. What is the width (a) of a rectangular waveguide whose characteristic impedance Z_0 is 408 ohms and which is required to propagate a 9.6 GHz signal in the TE_{10} mode?

Solution:

$$Z_0 = \frac{k_0 \eta}{\beta}$$

$$\beta = \frac{377}{408} k_0 \; ; \; k_0 = \frac{\omega}{c} = 2.01 \text{ cm}^{-1}$$

Also

$$\beta^2 = k_0^2 - \left(\frac{\pi}{a}\right)^2$$

$$\left(\frac{\pi}{a}\right)^2 = k_0^2 \left[1 - \left(\frac{377}{408}\right)^2\right] = 0.1462\, k_0^2$$

$$a^2 = \frac{\pi^2}{0.1462\, k_0^2} = 16.728$$

$$a = 4.09 \text{ cm.}$$

Example 2.4. A rectangular waveguide of cross-section 5 cm × 2 cm is used to propagate TM_{11} mode at 9 GHz. Determine the cut-off wavelength and wave impedance.

Solution: For TM_{11} mode, $m = 1$, $n = 1$ at $f = 9 \times 10^9$ Hz, $\lambda = 3.33$ cm

$$\lambda_c = \frac{2}{\sqrt{(m/a)^2 + (n/b)^2}} = \frac{2ab}{\sqrt{a^2 + b^2}} \quad \text{(from Eqn. 2.10}a\text{)}$$

$$= \frac{2 \times 5 \times 2}{\sqrt{5^2 + 2^2}} = 3.714 \text{ cm}$$

$$Z_0 = 120\pi\sqrt{1 - (\lambda/\lambda_c)^2} = 120\pi\sqrt{1 - \left(\frac{3.333}{3.714}\right)^2} = 325\ \Omega.$$

Example 2.5. A waveguide operating in TE_{10} mode has dimensions a = 2.26 cm and b = 1 cm. The measured guide wavelength is 4 cm. Find (*a*) the cut-off frequency of the propagating mode (*b*) the frequency of operation, (*c*) maximum frequency of propagation in this guide.

Solution:

(*a*) For TE_{10} mode, cut-off frequency f_c is

$$f_c = \frac{1}{2a\sqrt{\omega\varepsilon}} = \frac{c}{2a} = \frac{3 \times 10^{10}}{2 \times 2.26} = 6.64 \text{ GHz}$$

(*b*)

$$\beta = \frac{2\pi}{\lambda_g} = \frac{2\pi}{4}$$

$$\beta_{10} = \sqrt{\omega^2\mu\varepsilon - \left(\frac{\pi}{a}\right)^2}$$

$$\beta_{10}^2 + \left(\frac{\pi}{a}\right)^2 = \left(\frac{2\pi f}{c}\right)^2$$

$$f = \frac{c^2}{4\pi^2}\left[\beta^2 + \left(\frac{\pi}{a}\right)^2\right]$$

$$= 10 \text{ GHz}$$

(*c*) First higher order mode is TE_{20} mode

Here, $m = 2, n = 0$

$$f_{20} = \frac{c}{2\pi}\sqrt{\left(\frac{2\pi}{a}\right)^2} = \frac{c}{a} = 13.27 \text{ GHz}.$$

Example 2.6. A rectangular waveguide with dimensions a = 2.5 cm, b = 1 cm is to operate below 15.2 GHz. How many TE and TM modes can be propagated in the waveguide if the guide has the medium parameters given by $\sigma = 0$, $\varepsilon = 4\varepsilon_0$, $\mu_r = \mu_0$? (*a*) Calculate the cut-off frequencies of the modes. (*b*) Also calculate the phase constant, phase velocity and the wave impedance for TE_{10} mode operating at 15 GHz.

Solution:

(*a*) The cut-off frequency is given by

$$f_{c_{mn}} = \frac{c'}{2}\sqrt{\frac{m^2}{a^2} + \frac{n^2}{b^2}}$$

Given $a = 2.5$ cm, $b = 1$ cm, $\varepsilon_r = 4$

$$c' = \frac{1}{\sqrt{\mu\varepsilon}} = \frac{c}{\sqrt{\mu_r \varepsilon_r}} = \frac{c}{2}$$

Hence,

$$f_{c_{mn}} = \frac{c}{4a}\sqrt{m^2 + \frac{a^2}{b^2}n^2}$$

$$= \frac{3\times 10^8}{4(2.5\times 10^{-2})}\sqrt{m^2 + 6.25\,n^2}$$

$$= 3\sqrt{m^2 + 6.25\,n^2}\ \text{GHz}$$

To find all $f'_{c_{mn}} < 15$ GHz, we fix $m = 0$ and vary n, we obtain

for TE_{01} mode ($m = 0, n = 1$) $f_{c_{01}} = 7.5$ GHz

TE_{02} mode ($m = 0, n = 2$) $f_{c_{02}} = 15$ GHz

$f_{c_{mn}} < 15.2$ GHz, the maximum $n = 2$

Now fix n and increase m till $f_{c_{mn}} < 15.2$ GHz

For TE_{10} mode ($m = 1, n = 0$) $f_{c_{10}} = 3$ GHz

TE_{20} mode, $f_{c_{20}} = 6$ GHz

TE_{30} mode, $f_{c_{30}} = 9$ GHz

TE_{40} mode, $f_{c_{40}} = 12$ GHz

TE_{50} mode, $f_{c_{50}} = 15$ GHz

TE_{60} mode, $f_{c_{60}} = 18$ GHz

That is for $f_{c_{mn}} < 15.2$ GHz, the maximum $m = 5$. We have found the maximum m and n, other possible combinations in between these values of m and n are

For TE_{11}, TM_{11} (degenerate modes), $f_{c_{11}} = 8.078$ GHz

TE_{21}, TM_{21} $f_{c_{21}} = 9.6$ GHz

TE_{31}, TM_{31} $f_{c_{31}} = 11.72$ GHz

TE_{41}, TM_{41} $f_{c_{41}} = 14.14$ GHz

TE_{12}, TM_{12} $f_{c_{12}} = 15.3$ GHz

Hence, the modes for which the cut-off frequency is less than 15.2 GHz are eleven TE modes and four TM modes.

(*b*) For TE_{10} mode

$$\beta = \omega\sqrt{\mu\,\varepsilon}\sqrt{1-\left(\frac{f_c}{f}\right)^2}$$

$$= \frac{2\pi \times 5\times 10^9 \times 2}{3\times 10^8}\sqrt{1-\left(\frac{3}{15}\right)^2}$$

$$= 200\pi\sqrt{0.96} = 615.6 \text{ rad/m}$$

$$u_p = \frac{\omega}{\beta} = \frac{2\pi\times 15\times 10^9}{615.6} = 1.531\times 10^8 \text{ m/s}$$

$$\eta_{TE} = \sqrt{\frac{\mu}{\varepsilon}}\sqrt{1-\left(\frac{f_c}{f}\right)^2}$$

$$= 60\pi\sqrt{0.96} = 192.4\ \Omega.$$

Example 2.7. A rectangular waveguide has a = 3.0 cm, b = 1.5 cm, μ = 1 and ε = 2.25. Calculate (*a*) the cut-off wavelength and frequency for TE_{10}, TE_{20} and TM_{11} modes, (*b*) λ_g and Z_0 at 4.0 GHz.

Solution:

(*a*)

$$\lambda_c = \frac{1}{\sqrt{\left(\frac{m}{2a}\right)^2+\left(\frac{n}{2b}\right)^2}}$$

for

TE_{10}, $m = 1$, $n = 0$, $\lambda_c = 2a$

TE_{20}, $m = 2$, $n = 0$, $\lambda_c = a$

TE_{11}, $m = 1$, $n = 1$, $\lambda_c = \dfrac{2ab}{\sqrt{a^2+b^2}}$

Thus the cut-off wavelengths:

for

TE_{10} = 6 cm

TE_{20} = 3 cm

TM_{11} = 2.68 cm

The cut-off frequencies are functions of the dielectric properties of the insulating material as well as the guide dimensions.

$$f_c = \frac{c}{\lambda_c\sqrt{\varepsilon\mu}}$$

substitute for $\varepsilon = 2.25$, $\mu = 1$ and different value of λ_c. Then

$$f_{c_{10}} = 3.33 \text{ GHz}$$

$$f_{c_{20}} = 6.66 \text{ GHz}$$

$$f_{c_{11}} = 7.46 \text{ GHz}$$

(*b*) At 4.0 GHz, $\lambda_0 = 7.5$ cm

$$\lambda_g = \frac{\lambda_0}{\sqrt{\mu\varepsilon - (\lambda_0/\lambda_c)^2}}$$

In this case $\varepsilon = 2.25$, $\mu = 1$, hence

$$\lambda_g = \frac{7.5}{\sqrt{2.25 - (\lambda_0/\lambda_c)^2}}$$

$$= 9.05 \text{ cm.}$$

λ_g for TE_{10} mode is 9.05 cm.

The characteristic impedance of the guide is calculated from the relation

$$Z_0 = \frac{377\dfrac{b}{a}\sqrt{\dfrac{\mu}{\varepsilon}}}{\sqrt{1-\left(\dfrac{f_c}{f_0}\right)^2}} = 377\frac{b}{a}\sqrt{\frac{\mu}{\varepsilon}}\frac{\lambda_g}{\lambda}$$

$$= 377 \times \frac{1.5}{3.0}\sqrt{\frac{1}{2.25}} \times \frac{9.05}{7.5/\sqrt{2.25}} = 227 \text{ ohms.}$$

Example 2.8. A rectangular waveguide has a cross-section of 1.5 cm × 0.8 cm, $\sigma = 0$, $\mu = \mu_0$, and $\varepsilon = 4\,\varepsilon_0$. The magnetic field component is given as

$$H_x = 2\sin\left(\frac{\pi x}{a}\right)\cos\left(\frac{3\pi y}{b}\right)\sin\left(\pi \times 10^{11} t - \beta z\right) \text{A/m}$$

Determine

(*a*) The mode of operation.
(*b*) The cut-off frequency.
(*c*) The phase constant β.
(*d*) The propagation constant γ.
(*e*) The intrinsic wave impedance η.

Solution:

(*a*) From the expression for H_x, it is clear that $m = 1$, $n = 3$ that is the guide is operating in either TM_{13} or TE_{13} mode. We choose TE_{13} mode.

(*b*)
$$f_{c_{13}} = \frac{1}{2\sqrt{\mu\varepsilon}}\sqrt{\left(\frac{1}{a}\right)^2 + \left(\frac{3}{b}\right)^2}$$

$$\frac{1}{\sqrt{\mu\varepsilon}} = \frac{c}{\sqrt{\mu_r \varepsilon_r}} = \frac{c}{2}$$

Hence
$$f_{c_{13}} = \frac{c}{4}\sqrt{\left(\frac{1}{1.5\times10^{-2}}\right)^2 + \left(\frac{3}{0.8\times10^{-2}}\right)^2}$$

$$= \frac{3\times10^8}{4}\left(\sqrt{0.444+14.03}\right)\times10^2 = 28.57 \text{ GHz}$$

(*c*) Phase constant $\beta = \omega\sqrt{\mu\varepsilon}\sqrt{1-\left(\frac{f_c}{f}\right)^2}$

Here,
$$\omega = \pi\times10^{11} \text{ or } f = 50 \text{ GHz}$$

or
$$\beta = \frac{\pi\times10^{11}}{3\times10^8}\sqrt{1-\left(\frac{28.57}{50}\right)^2} = 1718.81 \text{ rad/m}$$

(*d*) Propagation constant, $\gamma = j\beta = j1718.81 \text{rad/m}$

$$\eta TE_{13} = \frac{\eta}{\sqrt{1-(f_c/f)^2}} = \frac{377}{\sqrt{\varepsilon_r}}\left[1-\left(\frac{28.57}{50}\right)^2\right]^{1/2}$$

$$= 229.69\ \Omega.$$

Example 2.9. An air-filled rectangular wave with cross-section of 4 cm × 2 cm transports energy in the dominant mode (TE_{10}) at a rate of 2 mW. If the frequency of operation is 10 GHz, determine the peak value of the electric field in the waveguide.

Solution:

The field expressions corresponding to TE_{10} mode are

$$E_x = 0,\ E_y = -\frac{j\omega\mu a c}{\pi}\sin\left(\frac{\pi x}{a}\right)e^{-j\beta z}$$

$$f_c = \frac{c}{2a} = \frac{3\times10^8}{2\left(4\times10^{-2}\right)} = 3.75 \text{ GHz}$$

$$\eta = \frac{\eta_0}{\sqrt{1-\left(\frac{f_c}{f_0}\right)^2}} = 406.7\ \Omega$$

The average power transmitted is

$$P_{\text{ave}} = \int_{y=0}^{b}\int_{x=0}^{a} \frac{|E_y|^2}{2\eta}\,dx\,dy$$

$$= \frac{E_s^2}{2\eta}\int_0^b dy \int_0^a \sin^2\left(\frac{\pi x}{a}\right)dx$$

where $E_0 = \frac{\omega\mu a}{\pi}c$

$$\therefore \qquad P_{\text{ave}} = \frac{E_0^2 ab}{4\eta}$$

Therefore, $$E_0^2 = \frac{4\eta P_{\text{ave}}}{ab} = 4(406.7)\times 2\times\frac{10^{-3}}{8\times 10^{-4}}$$

$$= 4067$$

or $$E_0 = 63.77 \text{ V/m}.$$

Example 2.10. A copper waveguide has dimensions a = 2.286 cm, b = 1.016 cm and is operating at a frequency f = 10 GHz in TE_{10} mode. For a guide of 1 m length, find the attenuation in dB.

Solution:

At 10 GHz, the propagation constant of the TE_{10} mode is

$$\beta = \sqrt{k_0^2 - \left(\frac{\pi}{a}\right)^2}$$

$$k_0^2 = \omega^2\mu\varepsilon = \left(\frac{2\pi f}{c}\right)^2 = 209.44 \text{ m}^{-1}$$

$$a = 0.02286 \text{ m}$$

$$\beta = 158.05 \text{ m}^{-1}$$

For copper walls ($\sigma = 5.8 \times 10^7$ s/m).

Hence $$R_s = \sqrt{\frac{\omega\mu}{2\sigma}} = 0.026\ \Omega$$

So the attenuation constant from Eqn. (1.55), is

$$\alpha_c = \frac{R_s}{a^3 b\beta k_0\eta}\left(2b\pi^2 + a^3k_0^2\right) = 0.0125 \text{ Np/m}$$

$$\alpha_c(\text{dB}) = -20\log e^{-\alpha_c} = 0.11 \text{ dB/m}.$$

Example 2.11. A dielectric field (ε_r = 9) rectangular waveguide, which has a = 7 cm, b = 3.5 cm, operates in the dominant TE_{10} mode. Determine (*i*) Cut-off frequency, (*ii*) phase velocity in the guide at a frequency of 2 GHz and (*iii*) the guide wavelength λ_g at the same frequency.

Solution:

$a = 7$ cm, $b = 3.5$ cm, $\varepsilon_r = 9$, $f = 2$ GHz

(*i*) Cut-off frequency (Eqn. 2.10)

$$f_c = \frac{1}{2\pi\sqrt{\mu\varepsilon}}\sqrt{\left(\frac{m\pi}{a}\right)^2 + \left(\frac{n\pi}{b}\right)^2}$$

$$= \frac{c}{2\sqrt{\varepsilon_r}}\sqrt{\left(\frac{m}{a}\right)^2 + \left(\frac{n}{b}\right)^2}$$

$$= \frac{c}{2\sqrt{\varepsilon_r}}\frac{1}{a} = \frac{3\times10^8}{\left(2\sqrt{9}\times7\times10^{-2}\right)} \qquad (TE_{10} \rightarrow m = 1, n = 0)$$

$$= 0.714 \text{ GHz}$$

(*ii*) $\bar{\beta} = \omega\sqrt{\mu\varepsilon}\sqrt{1-(f_c/f)^2}$ [(Eq. 2.10*b*)]

$$\bar{v}_p = \frac{c}{\sqrt{\varepsilon_r}\sqrt{1-\left(\frac{f_c}{f}\right)^2}} = \frac{3\times10^8}{\sqrt{9}\sqrt{1-\left(\frac{0.714}{2}\right)^2}}$$

$$= 1.07\times10^8 \text{ m/s}$$

(*iii*) Guide wavelength λ_g is

$$\lambda_g = \frac{1.07\times10^8}{2\times10^9} = 0.0535 \text{ m} = 5.35 \text{ cm}.$$

Example 2.12. A rectangular air-filled waveguide has a cross-section of 45 × 90 mm. Find (*a*) cut-off wavelength λ_{0c} for the dominant mode and (*b*) the relative phase velocity in the guide at 1.6 times the cut-off frequency.

Solution:

(*a*) $\lambda_{0c} = 2 \times 90 = 180$ nm.

(*b*) $$v = \frac{c}{\sqrt{1-\left(\frac{\lambda_0}{\lambda_{0c}}\right)^2}} = \frac{c}{\sqrt{1-\left(\frac{f_c}{f}\right)^2}} = \frac{c}{\sqrt{1-\left(\frac{1}{1.6}\right)^2}} = 1.286.$$

Example 2.13. An air-filled rectangular waveguide is dimensions of $a = 6$ cm and $b = 4$ cm. The signal frequency is 3 GHz. Compute the following:

(*a*) Cut-off frequency for the TE_{10}, TE_{01}, TE_{11} and TM_{11} modes.

(*b*) Wavelength in the waveguide for the dominant mode.

(*c*) Phase constant and phase velocity in the waveguide for dominant mode.

(*d*) Group velocity and wave impedance in the waveguide for dominant mode.

Solution: Given a = 6 cm, b = 4 cm and f = 3 GHz

(*a*) Cut-off frequency is

$$f_c = \frac{c}{2}\sqrt{\frac{m^2}{a^2} + \frac{n^2}{b^2}}$$

For TE_{10} mode,

$$f_c = \frac{3\times10^8}{2}\sqrt{\left(\frac{1}{6\times10^{-2}}\right)^2} = 2.5 \text{ GHz}$$

For TE_{01} mode,

$$f_c = \frac{3\times10^8}{2}\sqrt{\left(\frac{1}{4\times10^{-2}}\right)^2} = 3.75 \text{ GHz}$$

For TE_{11} mode,

$$f_c = \frac{3\times10^8}{2}\sqrt{\left(\frac{1}{4\times10^{-2}}\right)^2 + \left(\frac{1}{6\times10^{-2}}\right)^2}$$

$$= 4.5 \text{ GHz}$$

For TM_{11} mode,

$$f_c = \frac{3\times10^8}{2}\sqrt{\left(\frac{1}{4\times10^{-2}}\right)^2 + \left(\frac{1}{6\times10^{-2}}\right)^2}$$

$$= 4.5 \text{ GHz}$$

(*b*) Wavelength in the waveguide for the TE_{10} mode.

$$\lambda_g = \frac{\lambda}{\sqrt{1-\left(\frac{f_c}{f}\right)^2}}$$

$$\lambda = c/f = \frac{3\times10^8}{3\times10^9} = 0.1$$

$$\lambda_g = \frac{0.1}{\sqrt{1-\left(\frac{2.5}{3}\right)^2}} = 0.1786 \text{ m}$$

$$= 17.86 \text{ cm}$$

(*c*) Phase constant,

$$B_g = \omega\sqrt{\mu_0\varepsilon_0}\sqrt{1-\left(\frac{f_c}{f}\right)^2}$$

$$= \frac{2\pi\times3\times10^9}{3\times10^8}\sqrt{1-\left(\frac{2.5}{3}\right)^2}$$

$$= 34.97 \text{ rads/m}$$

Phase velocity,

$$v_g = \frac{v_p}{\sqrt{1-\left(\frac{f_c}{f}\right)^2}}$$

$$= \frac{3\times10^8}{\sqrt{1-\left(\frac{2.5}{3}\right)^2}}$$

$$v_g = 5.36\times10^8 \text{ m/s}$$

(*d*) Group velocity,

$$v_{gr} = \frac{c^2}{v_g} = \frac{(3\times10^8)^2}{5.36\times10^8} = 1.68\times10^8 \text{ m/s}$$

Wave impedance,

$$Z_g = \frac{\omega\mu}{\beta_g} = \frac{\eta}{\sqrt{1-\left(\frac{f_c}{f}\right)^2}} = \frac{377}{\sqrt{1-\left(\frac{2.5}{3}\right)^2}}$$

$$= 673.21\ \Omega.$$

Example 2.14. The dominant mode TE_{10} is propagated is a rectangular waveguide of dimensions $a = 6$ cm and $b = 4$ cm. The distance between a maximum and a minimum is 4.47 cm. Determine the signal frequency of the dominant mode.

Solution: Given $a = 6$ cm, $b = 4$ cm and $d = 4.47$ cm

So cut-off frequency

$$f_c = \frac{c}{2}\sqrt{\frac{m^2}{a^2}+\frac{n^2}{b^2}}$$

For dominant mode

$$f_c = \frac{3\times 10^8}{2}\sqrt{\frac{1}{(6\times 10^{-2})^2}}$$
$$= 2.5 \text{ GHz}$$

The distance between a maximum and minimum $d = \frac{\lambda_g}{4}$

So $\lambda_g = 4 \times 4.47$
$= 17.88$ cm

or the wavelength in the waveguide

$$\lambda_g = \frac{\lambda}{\sqrt{1-\left(\frac{f_c}{f}\right)^2}}$$

where $\lambda = \frac{c}{f}$

$$17.88\times 10^{-2} = \frac{\frac{3\times 10^8}{f}}{\sqrt{1-\left(\frac{f_c}{f}\right)^2}}$$

After solving $f = 3$ GHz.

Example 2.15. A TE_{11} mode of 10 GHz is propagated in an air-filled rectangular waveguide. The magnetic field in the z-direction is given by

$$H_z = H_0 \cos\left(\frac{\pi x}{\sqrt{6}}\right)\cos\left(\frac{\pi y}{\sqrt{6}}\right) \text{ A/m}$$

The phase constant is $\beta = 1.0475$ rad/cm, the quantities x and y are expressed in centimeters and $a = b = \sqrt{6}$ are also in cm. Determine the cut-off frequency f_c, phase velocity v_g, guided wavelength λ_g, and the magnetic field intensity in the y-direction.

Solution: Cut-off frequency

$$f_c = \frac{c}{2}\sqrt{\frac{m^2}{a^2}+\frac{n^2}{b^2}}$$

$$= \frac{3\times 10^{10}}{2}\sqrt{\frac{1}{6}+\frac{1}{6}} = 8.66 \text{ GHz}$$

The phase velocity v_g is,

$$v_g = \frac{\omega}{\beta_g} = \frac{2\pi\times10\times10^9}{1.0475\times10^2}$$

$$= 6.06 \times 10^8 \text{ m/s}$$

The guided wavelength λ_g is,

$$\lambda_g = \frac{v_g}{f} = \frac{6.06\times10^8}{10^{10}}$$

$$= 6.06 \times 10^{-2} \text{ m}$$

The magnitude field intensity in the y-direction is,

$$H_y = H_0 \frac{\partial H_z}{\partial y} = \frac{\pi}{\sqrt{6}} H_0 \cos\frac{\pi x}{\sqrt{6}} \sin\frac{\pi y}{\sqrt{6}} \text{ A/m.}$$

Example 2.16. A rectangular waveguide is designed to propagate the dominant mode TE_{10} at a frequency of 5 GHz. The cut-off frequency is 0.8 of the signal frequency. The ratio of the guide height to width is 2. The time average power flowing through the guide is 1 kW. Determine the magnitude of electric and magnetic intensities in the guide and indicate where these occur in the guide.

Solution: Given, $f = 5$ GHz, $f_c = 4$ GHz, $a/b = 2.0$

find a and b,

$$f_c = \frac{c}{2}\sqrt{\frac{m^2}{a^2} + \frac{n^2}{b^2}}$$

$$4\times10^9 = \frac{3\times10^8}{2}\frac{1}{a}$$

Then $a = 0.0375$ m and $b = 0.0187$ m

The phase constant is,

$$\beta_g = \omega\sqrt{\mu_0\varepsilon_0}\sqrt{1-\left(\frac{f_c}{f}\right)^2}$$

$$= \frac{2\pi\times5\times10^9}{3\times10^8}\sqrt{1-\left(\frac{4\times10^9}{5\times10^9}\right)^2}$$

$$= 20\pi \text{ rad/m}$$

The impedance is $$Z_g = \frac{\omega\mu}{\beta_g}$$

$$= \frac{2\pi\times5\times10^9\times4\pi\times10^{-7}}{20\pi}$$

$$Z_g = 628.3\ \Omega$$

The field equations for TE_{10} mode are

$$E_x = 0 \qquad H_x = H_{0x} \sin\frac{\pi x}{a} e^{j(\omega t - \beta_g z)}$$

$$E_y = E_{0y} \sin\frac{\pi x}{a} e^{j(\omega t - \beta_g z)} \qquad H_y = 0$$

$$E_z = 0 \qquad H_z = H_{0z} \cos\frac{\pi x}{a} e^{j(\omega t - \beta_g z)}$$

From Eq. (2.16):

$$P_{tr} = \frac{1}{2} Re \int_{x=0}^{a} \int_{y=0}^{b} E \times H \cdot Z \cdot dx\, dy$$

For lossless dielectric, the time-average power flow through a rectangular guide is given by

$$P_{tr} = \frac{1}{4Z_g} |E_{0y}|^2 ab = \frac{Z_g}{4} |H_{0x}|^2 ab$$

$$\therefore \qquad E_{0y} = \sqrt{\frac{4Z_g \cdot P_{tr}}{ab}}$$

$$= \sqrt{\frac{4 \times 628.3 \times 10^3}{3.15 \times 10^{-2} \times 1.87 \times 10^{-2}}} = 59.88 \text{ kV/m} \qquad \left(\text{at } z = 0 \text{ and } x = \frac{a}{2}\right)$$

$$H_{0x} = \sqrt{\frac{4 P_{tr}}{Z_g ab}} = \sqrt{\frac{4 \times 10^3}{628.3 \times 3.15 \times 10^{-2} \times 1.87 \times 10^{-2}}} = 9.53 \text{ A/m} \qquad (\text{at } z = 0 \text{ and } x = a/2)$$

and $\qquad H_{0z} = 9.53$ A/m at $z = 0$ and $x = 0$ or a.

Example 2.17. An air-filled rectangular waveguide has dimensions of $a = 6$ cm and $b = 4$ cm. The guide transports energy in the dominant mode TE_{10} at a rate of 1 horsepower (746 J). If the frequency is 20 GHz, what is the peak value of electric field occurring in the guide?

Solution: Given $f = 20$ GHz, TE_{10} mode, $a = 6$ cm and $b = 4$ cm

Cut-off frequency,

$$f_c = \frac{c}{2}\sqrt{\frac{m^2}{a^2} + \frac{n^2}{b^2}}$$

$$= \frac{3 \times 10^8}{2} \sqrt{\frac{1}{(6 \times 10^{-2})^2}} = 2.5 \text{ GHz}$$

Phase constant,

$$\beta_g = \omega\sqrt{\mu_0 \varepsilon_0}\sqrt{1 - \left(\frac{f_c}{f}\right)^2}$$

$$= \frac{2\pi\times 20\times 10^9}{3\times 10^8} = \sqrt{1-\left(\frac{2.5}{20}\right)^2}$$

$$\beta_g = 414.8 \text{ rad/sec.}$$

Wave impedance,

$$Z_g = \frac{\omega\mu}{\beta_g} = \frac{2\pi\times 20\times 10^9 \times 4\pi\times 10^{-7}}{414.8} = 381\ \Omega$$

Peak value of electric field

$$E_{0y} = \sqrt{\frac{4Z_g P_{tr}}{ab}}$$

$$= \sqrt{\frac{4\times 381\times 746}{4\times 10^{-2}\times 6\times 10^{-2}}}$$

$$E_{0y} = 53.31 \text{ kV/m.}$$

Example 2.18. A rectangular waveguide is filled by dielectric material of $\varepsilon_r = 9$ and has inside dimensions of 7 × 3.5 cm. It operates in the dominant TE_{10} mode.

(*a*) Determine the cut-off frequency.

(*b*) Find the phase velocity in the guide at a frequency of 2 GHz.

(*c*) Find the guided wavelength λ_g at the same frequency.

Solution: Given TE_{10} mode at f = 2 GHz, a = 7 cm, b = 3.5 cm and ε = 9

(*a*) Cut-off frequency is

$$f_c = \frac{1}{2\sqrt{\mu\varepsilon}}\sqrt{\frac{m^2}{a^2}+\frac{n^2}{b^2}}$$

$$= \frac{c}{2\sqrt{\varepsilon_r}}\cdot\frac{1}{a} = \frac{3\times 10^8}{2\sqrt{9}\cdot 7\times 10^{-2}}$$

$$= 0.714 \text{ GHz}$$

(*b*) Phase constant,

$$\beta_g = \omega\sqrt{\mu\varepsilon}\sqrt{1-\left(\frac{f_c}{f}\right)^2}$$

and phase velocity

$$v_p = \frac{\omega}{\beta_g} = \frac{c}{\sqrt{\varepsilon_r}\sqrt{1-\left(\frac{f_c}{f}\right)^2}}$$

$$= \frac{3 \times 10^8}{\sqrt{9}\sqrt{1-\left(\frac{0.714}{2}\right)^2}}$$

$$= 1.07 \times 10^8 \text{ m/s}$$

(*c*) The guided wavelength λ_g is

$$\lambda_g = \frac{v_p}{f} = \frac{1.07 \times 10^8}{2 \times 10^9} = 0.0535 \text{ m}$$

$$= 5.35 \text{ cm.}$$

CHAPTER 3

Microwave Transmission Lines II: Circular Waveguides and Microstrip Lines

3.1 CIRCULAR WAVEGUIDE

A hollow metal tube of circular cross-section also supports TE and TM waveguide modes. Figure 3.1 shows a circular waveguide of inner radius *a*. We employ cylindrical coordinates when dealing with cylindrical waveguide with circular cross-section. The wave equation for this geometry is written as

$$\frac{\partial^2 \psi}{\partial \rho^2} + \frac{1}{\rho}\frac{\partial \psi}{\partial \rho} + \frac{1}{\rho^2}\frac{\partial^2 \psi}{\partial \phi^2} + \frac{\partial^2 \psi}{\partial z^2} + \omega^2 \mu \varepsilon \psi = 0$$

For TM modes ψ stands for E_z and for TE modes as H_z.

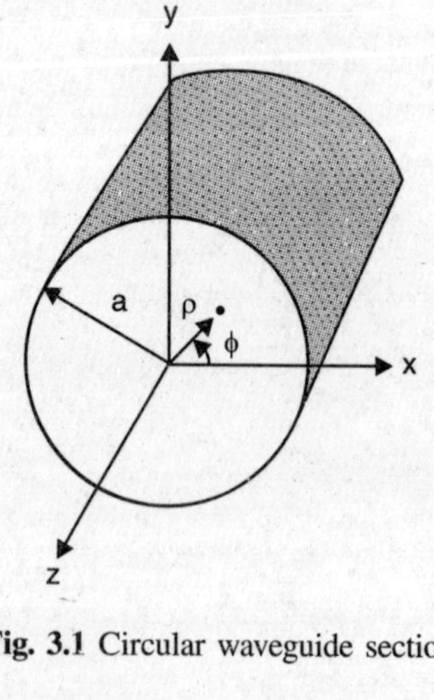

Fig. 3.1 Circular waveguide section

3.1.1 TM Wave Equations in Cylindrical Coordinates

In this case $H_z = 0$, the wave equation becomes

$$\frac{\partial^2 E_z}{\partial \rho^2}+\frac{1}{\rho^2}\frac{\partial^2 E_z}{\partial \phi^2}+\frac{\partial^2 E_z}{\partial z^2}+\frac{1}{\rho}\frac{\partial E_z}{\partial \phi}+\omega^2\mu\varepsilon E_z=0 \qquad ...(3.1)$$

A solution can be derived using the method of separation of variables as used for rectangular waveguides.

Hence $$E_z=P(\rho)Q(\phi)e^{-\bar{\gamma}z}=E_z^0 e^{-\bar{\gamma}z} \qquad ...(3.2)$$

We obtain

$$Q\frac{d^2P}{d\rho^2}+\frac{1}{\rho P}\frac{dP}{d\rho}+\frac{1}{Q\rho^2}\frac{d^2Q}{d\phi^2}+h^2=0 \qquad ...(3.3)$$

Equation (3.3) can be broken up into two differential equations.

$$\frac{d^2Q}{d\phi^2}=-n^2Q \qquad ...(3.4)$$

$$\frac{d^2P}{d(\rho h)}+\frac{1}{(\rho h)}\frac{dP}{d(\rho h)}+\left[1-\frac{n^2}{(\rho h)^2}\right]P=0 \qquad ...(3.5)$$

Equation (3.4) has a solution

$$Q=(A\sin n\phi+B\cos n\phi) \qquad ...(3.6)$$

Equation (3.6) is a standard form of Bessel's equation in terms of (ρh). The second order equation as usual will have two solutions $J_n(\rho h)$ and $N_h(\rho h)$, called Bessel functions of first and second kind respectively. For the present problem solution that is finite at $(\rho h) = 0$ is only acceptable, since $N_h(\rho h)$ becomes infinite at $\rho = 0$.

Hence, $$P(\rho h)=J_n(\rho h) \qquad ...(3.7)$$

where $J_n(\rho h)$ is Bessel's function of the first kind of order n. The value of E_z is obtained on substituting Eqns. (3.6) and (3.7) in Eqn. (3.2) as

$$E_z=J_n(\rho h)(A\sin n\phi+B\cos n\phi)e^{-\bar{\gamma}z} \qquad ...(3.8)$$

Similar solution will be obtained for H_z (TE) mode.

$$H_z=J_n(\rho h)(C\cos n\phi+D\sin n\phi)e^{-\bar{\gamma}z} \qquad ...(3.9)$$

3.1.2 TM Modes in Circular Waveguides

$$E_z(\rho,\phi)=(A\sin n\phi+B\cos n\phi)J_n(h\rho)$$

For the TM mode is that the boundary conditions can now be applied directly to E_z, since

$$E_z(\rho,\phi)=0 \text{ at } \rho=a \qquad ...(3.10a)$$

Thus, we must have

$$J_n(ha)=0 \qquad ...(3.10b)$$

or $$h_{nm} = \frac{P_{nm}}{a} \quad \text{...(3.11)}$$

where P_{nm} is the mth root of $J_n(x)$, that is $J_n(P_{nm}) = 0$. Table 3.1 gives first few values of P_{nm}.

Table 3.1 Values of P_{nm} for TM modes of a circular guide

n	P_{n1}	P_{n2}	P_{n3}
0	2.405	5.520	8.654
1	3.832	7.016	10.174
2	5.135	8.417	11.620

The propagation constant of TM_{nm} mode is

$$\bar{\beta}_{nm} = \sqrt{k_0^2 - h_{nm}^2} = \sqrt{k_0^2 - (P_{nm}/a)^2} \quad \text{...(3.12}a\text{)}$$

The cut-off frequency is

$$\bar{f}_{c_{nm}} = \frac{h_{nm}}{2\pi\sqrt{\mu\varepsilon}} = \frac{P_{nm}}{2\pi a\sqrt{\mu\varepsilon}} \quad \text{...(3.12}b\text{)}$$

The first TM mode to propagate is TM_{01} with $P_{01} = 2.405$.

This is greater than $P'_{11} = 1.81$. Hence TE_{11} is the dominant mode of circular waveguide. The transverse field components for TM_{01} are given as

$$E_\rho = \frac{-j\bar{\beta}}{h_{nm}} A \sin n\phi \, J_1'(h\rho) e^{-j\bar{\beta}z} \quad \text{...(3.13}a\text{)}$$

$$E_\phi = \frac{-j\bar{\beta}n}{h_{nm}^2 \rho} A \cos n\phi \, J_1(h\rho) e^{-j\bar{\beta}z} \quad \text{...(3.13}b\text{)}$$

$$H_\rho = \frac{j\omega\varepsilon n}{h_{nm}^2 \rho} A \cos n\phi \, J_1(h\rho) e^{-j\bar{\beta}z} \quad \text{...(3.13}c\text{)}$$

$$H_\phi = \frac{-j\omega\varepsilon}{h_{nm}} A \sin n\phi \, J_1(h\rho) e^{-j\bar{\beta}z} \quad \text{...(3.13}d\text{)}$$

3.1.3 TE Modes in Circular Waveguides

The boundary conditions to be met for TE waves are that $E_\phi = 0$ at $\rho = a$. But E_ϕ is proportional to $\frac{\partial H_z}{\partial \rho}$, and therefore to $J'_n(h\rho)$, where the prime denotes the derivative with respect to $(h\rho)$. Hence, for TE modes the boundary conditions required are

$$\frac{\partial J_n(ha)}{\partial a} = 0 = J'_n(ha) = 0 \quad \text{...(3.14)}$$

If the roots of $J_n'(x)$ are defined as P'_{nm}, so that $J_n'(P'_{nm}) = 0$, where $P'_{nm} = 0$ is the mth root of J_n', then h_{nm} must have the value

$$h_{nm} = \frac{P'_{nm}}{a} \qquad \text{...(3.15)}$$

First few values of P'_{nm} are given in Table 3.2.

Table 3.2 Values of P'_{nm} for TE modes of a circular waveguide

n	P'_{n1}	P'_{n2}	P'_{n3}
0	3.832	7.016	10.174
1	1.841	5.331	8.536
2	3.054	6.706	9.970

The TE_{nm} modes are thus defined by the cut-off wave number $h_{nm} = P'_{nm}/a$ where n refers to number of full wave variation in the circumferential (Φ) direction and the second number (m) refers to the Bessel function variation in the radial direction. The propagation constant of the TE_{mn} mode is

$$\bar{\beta}_{nm} = \sqrt{\omega^2\mu\varepsilon - h^2} = \sqrt{k_0^2 - h_{nm}^2}$$

$$= \sqrt{k_0^2 - \left(\frac{P'_{nm}}{a}\right)^2}$$

The cut-off frequency is

$$f_{c_{nm}} = \frac{h_{nm}}{2\pi\sqrt{\mu\varepsilon}} = \frac{P'_{nm}}{2\pi a\sqrt{\mu\varepsilon}} \qquad \text{...(3.16)}$$

The first mode (dominant mode) to propagate is the mode with the smallest P'_{nm} which from the table is TE_{11} mode. There is no TE_{10} mode in the circular waveguide but there exists a TE_{01} mode.

The transverse field components for TE_{11} mode are

$$H_z = D\sin\phi J_1(h\rho)e^{-j\bar{\beta}z} \qquad \text{...(3.17}a\text{)}$$

$$E_\rho = \frac{-j\omega\mu}{h_{nm}^2\rho} D\cos\phi J_1(h\rho)e^{-j\bar{\beta}z} \qquad \text{...(3.17}b\text{)}$$

$$E_\phi = \frac{j\omega\mu}{h_{nm}} D\sin\phi J_1'(h\rho)e^{-j\bar{\beta}z} \qquad \text{...(3.17}c\text{)}$$

$$H_\rho = -\frac{j\beta}{h_{nm}} D\sin\phi J_1'(h\rho)e^{-j\bar{\beta}z} \qquad \text{...(3.17}d\text{)}$$

$$H_\phi = \frac{-j\beta}{h_{mn}^2\rho} D\cos\phi J_1(h\rho)e^{-j\bar{\beta}z} \qquad \text{...(3.17}e\text{)}$$

$$H_\rho = \frac{j\omega\varepsilon n}{h_{mn}^2 \rho} A \cos n\phi \, J_1(h\rho) e^{-j\bar{\beta}z} \qquad ...(3.17f)$$

$$E_z = 0 \qquad ...(3.17g)$$

Here we have assumed that the coordinate system has been rotated about z axis to give H_z a sin ϕ variation hence, cos ϕ variation is equated to zero.

The wave impedance is

$$Z_{TE} = \frac{H_\rho}{H_\phi} = \frac{-E_\phi}{H_\rho} = \frac{\eta k_0}{\beta} = \frac{\eta}{\sqrt{1-(f_c/f)^2}} \qquad ...(3.18a)$$

Field patterns for some low order modes in a circular waveguide are shown in Fig. 3.2*a*. The wavelength for TE modes in a circular guide is given by (guide wavelength)

$$\lambda_g = \frac{\lambda}{\sqrt{1-(f_c/f)^2}} \qquad ...(3.18b)$$

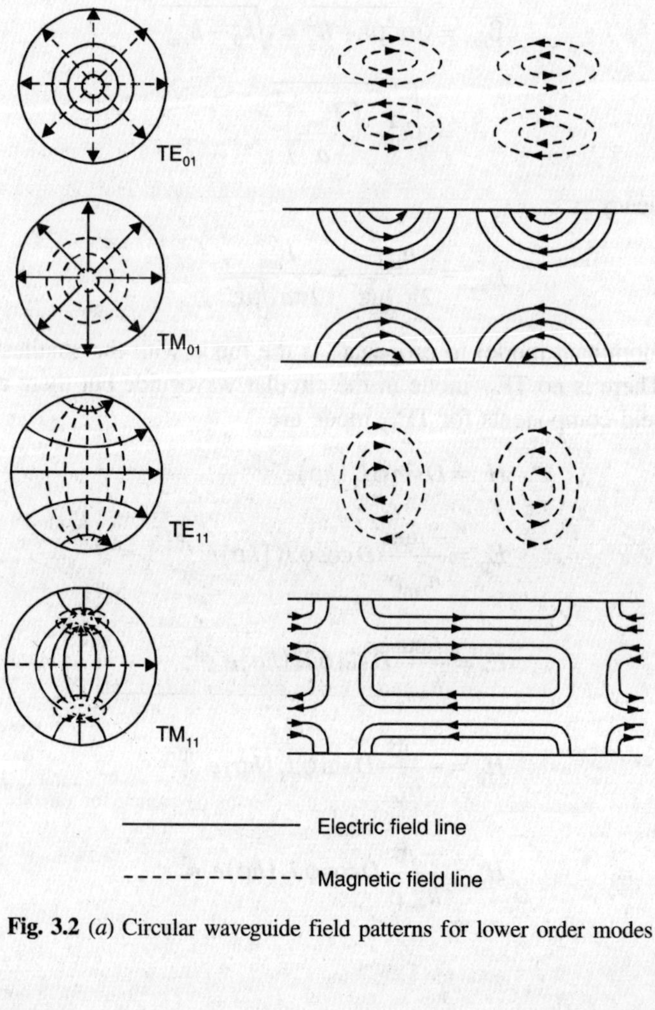

Fig. 3.2 (*a*) Circular waveguide field patterns for lower order modes

3.2 ATTENUATION IN CIRCULAR GUIDES

For TE modes the power flow down the guide can be computed as follows:

$$P_0 = \frac{1}{2}\text{Re}\int_{\rho=0}^{a}\int_{\theta=0}^{2\pi} E\times H^*\cdot\hat{z}\rho\, d\phi\, d\rho$$

$$= \frac{1}{2}\text{Re}\int_{\rho=0}^{a}\int_{\phi=0}^{2\pi}\left[E\rho H_\phi^* - E_\phi H_\rho^*\right]\rho d\phi d\rho$$

On substituting the values of required field components, we obtain

$$P_0 = \frac{\pi\omega\mu|D|^2\, Re(\bar{\beta})}{4h^4}\left[\left(P'_{11}\right)^2 - 1\right]J_1^2(ha) \quad \text{...(3.19}a\text{)}$$

Attenuation due to lossy conductor is given by

$$P_l = \frac{R_s}{2}\int_{\phi=0}^{2\pi}\left|J_s^2\right| a\, d\phi$$

$$= \frac{R_s}{2}\int_{\phi=0}^{2\pi}\left[H_\phi^2 + H_\rho^2\right] a\, d\phi$$

On substituting for H_ϕ and H_P and carrying out the integration P_l is found to be

$$P_l = \frac{\pi\left|D^2\right| R_s a}{2}\left(1 + \frac{\bar{\beta}^2}{h_{nm}^4 a^2}\right)J_1^2(ha) \quad \text{...(3.19}b\text{)}$$

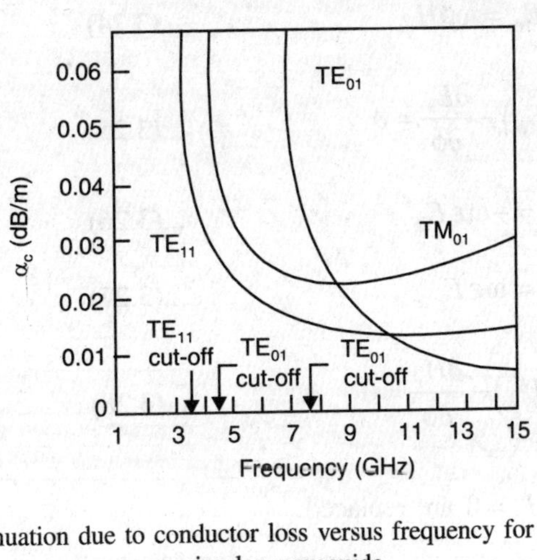

Fig. 3.2 (*b*) Attenuation due to conductor loss versus frequency for various modes of a circular waveguide

Substitute for P_l and P_0 to obtain α.

$$\alpha_c = P_l/2P_0$$

$$\alpha_c = \frac{R_s}{a\omega\mu\beta}\left(h_{nm}^2 + \frac{\omega^2\mu\varepsilon}{(P'_{11})_{11}^2 - 1}\right) \text{ Np/m} \quad ...(3.20)$$

Attenuation for TM waves in circular waveguide can be obtained similarly. Figure 3.2*b* shows the attenuation due to conductor loss versus frequency for various modes of a circular waveguide. From the figure it is observed that the attenuation of the TE_{01} mode decreases to a very small value with increasing frequency. Because of this property of TE_{01} mode it is of interest for low loss transmission over long distance. Note this mode is not the dominant mode of the circular waveguide, hence when used for signal propagation power can be lost from this mode (TE_{01}) to the lower order mode (TE_{11} or TM_{01}) of propagation.

3.3 IMPOSSIBILITY OF TEM MODE

The transverse electric and transverse magnetic (TEM) mode or transmission line modes are characterized by

$$E_z = H_z = 0$$

This means that the electric and magnetic fields are completely transverse to the direction of wave propagation. This mode cannot exist in hollow waveguides, since it requires two conductors, such as the coaxial transmission line and two-open-wire line. Figure 3.3 shows a coaxial line.

Maxwell's equation in cylindrical coordinates

$$\nabla \times E = -j\omega\mu H \quad ...(3.21)$$

$$\nabla \times H = j\omega\varepsilon E \quad ...(3.22)$$

Become

$$\beta_g E_r = \omega\mu H_\phi \quad ...(3.23)$$

$$\beta_g E_\phi = \omega\mu H_r \quad ...(3.24)$$

$$\frac{\partial}{\partial r}(rE_\phi) - \frac{\partial E_r}{\partial \phi} = 0 \quad ...(3.25)$$

$$\beta_g H_r = -\omega\varepsilon E_\phi \quad ...(3.26)$$

$$\beta_g H_\phi = \omega\varepsilon E_r \quad ...(3.27)$$

$$\frac{\partial}{\partial r}(rH_\phi) - \frac{\partial H_r}{\partial \phi} = 0 \quad ...(3.28)$$

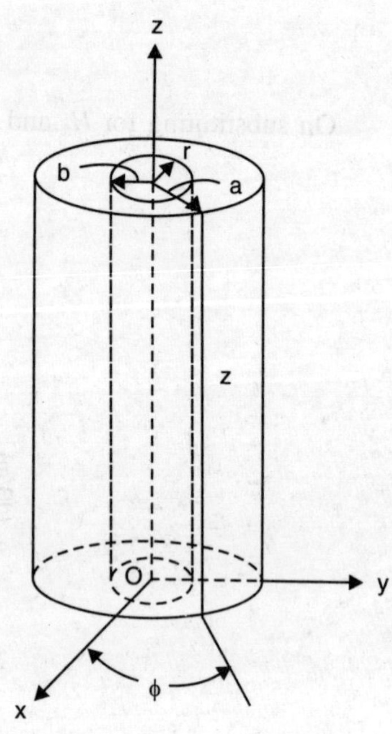

Fig. 3.3 Coordinates of *a*

where $\frac{\partial}{\partial r} = -j\beta_g$ and $E_z = H_z = 0$ are replaced.

Substitution of Eqn. (3.24) in (3.26), the propagation constant of the TEM mode in a coaxial line:

$$\beta_g = \omega\sqrt{\mu\varepsilon} \quad ...(3.29)$$

which is the phase constant of the wave in a lossless transmission line with a dielectric. In comparing the preceding equation with characteristic equation of the cylindrical coordinates, it is found that the cutoff frequency of the TEM mode in a coaxial line is zero, which is the same as in ordinary transmission lines.

The phase velocity of the TEM mode can be expressed from Eqn. (3.29) as

$$v_p = \frac{\omega}{\beta_g} = \frac{1}{\sqrt{\mu\varepsilon}} \qquad ...(3.30)$$

which is the velocity of light in an unbounded dielectric.

The wave impedance of the TEM mode is

$$\eta(\text{TEM}) = \sqrt{\frac{\mu}{\varepsilon}} \qquad ...(3.31)$$

which is the wave impedance of a lossless transmission line in a dielectric.

The properties of TEM modes in a lossless medium are as follows:

1. The cutoff frequency is zero.
2. Its transmission line is a two-conductor system.
3. Its wave impedance is the impedance in an unbounded dielectric.
4. Its propagation constant is the constant in an unbounded dielectric.
5. Its phase velocity is the velocity of light in an unbounded dielectric.

Example 3.1. A TE_{11} wave is propagated through a air filled circular waveguide of diameter 10 cm. Determine

(i) the cut-off frequency.

(ii) guide wavelength λ_g for a frequency of 3 GHz.

(iii) wave impedance of the guide.

Solution:

(i) TE_{11} mode and a = 5 cm

From Eqn. (1.75),

$$f_{c_{nm}} = \frac{P'_{nm}}{2\pi a\sqrt{\mu\varepsilon}}, \; n=1, \; m=1$$

The value of $P'_{11} = 1.841$, (Table 3.2)

$$f_{c11} = \frac{1.841\times 3\times 10^8}{6.28\times 5\times 10^{-2}} = 1.76 \text{ GHz}$$

(ii) f = 3 GHz, *i.e.*, λ_0 = 10 cm

$$\lambda_g = \frac{\lambda_0}{\sqrt{1-(f_c/f)^2}} = \frac{10}{\sqrt{1-\left(\frac{1.76}{3}\right)^2}}$$

$$= \frac{10}{0.81} = 12.35 \text{ cm}$$

(*iii*) Wave impedance
$$Z_{TE} = \frac{\eta}{\sqrt{\left(1-(f_c/f)^2\right)}} = \frac{377}{0.81} = 465.43\ \Omega.$$

Example 3.2. An air-filled circular waveguide has a radius of 2 cm and is to carry energy at frequency of 10 GHz. Determine all TE_{np} modes for which transmission is possible.

Solution:

$$a = 2 \text{ cm and } f = 10 \text{ GHz}$$

$h_{nm}a$ must be constant
$$h_{nm} = \frac{P'_{nm}}{a} = \omega_c\sqrt{\mu\varepsilon}$$

$$h_{nm}a = \omega_c\sqrt{\mu\varepsilon}\, a$$

$$= \frac{2\pi\times10^{10}\times2\times10^{2}}{3\times10^{8}} = 4.19$$

For all impossible modes from Tables 3.1 and 3.2.

Table 3.1	TE_{11} (1.841) TE_{21} (3.054) TE_{01} (3.832)	TM_{01} (2.45) TM_{11} (3.832)	Table 3.2

Example 3.3. A circular waveguide has a cut-off frequency of 9 GHz in dominant mode. (*i*) Calculate the inside radius of the guide if it is air-filled and (*ii*) calculate the inside radius of the guide if it is dielectric filled ($\varepsilon = 4$).

Solution:

For TE_{11}, which is dominant mode,
$P'_{11} = 1.841$ (Table 3.2)

(*i*)
$$a = \frac{P'_{11}}{2\pi f_c\sqrt{\mu\varepsilon}} = \frac{1.841\times3\times10^{10}}{6.28\times9\times10^{9}} = 0.98 \text{ cm}$$

(*ii*)
$$a_\varepsilon = \frac{a_{air}}{\sqrt{\varepsilon_r}} = \frac{0.98}{\sqrt{4}} = 0.49 \text{ cm}$$

i.e., the radius of the dielectric loaded guide is 0.49 cm.

Example 3.4. An air-filled circular waveguide is to be operated at a frequency of 6 GHz and is to have dimensions such that $f_c = 0.8\, f$ for the dominant mode. Determine

(*a*) the diameter of the guide.

(*b*) the wavelength λ_g and the phase velocity v_g in the guide.

Solution: Given dominant mode TE_{11} at $f = 6$ Hz, $f_c = 0.8 f = 4.8$ GHz

(*a*) From Eqn. (3.18),

$$f_{c_{nm}} = \frac{P'_{nm}}{2\pi a\sqrt{\mu\varepsilon}}$$

$n = 1, m = 1, P'_{11} = 1.841$ (From Table 3.2), $f_c = 4.8 \times 10^9$

$$\therefore \qquad a = \frac{1.841 \times 3 \times 10^8}{2\pi \times 4.8 \times 10^9}$$

$$a = 1.83 \text{ cm}$$

The diameter is 3.66 cm.

(*b*) The phase velocity v_g is

$$v_g = \frac{v_p}{\sqrt{1-\left(\frac{f_c}{f}\right)^2}} = \frac{3 \times 10^8}{\sqrt{1-\left(\frac{4.8}{6}\right)^2}} = 5 \times 10^8 \text{ m/s}$$

The wavelength is,

$$\lambda_g = \frac{\lambda}{\sqrt{1-\left(\frac{f_c}{f}\right)^2}} = \frac{\frac{3 \times 10^8}{6 \times 10^9}}{\sqrt{1-\left(\frac{4.8}{6}\right)^2}}$$

$$= 8.33 \text{ cm.}$$

Example 3.5. **An air-filled circular waveguide of 2 cm inside radius is operated in the TE_{01} mode.**

(*a*) Compute the cut-off frequency.

(*b*) If the guide is to be filled with a dielectric material of $\varepsilon_r = 2.25$, to what value must its radius be changed in order to maintain the cut-off frequency at its original value?

Solution: Given TE_{01} mode at $a = 2$ cm and $P'_{nm} = 3.832$ (From Table 3.2).

(*a*) The cut-off frequency is

$$f_c = \frac{P'_{nm}}{2\pi a\sqrt{\mu_0\varepsilon_0}} = \frac{3.832 \times 3 \times 10^8}{2\pi \times 2 \times 10^{-2}}$$

$$= 9.15 \text{ GHz.}$$

(*b*) $\varepsilon_r = 2.25$

$$a_\varepsilon = \frac{a}{\sqrt{\varepsilon_r}} = \frac{2}{\sqrt{2.25}} = \frac{2}{1.5}$$

$$= 1.33 \text{ cm}$$

The new radius is 1.33 cm with dielectric loading.

Example 3.6. An air-filled circular waveguide has a radius 1.5 cm and is to carry energy at a frequency of 10 GHz. Find all TE and TM modes for which transmission is possible.

Solution: Given f = 10 GHz and a = 1.5 cm.

For all the TE and TM modes the value of a h_{nm} must be constant

$$h_{nm} \leq \omega_c \sqrt{\mu_0 \varepsilon_0}\, a \leq \frac{2\pi \times 10^{10} \times 1.5 \times 10^{-2}}{3 \times 10^{8}} = 3.1416$$

For all possible modes from Tables 3.1 and 3.2.

$$\text{Table 3.1} \left| \begin{array}{l} TE_{11}\,(1.841) \\ TE_{21}\,(3.054) \end{array} \right. \qquad \text{Table 3.2} \left| TM_{01}\,(2.405) \right.$$

3.4 STRIP AND MICROSTRIP TRANSMISSION LINES

Widely used in the microwave bands, strip lines mainly serve to interconnect a multiplicity of microwave devices into a branched microwave system.

A common types of strip line (Fig. 3.4*a*) consists of a thin central conductor of rectangular cross-section situated midway between two parallel metal plates ("ground planes") with the long dimension of the central strip conductor running parallel to the outer plates. The outer plates are somewhat greater in width than the central strip. The strip may be held in place by solid dielectric filling the line or, as is more common, on a thin dielectric sheet held in position by posts at the edges of the parallel plates.

Because the central strip is placed between two parallel plates located symmetrically about the strip, this form of transmission line is often called shielded, balanced, or symmetrical. There is another form, known as an open-strip line which has only one ground plane. Sometimes, it is referred to as an unshielded, unbalanced, or unsymmetrical strip. The more common usage is to call the former type simply as *strip lines* (Fig. 3.4*a*) and the latter type, *microstrip lines* (Fig. 3.4*b*). In recent years, with the advent of microwave integrated circuits, microstrip lines have been extensively used, as they provide one free and accessible surface on which solid-state devices can be placed.

Strip and microstrip lines are shown in Fig. 3.4 along with their field patterns over the cross-section.

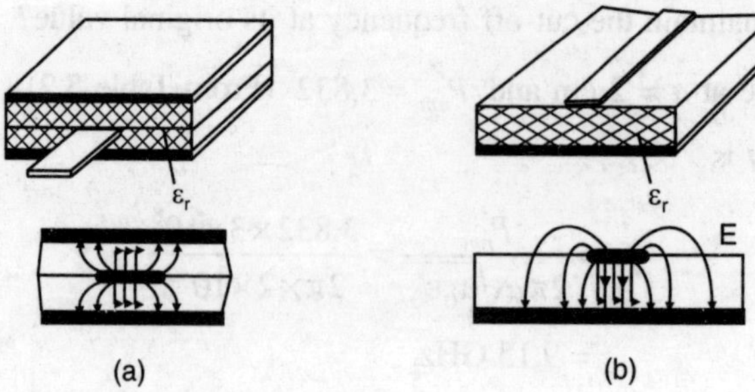

(a) (b)

Fig. 3.4 Microwave striplines (*a*) stripline; (*b*) microstrip line

3.4.1 Stripline

Stripline is fabricated by a photochemical process on dielectric substrate sheets (ε_r = 2 to 7, tan $\delta = 10^{-4}$ to 10^{-3}, and the substrate thickness is 1.5–5 mm), foil-cladded on either side. The foil thickness is 15 to 100 μm which is several times the skin depth. In the unbalanced design, one layer of foil serves as a shield, or ground plate, and the other is etched to form the line pattern. In the balanced design, the etched line pattern is covered by a mating board fully stripped of foil on the inner side. The outer "ground" planes in a stripline are reliably connected together around the periphery of the device and also at several points between the line pattern. The propagation mode in a stripline is nearly TEM mode, the wavelength in the line is $\lambda_0/\sqrt{\varepsilon_r}$.

3.4.2 Stripline Characteristics: Propagation Constant and Characteristic Impedance

We know that, the phase velocity of a TEM mode is given by

$$v_p = c/\sqrt{\varepsilon_r} \quad ...(3.32)$$

where c is the velocity of EM wave in free space ($c = 3 \times 10^8$ m/sec) and ε_r is the relative permittivity of the dielectric medium. Thus the propagation constant of the stripline is

$$\beta = \frac{\omega}{v_p} = \frac{\omega}{c}\sqrt{\varepsilon_r} \quad ...(3.33)$$

The characteristic impedance of a lossless transmission line is given by

$$Z_0 = \sqrt{\frac{L}{C}} = \frac{\sqrt{LC}}{C} = \frac{\omega}{C} \quad ...(3.34)$$

$$= \frac{1}{v_p C} = \frac{\sqrt{\varepsilon_r}}{c \times C} = \frac{\sqrt{\varepsilon_r}}{3\times 10^8\, C} \text{ ohms} \quad ...(3.35)$$

where L and C are the inductance and capacitance per unit length of the line. Thus, the determination of the high frequency characteristic impedance reduces to finding the capacitance per unit length of the stripline configuration.

For a symmetrical homogeneous stripline (Fig. 3.4*a*) having centre strip thickness $t \to 0$, an exact derivation of the capacitance per unit length can be performed by the use of conformal transformations. The characteristic impedance of the symmetric stripline having zero strip thickness can be expressed as follows:

For $W/b \le 0.5$

$$Z_0 = \frac{30}{\sqrt{\varepsilon_r}} \ln\left[\frac{2\left(1+\sqrt{k}\right)}{\left(1-\sqrt{k}\right)}\right] \quad ...(3.36)$$

For $W/b \ge 0.5$

$$Z_0 = \frac{30\pi^2}{\sqrt{\varepsilon_r}\ \ln\left[\frac{2\left(1+\sqrt{k'}\right)}{\left(1-\sqrt{k'}\right)}\right]} \text{ ohms} \quad ...(3.37)$$

where $k = \sec\dfrac{h\pi W}{2b}$ ...(3.38)

$$k' = \tan\frac{h\pi W}{2b} \quad ...(3.39)$$

Here W = the strip width

b = ground plane spacing

For $W \geq b$ the formula for Z_0 reduces to the simple form

$$Z_0 = \frac{120\pi^2}{8\sqrt{\varepsilon_r}\,(\ln 2 + \pi W/2b)}, \quad (W \geq b) \qquad ...(3.40)$$

For a very narrow strip

$$Z_0 = \frac{60}{\sqrt{\varepsilon_r}} \ln \frac{8b}{\pi W}, \qquad (W \leq 0.2b) \qquad ...(3.41)$$

These formulas assume a zero strip thickness, and are accurate to about 1% of the exact results. It is seen from Eqns. (3.40) to (3.41) that **the characteristic impedance decreases as the strip width W increases.**

When the strip thickness is not negligible there are variety of close form approximate solutions available in the literature.

Since the stripline is usually fabricated using a pair of printed circuit boards, the thickness t is typically a few thousandths of an inch, while b ranges from 1/16 to 1/4 of an inch. For such structure, practical values of Z_0 range from 10 to 100 ohms for most dielectric materials.

3.4.3 Attenuation/Losses in Striplines

Attenuating in a stripline can be caused by (*i*) dielectric loss and (*ii*) conductor loss. If α_d is the attenuation constant due to dielectric loss, and α_c in the attenuation constant due to conductor loss, then the total attenuation constant is

$$\alpha = \alpha_d + \alpha_c \qquad ...(3.42)$$

The attenuation constant due to dielectric loss is given by

$$\alpha_d = \frac{\pi}{\lambda_0}\sqrt{\varepsilon_r}\,\tan\delta \quad \text{Np/length} \qquad ...(3.43)$$

$$= 27.3\frac{\sqrt{\varepsilon_r}}{\lambda_0}\tan\delta \quad \text{dB/length}$$

where $\tan\delta$ is loss tangent of the dielectric and α_c is given by

$$\alpha_c = \bar{\alpha}_c \frac{\sqrt{f\varepsilon_r}}{b} \quad \text{dB/length} \qquad ...(3.44)$$

where the frequency f is in GHz and b is ground-plate spacing.

3.5 MICROSTRIP LINES

Microstrip lines are most exclusively built on dielectric substrates with a high dielectric constant (ε_r = 10 to 15, $\tan\delta = 1 \times 10^{-4}$ to 5×10^{-4}, with a thickness of 0.5 to 1.0 mm). Their size is about $1/\sqrt{\varepsilon_r}$ that of air-filled rigid strip lines, so they are admirably adapted to application in the technology of hybrid integrated microcircuits.

Substrate plates for microstrip lines come in standard sizes, fabricated to a high degree of precision and polished. The desired line pattern is applied to a substrate by vacuum sputtering through a mask, with the conductors then built up to the required thickness by electroplating. The most commonly used form of microstrip lines is shown in Fig. 3.4(*b*).

In microstrip lines, the lowest mode is a hybrid quasi-TEM mode for which the phase velocity stands midway between that for electromagnetic energy propagating in a dielectric and in free space.

Microstrip line characteristics: Characteristic impedance effective dielectric constant (permittivity) and propagation constant.

3.5.1 Characteristic Impedance

For any TEM type transmission line the characteristic impedance at high frequency may be expressed as

$$Z_0 = \sqrt{L/C} \quad \text{or} \quad Z_0 = v_p L \quad \text{or} \quad Z_0 = 1/v_p C \qquad \text{...(3.45 } a,b,c)$$

where v_p is the velocity of the wave travelling along the line.

The phase velocity is also given as

$$v_p = \frac{1}{\sqrt{LC}} \qquad \text{...(3.46)}$$

If we assume that the substrate of the microstrip line is removed (effectively) then we will have an air filled line along which the wave will travel with the velocity of light ($c = 3 \times 10^8$ m/s) in free space. The characteristic impedance of this air filled 'microstrip' line is, say Z_{01} where L remains unaltered and C_1 is the capacitance per unit length for this structure.

Since $v_p = c$,

$$Z_{01} = \sqrt{\frac{L}{C_1}} \qquad \text{...(3.47)}$$

or

$$Z_{01} = cL \text{ or } Z_{01} = \frac{1}{cC_1} \qquad \text{...(3.48)}$$

Combining Eqns. (3.45*a*), (3.45*b*) and (3.45*c*) yield a very significant result

$$Z_0 = \frac{1}{c\sqrt{CC_1}} \qquad \text{...(3.49)}$$

This means we can evaluate the characteristic impedance, if only we can evaluate the capacitance per unit length of the structure with and without the presence of the dielectric substrate.

3.5.2 The Effective Dielectric Constant (Permittivity) ε_{eff}

Since some of the field lines are in the dielectric region and some are in air, one might suspect that the actual value of dielectric constant (permittivity) of the microstrip line must lie somewhere between 1 and ε_r which is termed as the effective dielectric constant (permittivity) ε_{eff} of the microstrip line, and satisfies the relation

$$1 < \varepsilon_{eff} < \varepsilon_r \qquad \text{...(3.50)}$$

The effective dielectric constant (permittivity) can be interpreted as the permittivity of a homogenous medium that replaces the air and dielectric region of the microstrip as shown in Fig. 3.5.

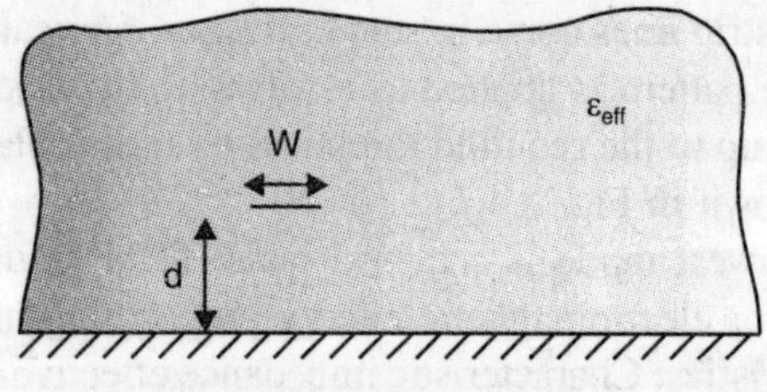

Fig. 3.5 Equivalent geometry of microstrip line (quasi-TEM mode)

It is defined as the ratio of capacitance per unit length for the microstrip line with and without the dielectric present (*i.e.*, air is dielectric):

$$\varepsilon_{eff} = \frac{C(\varepsilon_r)}{C_1(\varepsilon_r = 1)} \qquad ...(3.51)$$

For the very wide strip width ($W/h >> 1$ Fig. 3.6*a*) most of the electric fields in the dielectric and ε_{eff} approach the permittivity of the material ε_r (*i.e.*, $\varepsilon_{eff} \rightarrow \varepsilon_r$). In the case of very narrow strip width ($W/h << 1$, Fig. 3.6*b*) the field is almost equally shared by the ($\varepsilon_r = 1$) and the dielectric substrate so that ε_{eff} approaches $\varepsilon_{eff} \approx \frac{1}{2}(\varepsilon_r + 1)$.

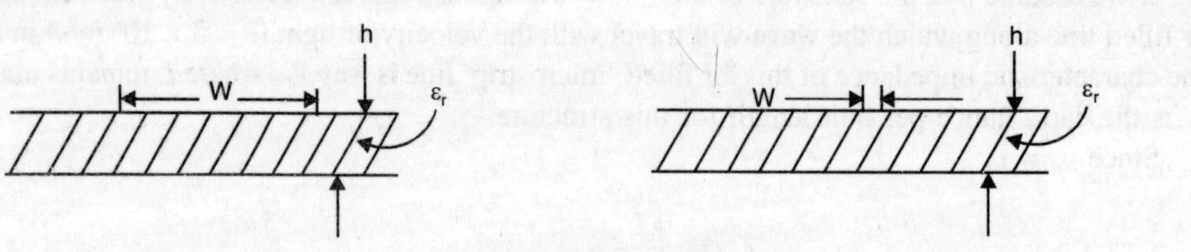

Fig. 3.6 (*a*) Extremely wide ($W >> h$) and (*b*) extremely narrow ($W << h$) microstrip lines

3.5.3 Phase Velocity and Propagation Constant

The phase velocity and propagation constant can be expressed as

Phase velocity
$$v_p = \frac{C}{\sqrt{\varepsilon_{eff}}} \qquad ...(3.52)$$

Propagation
$$\beta = \frac{\omega}{v_p} = \frac{\omega}{c}\sqrt{\varepsilon_{eff}} \qquad ...(3.53)$$

We can write from $\frac{v_p}{c} = \frac{\lambda_g}{\lambda_0} = \frac{1}{\sqrt{\varepsilon_{eff}}}$

$$\varepsilon_{eff} = \left(\frac{c}{v}\right)^2 \qquad ...(3.54a)$$

A useful relationship between Z_0, Z_{01} and ε_{eff} can be obtained as

$$Z_0 = \frac{Z_{01}}{\sqrt{\varepsilon_{eff}}}$$

or $$Z_{01} = Z_0\sqrt{\varepsilon_{eff}} \qquad ...(3.54b)$$

3.5.4 Approximate Formulas for Characteristic Impedance and Effective Dielectric Constant

There are many relations which can be used to calculate characteristic impedance and effective microstrip permittivity as a function of the dimensions and substrate permittivity.

(*A*) The empirical equation is expressed as

$$\varepsilon_{eff} = 0.475\varepsilon_r + 0.67 \qquad ...(3.55)$$

where ε_r is the relative permittivity of the substrate.

(*B*) Characteristic impedance for a narrow $\left(\frac{h}{W} < 0.8\right)$ microstripline

The cross-section of microstrip line (Fig. 3.7*a*) is rectangular. The rectangular conductor of width *W* and thickness *t* (Fig. 3.7*a*) can be transformed into an equivalent circular conductor by an empirical equation (with reference of Fig. 3.7),

$$d = 0.67W\left(0.8 + \frac{t}{W}\right) \qquad ...(3.56)$$

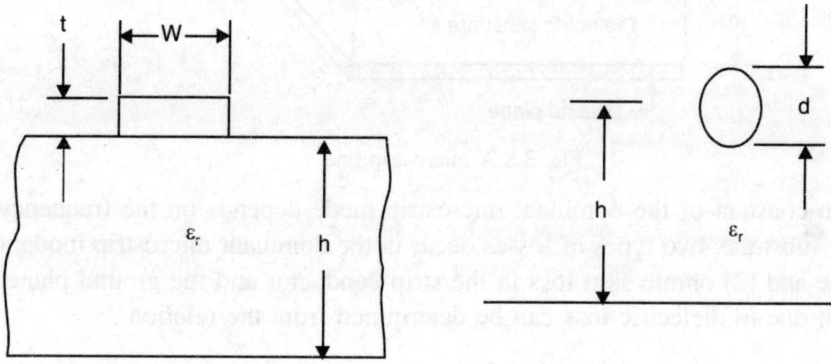

Fig. 3.7 (*a*) Cross-section of microstrip line and (*b*) a wire-over-ground line

The characteristic impedance of a wire-over-ground (Fig. 3.7*b*) transmission line, is given by

$$Z_0 = \frac{60}{\sqrt{\varepsilon_r}} \ln\frac{4h}{d} \quad \text{for } h >> d \qquad ...(3.57)$$

Substituting Eqn. (3.55) for the dielectric constant and Eqn. (3.56) for the equivalent diameter in Eqn. (3.57) yields.

$$Z_0 = \frac{87}{\sqrt{\varepsilon_r + 1.41}} \ln\left[\frac{5.98h}{0.8W + t}\right], \qquad (h < 0.8W) \qquad ...(3.58)$$

where t = the thickness of the microstrip line

W = width of the microstrip line

h = height from the microstrip line to the ground (thickness of the dielectric)

ε_r = relative dielectric constant of the dielectric substrate

and the characteristic impedance for a wide microstrip line ($W >> h$) is expressed by

$$Z_0 = \frac{h}{W}\frac{377}{\sqrt{\varepsilon_r}}, \quad (W >> h) \qquad ...(3.59)$$

The guide wavelength for the propagation of quasi-TEM mode is given by

$$\lambda_g = \lambda_0 / \sqrt{\varepsilon_{eff}}\,.$$

3.5.5 Losses in Microstrip

Microstrip transmission lines consisting of a conducting ribbon attached to a dielectric sheet with conductive backing are widely used in both microwave and computer technology (see Fig. 3.8).

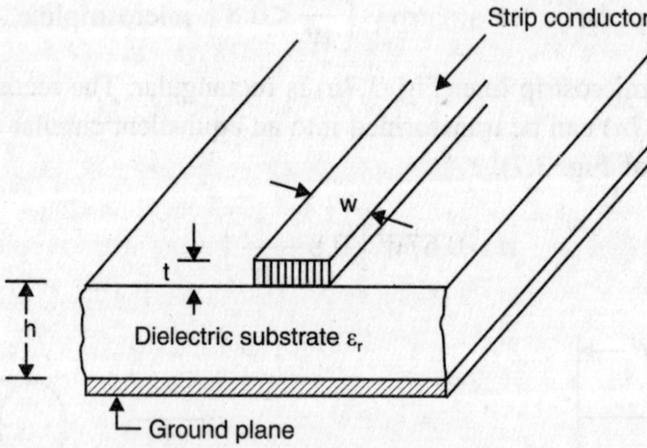

Fig. 3.8 A microstrip line

The attenuation constant of the dominant microstrip mode depends on the frequency. For a non-magnetic dielectric substrate, two types of losses occur in the dominant microstrip mode: (1) dielectric loss in the substrate and (2) ohmic skin loss in the strip conductor and the ground plane.

The attenuation due to dielectric loss can be determined from the relation

$$\alpha_d = 27.3\left[\varepsilon_r / \sqrt{\varepsilon_{eff}(f)}\right]\left\{\frac{\varepsilon_{eff}(f)-1}{\varepsilon_r - 1}\right\}\tan\delta/\lambda_g \quad \text{dB/unit length} \qquad ...(3.60)$$

where $\tan\delta$ is the loss tangent of the dielectric and $\left(\frac{\varepsilon_{eff}-1}{\varepsilon_r-1}\right)$ is a filling factor which accounts for the fact that the fields around the microstrip line are partly in air (lossless) and partly in the dielectric. The other radiation which is also used as

$$\alpha_d = \frac{g\sigma_\alpha}{\sqrt{\varepsilon_{eff}}}\sqrt{\frac{\mu_0}{\varepsilon_0}}$$

$$= 1.637 \times 10^7 \frac{g\sigma_\alpha}{\sqrt{\varepsilon_{eff}}} \quad \text{dB/cm} \qquad \text{...(3.61)}$$

In a microstrip line over a low-loss dielectric substrate, the predominant sources of losses at microwave frequencies are the nonperfect conductors. The microstrip conductor contributes the major part of the ohmic loss. The conducting attenuation constant of a wide microstrip line is given by

$$\alpha_c \cong \frac{8.686 R_s}{Z_0 w} \quad \text{dB/cm} \quad \text{for} \quad \frac{w}{h} > 1 \qquad \text{...(3.62)}$$

where $R_s = \sqrt{\frac{\pi f \mu}{\sigma}}$ is the surface skin resistance in Ω/square,

$R_s = \frac{1}{\delta\sigma}$ is Ω/square

$\delta = \frac{1}{\sqrt{\pi f \mu \sigma}}$ is the skin depth in cm.

In addition to the conductor and dielectric losses, microstrip line also has radiation losses. The radiation loss depends on the substrate's thickness and dielectric constant.

The ratio of radiated power to total dissipated power for an open-circuited microstrip line is

$$\frac{P_{rad}}{P_t} = 240\pi^2 \left(\frac{h}{\lambda_0}\right)^2 \frac{F(\varepsilon_{re})}{Z_0} \qquad \text{...(3.63)}$$

where $F(\varepsilon_{re})$ is a radiation factor.

The radiation factor decreases with increasing substrate dielectric constant. So it can be expressed as

$$\frac{P_{rad}}{P_t} = \frac{R_r}{Z_0} \qquad \text{...(3.64)}$$

where R_r is the radiation resistance of an open-circuited microstrip.

3.6 QUALITY FACTOR Q OF MICROSTRIP LINES

The quality factor Q of a microstrip lines is very high, but it is limited by the radiation losses of the substrates and with low dielectric constant.

The ohmic attenuation constant of a wide microstrip line is given by Eqn. (3.62) as

$$\alpha_c = \frac{8.686 R_s}{Z_0 w} \quad \text{dB/cm}$$

and that the characteristic impedance of a wide microstrip line, as Eqn. (3.40) is

$$Z_0 = \frac{h}{w}\sqrt{\frac{\mu}{\varepsilon}} = \frac{377 h}{\sqrt{\varepsilon_r}\, w} \quad \Omega$$

The wavelength in the microstrip line is

$$\lambda_g = \frac{30}{f\sqrt{\varepsilon_r}} \text{ cm} \qquad ...(3.65)$$

where f is the frequency in GHz.

The quality factor Q_c of a wide microstrip line is

$$Q_c = 0.63h\sqrt{\sigma f \text{ GHz}} \qquad ...(3.66)$$

A quality factor Q_d is related to the dielectric attenuation constant:

$$Q_d = \frac{27.3}{\alpha_d} \qquad ...(3.67)$$

where α_d is in $dB/\lambda g$.

3.7 SLOT AND COPLANAR MICROSTRIP LINE

Also classed with stripline are slotted and coplanar lines. Among their advantages are the possibility of connecting discrete and semiconductor devices in both series and shunt, ease of joining to other types of striplines, and coaxial cables, and usability as components for non-reciprocal ferrite devices.

A slotted microstrip line has a narrow slot in the conductor applied to one side of a dielectric board or sheet of high dielectric constant (Fig. 3.9). A slotted microstrip line sustains a slowed-down TE wave the electromagnetic field of which is concentrated near the slot. The cut-off frequency of the dominant mode in a slotted microstrip line is zero, but the mode shows a considerable dispersion. The characteristic

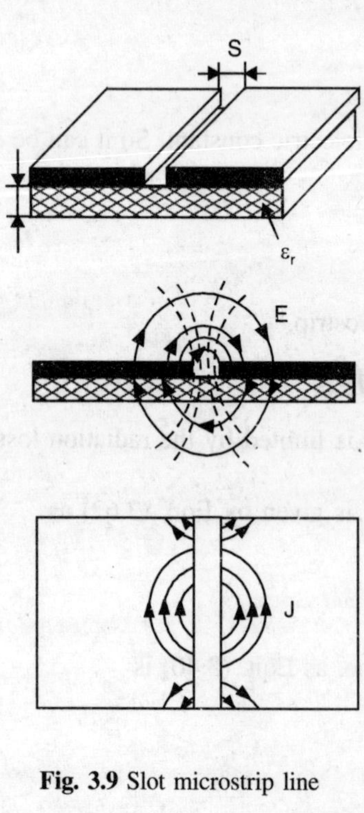

Fig. 3.9 Slot microstrip line

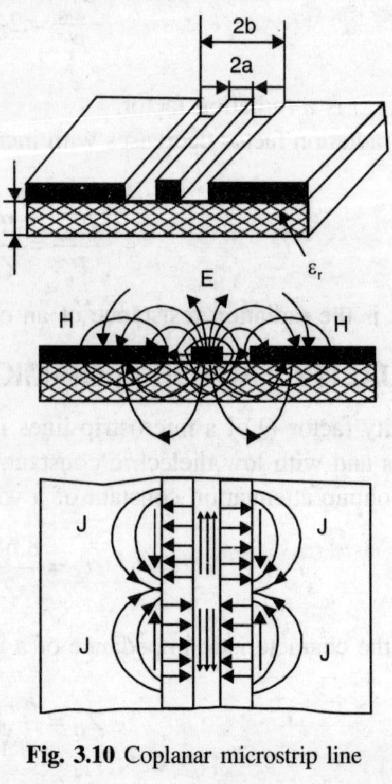

Fig. 3.10 Coplanar microstrip line

impedance of a slotted microstrip line is arbitrarily expressed in terms of the voltage at the slot and the transmitted power, but existent evidence needs experimental verification.

A coplanar microstrip line (Fig. 3.10) is a three-conductor microstrip line formed by two parallel, closely spaced narrow slots in a metal sheet on one side of a dielectric substrate. As in a slotted line, the substrate of a coplanar line has a high dielectric constant ($\varepsilon_r \approx 10$) which results in a marked reduction in the wavelength and in the concentration of the field near the central strip. The dominant mode that can be supported by a coplanar line is the slowed-down TE mode. This mode is the display dispersion, but its cut-off frequency is zero. Its characteristic impedance is defined in arbitrary terms and is usually calculated on a computer, using a special program.

3.8 COUPLED LINE DIRECTIONAL COUPLER

In this section first coupled strip and microstrip lines are discussed. Then parallel coupled and branch line direction coupler are described.

Coupled line: Figure 3.11, shows various coupled transmission line geometries. In these line power can be coupled between the lines due to interaction of the electromagnetic field of each line. These coupled lines are usually assumed to operate in the TEM mode, which is rigorously valid for stripline structures and approximately valid (quasi TEM mode) for microstrip lines. In general these coupled lines (Fig. 3.11) can support two distinct propagation modes. This feature can be used for designing directional coupler, hybrids, and filters.

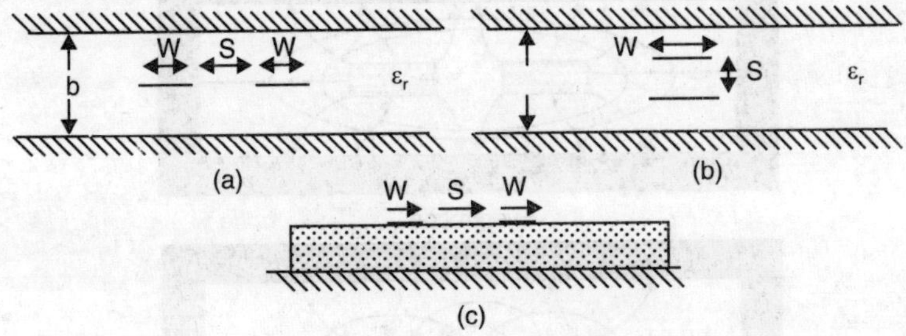

Fig. 3.11 Various coupled lines

Even and odd mode excitations: Even and odd mode excitation for a coupled line, and the resulting equivalent-capacitance networks are shown in Fig. 3.12(*a*).

Even-mode excitation is the excitation in which the currents in the strip conductors are equal in magnitude and in the same direction (Fig. 3.12*a*). The two conductors have the same potential.

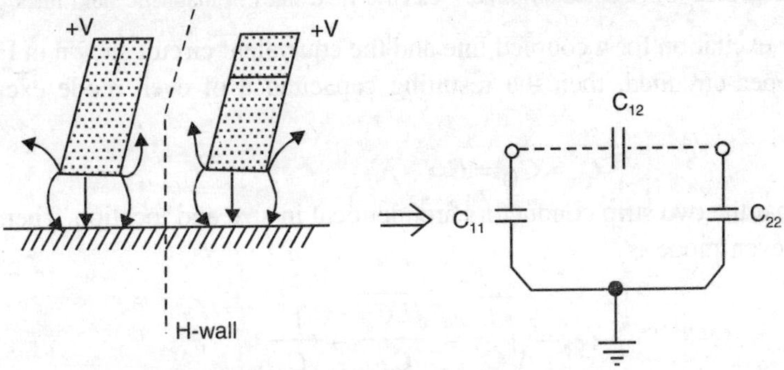

Fig. 3.12 (*a*) Even-mode excitation for a coupled line (microstrip), and the resulting equivalent

Odd-mode excitation is the excitation in which the currents in the strip conductors are equal in amplitude but in opposite direction (Fig. 3.12*b*). The two conductors have opposite potential.

The electric field lines for these two cases are sketched in Fig. 3.12 (*a*) and (*b*).

For the even mode, the electric field has even symmetry about the center line, and no curent flows between the two strip conductors.

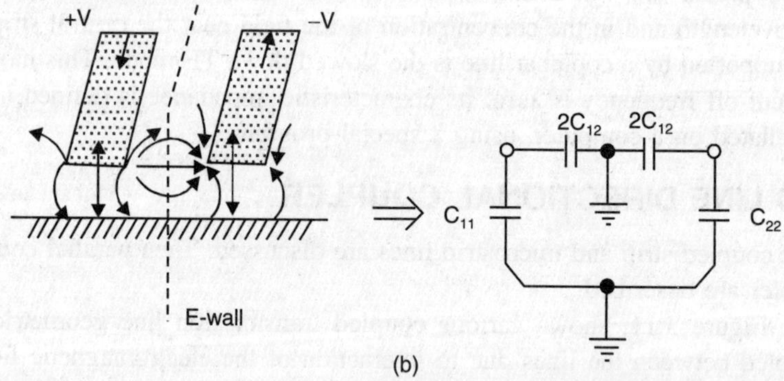

Fig. 3.12 (*b*) Odd-mode excitations for a coupled line (microstrip), and the resulting equivalent

In Fig. 3.12*c*, electric and magnetic field lines for coupled stripline are sketched (both for even and odd mode excitation).

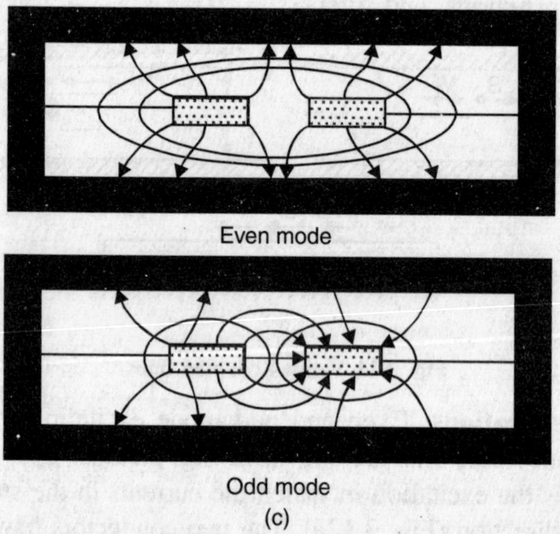

Fig. 3.12 (*c*) Coupled stripline:—electric field lines ... magnetic field lines

The even mode excitation for a coupled line and the equivalent circuit shown in Fig. (3.129), where C_{12} is effectively open-circuited, then the resulting capacitance of even mode excitation of lines to ground is

$$C_e = C_{11} = C_{22} \qquad \text{...(3.68)}$$

It is assumed that the two strip conductors are identical in size and location. Then the characteristic impedance for the even mode is

$$Z_{0e} = \sqrt{\frac{L}{C_e}} = \frac{\sqrt{LC_e}}{C_e} = \frac{1}{vC_e} \qquad \text{...(3.69)}$$

where v is the velocity of propagation on the line.

For the odd-mode, the electric field lines have an odd symmetry about the center line, and a voltage null exists between the two strip conductors. We can imagine this as a ground plane through the middle of C_{12}, which leads to the equivalent circuit as shown. In this case, the effective capacitance between either strip conductor and ground is

$$C_0 = C_{11} + 2C_{12}$$
$$= C_{22} + 2C_{12} \quad \text{...(3.70)}$$

and the characteristic impedance for the odd-mode is

$$Z_{0e} = \frac{1}{vC_0} \quad \text{...(3.71)}$$

In words, Z_{0e} is the characteristic impedance of one of the strip conductors relative to ground when the coupled line is operated in the even (odd) mode. An arbitrary excitation of a coupled line can always be treated as a superposition of appropriate amplitudes of even and odd modes.

If the coupled line is purely TEM, such as coaxial, parallel plate, or stripline, analytical techniques such as conformal mapping can be used to evaluate the capacitances per unit length of line, and the even and odd mode characteristic impedances can then be determined. For quasi-TEM lines, such as microstrip, these results can be obtained numerically or by approximate quasi-static techniques.

3.9 COUPLED LINE DIRECTIONAL COUPLER (PARALLEL COUPLED)

Single section coupled line coupler: When two microstrip or strip transmission lines are coupled over a certain length, we have a TEM directional coupler. Figure 3.13 shows a single section coupled line coupler and port designation. The theory of TEM directional couplers is cumbersome and as such not discussed. However, the major design equations are given below which were obtained applying the even-odd mode technique in conjunction with the input impedance of the line.

The analysis shows the: (*i*) the coupled voltages at parts 2 and 3 are out of phase by $\pi/2$, (*ii*) coupling is a function of $\theta = \beta l$, the electrical length of the parallel coupled section of length l (Fig. 3.13) of a loss-less ($\alpha = 0$) structure uncoupled voltage at port 3 is maximum when θ = odd multiple of $\pi/2$. This corresponds to a minimum length of $l = \lambda_g/4$ at the midband of frequency *i.e.*, the coupling is strongest when the length of the coupler is a quarter wavelength. Strictly speaking a coupler like this only works at one frequency, namely that for which the length is one-quarter length. In practice it is also usable in the frequency band around it, with special tricks. If C_0 is the midband coupling coefficient ($\theta = \pi/2$).

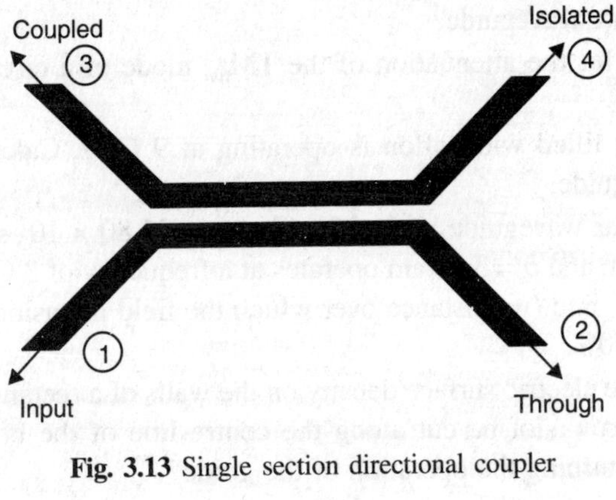

Fig. 3.13 Single section directional coupler

$$C_0 = \frac{V_3}{V_1} = \frac{Z_{0e} - Z_{0o}}{Z_{0e} + Z_{0o}}, \quad (\text{Directivity} = 20 \log C_0)$$

Then the characteristic impedances required are

$$Z_{0e} = Z_0 \sqrt{\frac{1+C_0}{1-C_0}} \qquad \text{...(3.72)}$$

$$Z_{0o} = Z_0 \sqrt{\frac{1-C_0}{1+C_0}} \qquad \text{...(3.73)}$$

In the above analysis, it was assumed that the even and odd modes of the coupled line structure have the same velocities of propagation, so that the line has the same electrical length for both modes. For a coupled microstrip, or other non-TEM, line this condition will generally not be statisfied, and the coupler will have poor directivity because of unequal even and odd mode phase velocities. Techniques for compensating coupled microstrip lines to achieve equal even and odd mode phase velocities include the use of dielectric overlays and anistropic substrates.

This type of coupler is best suited for weak couplings, as tight coupling requires lines that are too close together to be practical, or a combination of even and odd mode characteristic impedances that is non-realizable.

REVIEW QUESTIONS AND PROBLEMS

1. What is meant by a cut-off frequency of a waveguide?
2. Explain why single conductor hollow or dielectric filled waveguide cannot support TEM waves.
3. Can a waveguide have more than one cut-off frequency? On what factors does the cut-off frequency of a waveguide depend?
4. Is the guide wavelength of a propagating wave in a waveguide longer or shorter than the wavelength in the corresponding unbounded dielectric medium?
5. What is meant by the dominant mode of a waveguide? What is the dominant mode of a parallel plate waveguide?
6. Does the attenuation constant due to dielectric losses increase with frequency for TM and TE modes in a parallel plate waveguide?
7. Derive the expression for the attenuation of the TM_{nm} mode of a circular waveguide with finite conductivity.
8. An *X* band waveguide filled with teflon is operating at 9 GHz. Calculate the phase and group velocities in the waveguide.
9. An air filled rectangular waveguide made of copper ($\sigma = 5.80 \times 10^7$ s/m) and having transverse dimensions $a = 7.20$ cm and $b = 3.40$ cm operates at a frequency of 3 GHz in the dominant mode. Find (*i*) f_c (*ii*) λ_g (*iii*) λ_c and (*iv*) distance over which the field intensities of the propagating wave will be attenuated by 50%.
10. Find expressions for the electric surface density on the walls of a rectangular waveguide for a TE_{10} mode. Why can a narrow slot be cut along the centre line of the broad wall of a rectangular waveguide without perturbing the operation of the guide?

11. Write Maxwell's equation for propagation in free space and derive the wave equation in rectangular coordinates. What is the effect of lossy medium on the wave equation?

12. Use Maxwell's equations to show that it is impossible for the TEM wave to exist within any single conductor waveguide such as an ordinary rectangular or circular guide.

13. Derive an expression for the attenuation factor for the TM_{10} were between parallel conducting planes.

14. Obtain an expression for power in a TE_{11} mode in a circular guide.

15. In free space,

$$H(z, t) = 1.33 \times 10^{-1} \cos(4 \times 10^7 t - Bz) a_x \text{ (A/m)}$$

obtain an expression per $E(z, t)$. Find β and λ.

16. Does the attenuation constant due to dielectric losses increase or decrease with frequency for TM and TE modes in a parallel-plate waveguide?

17. State the boundary conditions to be satisfied by E_z for TM waves in a rectangular waveguide.

18. Which mode is the dominant mode in a rectangle waveguide if *(a)* $a > b$, *(b)* $a < b$, and *(c)* $a = b$?

SELECTED BIBLIOGRAPHY

1. Ramo, S., Whinnery, J.R. and Van Duzer, T., *Fields and Waves in Communication Electronics,* John Wiley & Sons, 1965.
2. Jordan, E.C., and Balmain, K.G., *Electromagnetic Waves and Radiating Systems,* Prentice Hall Inc., 1968.
3. Collen, R.E., *Foundations for Microwave Engineering*, McGraw-Hill Inc., New York, 1992.
4. Sisodia, M.L., Raghuvanshi, G.S., *Microwave Circuits and Passive Devices,* New Age International (P) Ltd. Publishers, 2001.

CHAPTER 4

Microwave Cavity Resonators

The simplest waveguide resonator (cavity) is a length of circular or rectangular waveguide short-circuited at both ends of the waveguide, this forms a closed box or cavity. Electric and magnetic energy is stored within the cavity, and power can be dissipated in the metallic walls of the cavity and the dielectric filling of the cavity. Resonators of the type find extensive use as resonant circuit at microwave frequencies. Their behaviours is analogous to that of coil and capacitor combinations. However at microwave frequency cavity resonators are (*i*) simple (*ii*) reasonable dimensions (*iii*) high Q and (*iv*) very high shunt impedance. The resonant frequencies for a general TE or TM mode can be derived using the wave equation. The method of separation of variables is used to solve for the electric and magnetic fields that satisfy the boundary conditions of the cavity.

4.1 RECTANGULAR WAVEGUIDE CAVITY

The geometry of a rectangular cavity is shown in Fig. 4.1. It consists of a rectangular waveguide of length d which is shorted at both ends (z = 0, d).

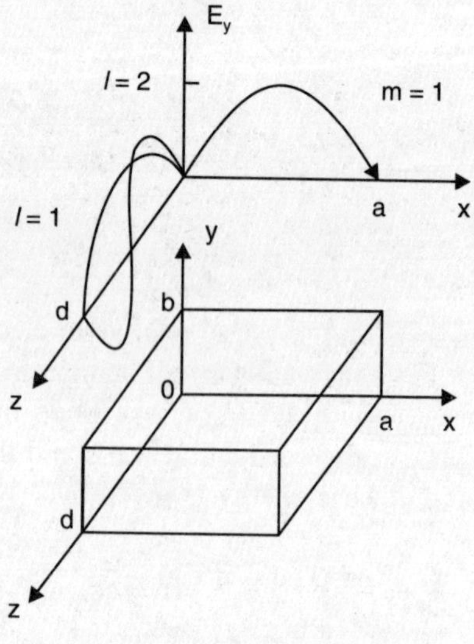

Fig. 4.1 A rectangular resonant cavity, and the electric field distribution for the TE_{101} and TE_{102} resonant modes

The dimension d of the cavity is generally a multiple of half guide wavelength at the frequency of operation (f). The resulting standing wave pattern is such that the x and y field components of the electric field are zero at $z = 0$. The cavity geometry reveals that field solutions can be obtained directly from the corresponding waveguide solutions. For the TE_{mn} or TM_{mn} mode, the propagation constant is given by

$$\overline{\beta}_{mn}^2 = \omega^2\mu\varepsilon - \left(\frac{m\pi}{a}\right)^2 - \left(\frac{n\pi}{b}\right)^2 \qquad \text{...(4.1)}$$

We require $\overline{\beta}_{mn}\, d = l\,\pi$, where l is an integer, in order for cavity to be a multiple of half guide wavelength long. Thus when l is specified as

$$\overline{\beta}_{mn} = \frac{l\,\pi}{d}, \qquad l = 1, 2, \ldots \qquad \text{...(4.2)}$$

If we put $k^2 = \omega^2\mu\,\varepsilon$ and define a cut-off wave number for the rectangular cavity as

$$k_{mnl} = \left[\left(\frac{l\pi}{d}\right)^2 + \left(\frac{m\pi}{a}\right)^2 + \left(\frac{n\pi}{b}\right)^2\right]^{1/2} \qquad \text{...(4.3)}$$

Hence, the resonance frequency of the cavity is given by

$$f_{mnl} = \frac{ck_{mnl}}{2\pi} = c\left[\left(\frac{l}{2d}\right)^2 + \left(\frac{m}{2a}\right)^2 + \left(\frac{n}{2b}\right)^2\right]^{1/2} \qquad \text{...(4.4)}$$

where c is the velocity of light. It can be observed from Eqn. 4.4 that there is a triply infinite number of resonant frequencies corresponding to different field distributions. Also note that there is more than one field distribution since Eqn. 4.4 holds for both TE and TM modes.

For a TE_{101} resonant mode the field expression are given as follows:

$$E_y = E_0 \sin\left(\frac{\pi x}{a}\right)\sin\left(\frac{l\pi}{d}z\right) \qquad \text{...(4.5}a\text{)}$$

$$H_x = -j\frac{E_0}{Z_{TE}}\sin\left(\frac{\pi x}{a}\right)\cos\left(\frac{l\pi}{d}z\right) \qquad \text{...(4.5}b\text{)}$$

$$H_z = \frac{j\pi E_0}{k\eta a}\cos\frac{\pi x}{a}\sin\left(\frac{l\pi}{d}z\right) \qquad \text{...(4.5}c\text{)}$$

Note that the magnetic field is ± 90° out of phase relative to electric field. This is always the case in a lossless cavity and corresponds to voltage and current being ± 90° out of phase with each other in a lossless LC circuit. We can now compute the Q of this mode by finding the stored electric and magnetic energies, and the power lost in the conducting walls and the dielectric.

At resonance, the average stored electric energy is given by

$$W_e = \frac{\varepsilon_0}{4}\int_0^a\int_0^b\int_0^d E_y E_y^* \,dx\,dy\,dz$$

$$= \frac{\varepsilon_0 abd}{16} E_0^2 \qquad \text{...(4.6)}$$

$$W_m = \frac{\mu}{4} \int_v \left(H_x H_x^* + H_z H_z^* \right) dx\, dy\, dz = W_e$$

This shows that at resonance the stored electric and magnetic energies are equal, which is analogous to the LC resonant circuits.

For small losses we can find the power dissipated in the cavity walls by

$$P_c = \frac{R_s}{2} \int_{walls} J_s \cdot J_s^* \, ds = \frac{R_s}{2} \int_{walls} |H_t|^2 \, ds$$

where $R_s = \sqrt{\omega\mu/2\sigma}$ is the surface resistivity of the metallic walls, and H_t is the tangential magnetic field at the surface of the walls. Substituting from eqn. 4.5*b*, *c* we get

$$P_c = \frac{R_s E_0^2 \lambda^2}{8\eta^2} \left(\frac{l^2 ab}{d^2} + \frac{bd}{a^2} + \frac{l^2 a}{2d} + \frac{d}{2a} \right) \qquad \text{...(4.7)}$$

For arriving to this result the symmetry property of the cavity has been used and the contributions from the walls at $x = 0$, $y = 0$ and $z = 0$ doubled, to account for the contributions from $x = a$, $y = b$ and $z = d$ respectively. We can now find the Q of the cavity with lossy conducting walls but lossless dielectric as

$$Q_c = \frac{2\omega_0 W_e}{P_c}$$

$$= \frac{(kad)^3 b\eta}{2\pi^2 R_s} \frac{1}{\left[2l^2 a^3 b + 2bd^3 + l^2 a^3 d + ad^3 \right]} \qquad \text{...(4.8)}$$

If we have to compute power loss in the dielectric, consider a lossy dielectric which is complex, the imaginary part will contribute to power loss, *i.e.*,

$$\varepsilon = \varepsilon' - j\varepsilon'' = \varepsilon_r \varepsilon_0 (1 - j \tan\delta)$$

Then the power dissipated in the dielectric is

$$P_d = \frac{1}{2} \int_v J \cdot E^* \, dv = \frac{\omega\varepsilon''}{2} \int_v |E^2| \, dv = \frac{abd\omega\varepsilon''}{8} |E_0|^2 \qquad \text{...(4.9)}$$

Hence the Q of the cavity with a lossy dielectric inside the cavity, but with perfectly conducting walls is

$$Q_d = \frac{2\omega W_e}{P_d} = \frac{\varepsilon'}{\varepsilon''} = \frac{1}{\tan\delta} \qquad \text{...(4.10)}$$

When both wall losses and dielectric losses are present, the total power loss is

$$Q = \left(\frac{1}{Q_c} + \frac{1}{Q_d} \right)^{-1} \qquad \text{...(4.11)}$$

4.2 CYLINDRICAL CAVITY

The cylindrical cavity is obtained from a small section of circular waveguide of length d and radius a, whose ends are short-circuited. Since the dominant mode of a circular waveguide is TE_{11}, TE_{111} will be the dominant mode of cylindrical cavity. Figure 4.2 shows the geometry of a cylindrical cavity. This type of cavity is generally used for wave metres, which are used to measure frequency, because it provides very high Q and wide frequency range of operation. Having high Q cavity ensures high degree of resolution or accuracy in the measurement of an unknown frequency. When the cavity is tuned to the frequency of the unknown source, it absorbs a maximum power from the input waveguide. A detector coupled to the input line is used to indicate this dip in power at resonance.

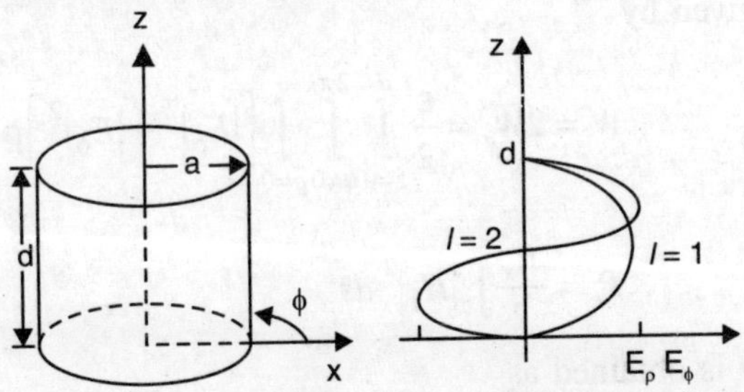

Fig. 4.2 A cylindrical resonant cavity, and the electric field distribution for resonant modes with $l = 1$ and $l = 2$

The fields in the cylindrical cavity are determined from the corresponding waveguide fields. The resonant frequency for the TE_{nml} mode is obtained from the relation,

$$f_{nml} = \frac{c}{2\pi\sqrt{\mu\varepsilon}}\sqrt{\left(\frac{P'_{nm}}{a}\right)^2 + \left(\frac{l\pi}{d}\right)^2} \qquad ...(4.12)$$

and the resonance frequency for the TM_{nml} mode is

$$f_{nml} = \frac{c}{2\pi\sqrt{\mu\varepsilon}}\sqrt{\left(\frac{P_{nm}}{a}\right)^2 + \left(\frac{l\pi}{d}\right)^2} \qquad ...(4.13)$$

For the TE_{nml} mode

$n = 0, 1, 2,$ is the number of field variation along the circumference (ϕ)

$m = 1, 2, 3 ...,$ is the number of zeroes of the field in radial direction

$l = 1, 2, 3 ...,$ is the number of half wave in z direction.

The fields for the TE_{111} mode are

$$H_z = H_0 J_1\left(\frac{P'_{11}\rho}{a}\right)\cos\phi \sin\frac{l\pi z}{d} \qquad ...(4.14a)$$

$$H_\rho = \frac{H_0\bar{\beta}a}{P'_{11}} J'_1\left(\frac{P'_{11}\rho}{a}\right)\cos\phi\cos\frac{l\pi z}{d} \qquad ...(4.14b)$$

$$H_\phi = \frac{-\bar{\beta}a^2 H_0}{(P'_{11})^2\rho^2} J_1\left(\frac{P'_{11}\rho}{a}\right)\sin\phi\cos\frac{l\pi z}{d} \qquad ...(4.14c)$$

$$E_\rho = \frac{jk_0 \eta a^2 H_0}{(P'_{11})^2 \rho} J_1\left(\frac{P'_{11}\rho}{a}\right) \sin\phi \sin\frac{l\pi z}{d} \quad \text{...(4.14}d\text{)}$$

$$E_\phi = \frac{jk_0 \eta a H_0}{P'_{11}} J'_1\left(\frac{P'_{11}\rho}{a}\right) \cos\phi \sin\frac{l\pi z}{d} \quad \text{...(4.14}e\text{)}$$

$$E_z = 0 \quad \text{...(4.14}f\text{)}$$

To find Q of the TE_{111} mode, the time average stored electric and magnetic energies are equal, and the total stored energy is given by

$$W = 2W_e = \frac{\varepsilon}{2} \int_{z=0}^{d} \int_{\phi=0}^{2\pi} \int_{\rho=0}^{a} \left[|E_\rho|^2 + |E_\phi|^2 \right] \rho \, d\rho \, d\phi \, dz$$

$$P_c = \frac{R_s}{2} \int_s |H_t|^2 \, ds$$

Finally Q of the cavity is obtained as

$$\frac{Q\delta_s}{\lambda_0} = \frac{\left[1 - \left(\frac{n}{P'_{nm}}\right)^2\right]\left[\left(P'^2_{nm}\right) + \left(\frac{\pi a l}{d}\right)^2\right]^{3/2}}{2\pi\left[\left(P'_{nm}\right)^2 + \frac{2a}{d}\left(\frac{l\pi a}{d}\right)^2 + \left(1 - \frac{2a}{d}\right)\left(\frac{\pi a n l}{(P'_{nm}) d}\right)^2\right]}$$

where $l = n = m = 1$.

From above relation it can be noted that right hand side is independent of frequency, hence Q varies as λ_0/δ_s, for any cavity and thus decreases as $f^{-1/2}$. For wave meters TE_{011} mode is significant because its Q is two to three times that of TE_{111} mode. Another advantage of the TE_{011} mode is that $H_\phi = 0$ and as such there are no axial currents present. Hence the end plates of the cavity can be freely moved to adjust the cavity length d for tuning purposes without introducing significant loss, since no currents flow across the gap, *i.e.*, the gap between the circular end plate and the cylinder wall is parallel to the current flow lines. Care should be taken when working with TE_{011} mode, since TE_{011} mode is not the lowest order mode, other mode should not be excited within the tuning range of the cavity. A mode chart is shown in Fig. 4.3 which shows what modes will resonate for a given value of $(2a/d)$ and frequency.

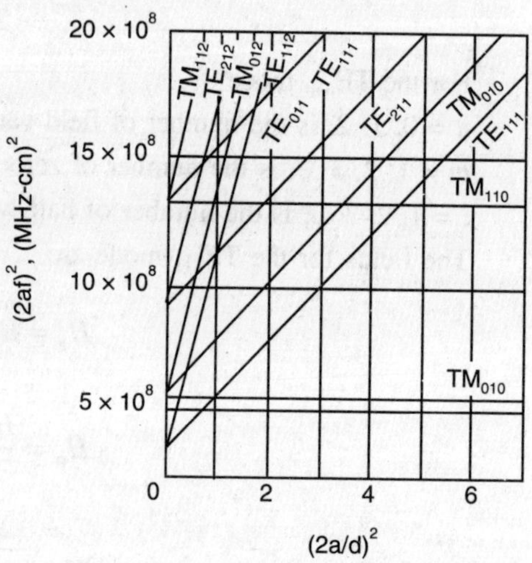

Fig. 4.3 Resonant frequencies for the first few modes of a circular cylindrical cavity

Such a chart is very useful for the design of cavity resonators, as it shows which modes can be excited at a given frequency for a given cavity size. For example, for $(2a/d)^2$ between 3 and 4, only TE_{011} mode can resonate in the frequency range corresponding to $(2af)^2$ between 12.5 to 14.5×10^8.

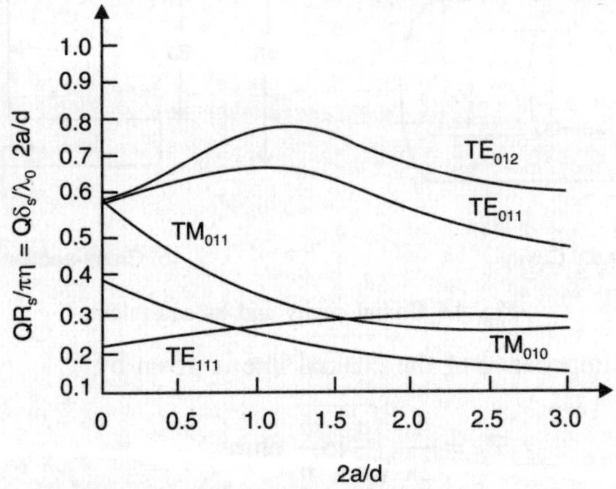

Fig. 4.4 Normalized Q for various cylindrical cavity modes

Figure 4.4 gives a plot of $Q\ \delta_s/\lambda_0$ versus $\left(\frac{2a}{d}\right)$ for several modes. It is evident from the figure that considerably higher value of Q *is* obtained for the TE_{011} mode compared to that for the TE_{111} mode. It is also clear that typical values of Q range from 10,000 to 40,000 or more.

Resonant cavities are employed mainly in the design of microwave tubes and in measuring equipment as cavity wave meters.

4.2.1 Reentrant Cavities

A reentrant cavity is one in which the metallic boundries extend into the interior of the cavity. When the operating frequency is increased to several tens of megahertz, both the inductance and capacitance must be reduced to a minimum in order to maintain resonance at the frequency. Therefore the reentrant cavities are design for use in klystrons and microwave triodes. Several types of reentrant cavities are shown in Fig. 4.5. One of the commonly used reentrant cavity is the coaxial cavity shown in Fig. 4.6.

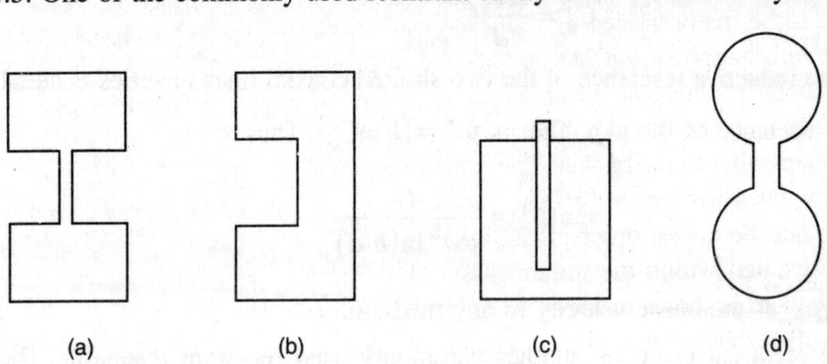

Fig. 4.5 Reentrant cavities (*a*) Coaxial cavity (*b*) Radial cavity (*c*) Toroidal cavity (*d*) Butterfly cavity.

It is clear from Fig. 4.6 that not only has the inductance been considerably decreased but the resistance losses are markedly reduced as well, and the shelf shielding enclosure prevents radiation

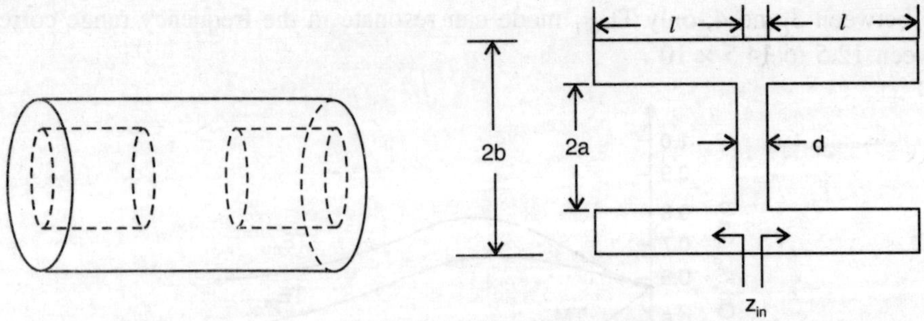

(a) Coaxial Cavity (b) Cross-section of Coaxial Cavity

Fig. 4.6 Coxial cavity and its equivalent.

losses. The characteristic impedance of the coaxial line is given by

$$Z_0 = \frac{1}{2\pi}\sqrt{\frac{\mu}{\varepsilon}}\ln\frac{b}{a} \text{ ohms.} \qquad ...(4.15)$$

The coaxial cavity is similar to a coaxial line shorted at two ends and joined at the center by a capacitor. The input impedance to each shorted coaxial line is given by

$$Z_{in} = jZ_0(\tan\beta l) \qquad ...(4.16)$$

where l is the length of the coaxial line.

Substitute Eqns. (4.15) in (4.16)

$$Z_{in} = j\frac{1}{2\pi}\sqrt{\frac{\mu}{\varepsilon}}\ln\frac{b}{a}\tan(\beta l) \qquad ...(4.17)$$

The inductance of the cavity is given by

$$L = \frac{2X_{in}}{\omega} = \frac{1}{\pi\omega}\sqrt{\frac{\mu}{\varepsilon}\ln\frac{b}{a}\tan(\beta l)} \qquad ...(4.18)$$

and the capacitance of the gap by

$$C_g = \frac{\varepsilon\pi a^2}{d} \qquad ...(4.19)$$

at resonance the inductive reactance of the two shorted coaxial lines in series is equal in magnitude to the capacitive reactance of the gap. That is, $\omega L = (1/\omega C_g)$. Thus,

$$\tan(\beta l) = \frac{d\nu}{\omega a^2 \ln(b/a)} \qquad ...(4.20)$$

where $\nu = 1/\sqrt{\mu\varepsilon}$ is the phase velocity in any medium.

The radial reentrant cavity is another commonly used reentrant resonator. The inductance and capacitance of a radial reentrant cavity is

$$L = \frac{\mu l}{2\pi}\ln\frac{b}{a} \qquad ...(4.21)$$

$$c = \varepsilon_0 \left[\frac{\pi a^2}{2} - 4a \ln \frac{0.765}{\sqrt{l^2 + (b-a)^2}} \right] \qquad ...(4.22)$$

and the resonant frequency is,

$$f_r = \frac{c}{2\pi\sqrt{\varepsilon_r}} \left\{ al \left[\frac{a}{2d} - \frac{2}{l} \ln \frac{0.765}{\sqrt{l^2 + (b-a)^2}} \right] \ln \frac{b}{a} \right\}^{1/2} \qquad ...(4.23)$$

where $C = 3 \times 10^8$ m/s is velocity of light in a vacuum.

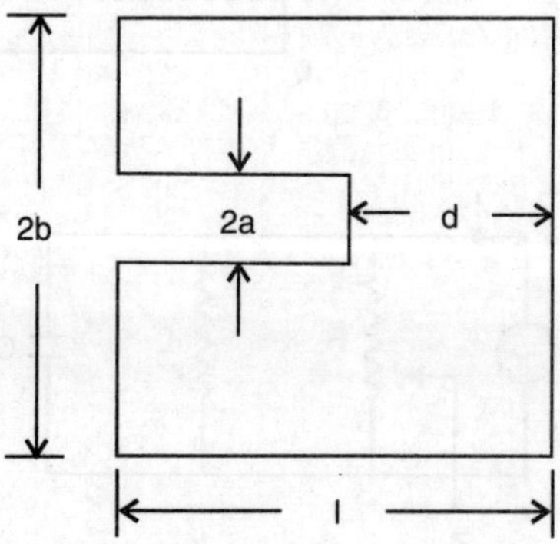

Fig. 4.7 Radial reentrant cavity

4.3 QUALITY FACTOR AND COUPLING COEFFICIENTS OF MICROWAVE RESONATORS

The operation of microwave resonators is very similar to that of lumped element resonators of circuit theory. A series resonator is shown in Fig. 4.8*a*, its input impedance versus frequency graph is shown in Fig. 4.8*b*. Its input impedance is given by

$$Z_{in} = R + j\omega L - j\frac{1}{\omega C} \qquad ...(4.15)$$

the complex power delivered to the resonator is given by

$$P_{in} = \frac{1}{2}VI^* = \frac{1}{2}Z_{in}|I|^2 = \frac{1}{2}Z_{in}\left|\frac{V}{Z_{in}}\right|^2 \qquad ...(4.16a)$$

$$= \frac{1}{2}|I|^2 \left(R + j\omega L - j\frac{1}{\omega C} \right) \qquad ...(4.16b)$$

The power dissipated by the resistor, R, is

$$P_{loss} = \frac{1}{2}|I|^2 R \qquad ...(4.17)$$

The average magnetic energy stored in the inductor, L, is given by

$$W_m = \frac{1}{4}|I|^2 L \qquad ...(4.18a)$$

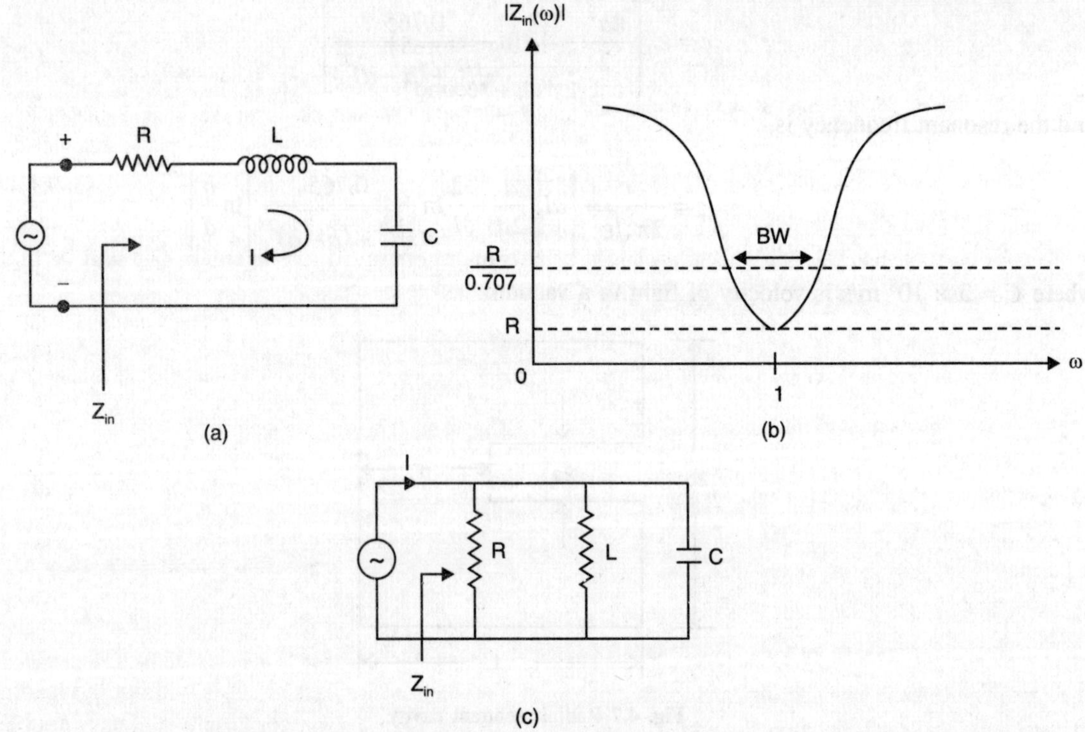

Fig. 4.8 A series RLC resonator and its response (*a*) The series RLC circuit (*b*) The input impedance magnitude versus frequency

and the average electric energy stored in the capacitor, *C*, is given by

$$W_e = \frac{1}{2}\left|V_c\right|^2 C = \frac{1}{4}\left|I\right|^2 \frac{1}{\omega^2 C} \qquad \text{...(4.18}b\text{)}$$

where V_c is the voltage across the capacitor. Then the complex power of Eqn. (4.16) can be written as

$$P_{in} = P_{loss} + 2j\omega\left(W_m - W_e\right) \qquad \text{...(4.19)}$$

and also the input impedance of Eqn. (4.15) can be written as

$$Z_{in} = \frac{2P_{in}}{|I|^2} = \frac{P_{loss} + 2j\omega\left(W_m - W_e\right)}{\frac{1}{2}|I|^2} \qquad \text{...(4.20)}$$

At resonance the average stored magnetic and electric energies are equal. Hence the input impedance can be rewritten

$$Z_{in} = P_{loss}\frac{1}{2}|I|^2 = R, \text{ which is purely real.} \qquad \text{...(4.21)}$$

When $W_m = W_e$, the resonant frequency ω_0 is defined as

$$\omega_0 = \frac{1}{\sqrt{LC}} \qquad \text{...(4.22)}$$

In this case the quality factor (Q) is defined as

$$Q = \omega \frac{(\text{average energy stored})}{(\text{energy loss/second})}$$

$$= \omega\left(\frac{W_m + W_e}{P_c}\right) \quad ...(4.23)$$

Hence we say that Q_{SR} is a measure of loss of a resonant circuit. If loss is small, Q_{SR} will be high. In the case of series resonator Q_{SR} is given by

$$Q_{SR} = \omega_0 \frac{2W_m}{P_{loss}} = \frac{\omega_0 L}{R} = \frac{1}{\omega_0 RC} \quad ...(4.24)$$

Similarly for a parallel resonant circuit shown in Fig. 4.8*c* The Q_{PR} is given by

$$Q_{PR} = \omega_0 \frac{2W_m}{P_{loss}} = \frac{R}{\omega_0 L} = \omega_0 RC \quad ...(4.25)$$

Since $W_m = W_e$, the resonance, this result shows that Q_{PR} of a parallel circuit increases as R increases.

4.3.1 Loaded and Unloaded *Q* of Resonators

The Q defined in the last section is a characteristic of a resonator circuit itself. It is without the loading effect of external circuitry, hence is called unloaded Q. In practice, a resonator circuit is always coupled to other circuitry, which will have the effect of lowering the overall Q of the resonator and is called the loaded Q and is represented by Q_L. Figure 4.9 shows a resonator coupled to an external resistor R_L. So the effective resistor in the case of series resonator becomes equal to ($R + R_L$), which for the parallel resonator is $RR_1/R + R_L$. And we define an external $Q(Q_e)$ as

$$Q_e = \begin{cases} \dfrac{\omega_0 L}{R_L} & \text{for } \textit{series circuits} \\[2ex] \dfrac{R_L}{\omega_0 L} & \text{for } \textit{parallel circuits} \end{cases} \quad ...(4.26)$$

We can define Q of a resonator circuit in terms of its stored energy loss at the resonant frequency *i.e.*,

$$Q = 2\pi \left.\frac{\textit{Energy Stored}}{\textit{Energy Loss per Cycle}}\right|_{at\,\omega_r} = \omega_r \left.\frac{\textit{Energy Stored}}{\textit{Average Power Loss}}\right|_{at\,\omega_r} \quad ...(4.27)$$

There are actually three Q factors that can be defined, depending upon which loss is being considered. Namely,

$$\text{Unloaded } Q: Q_U \equiv \omega_r \left.\frac{\textit{Energy Stored in the Resonator Circuit}}{\textit{Power Loss in the Resonator Circuit}}\right|_{at\,\omega_r}$$

$$\text{External } Q : Q_E \equiv \omega_r \left.\frac{\textit{Energy Stored in the Resonator Circuit}}{\textit{Power Loss in the External Circuit}}\right|_{at\,\omega_r} \qquad ...(4.28)$$

$$\text{Loaded } Q : Q_L \equiv \omega_r \left.\frac{\textit{Energy Stored in the Resonator Circuit}}{\textit{Total Power Loss}}\right|_{at\,\omega_r}$$

The unloaded Q is a measure of the quality of the resonant circuit itself. Infinite Q_u means that the resonant circuit is dissipationless. External Q is a measure of the degree to which the resonant circuit

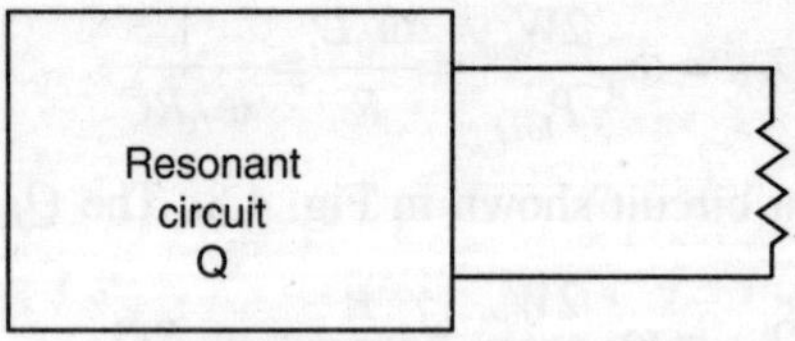

Fig. 4.9 A resonant circuit connected to an external load, R_L

is coupled to the external circuitry. A relationship among the three Q factors can be derived from Eqn. 4.28.

$$\frac{1}{Q_L} = \frac{1}{Q_E} + \frac{1}{Q_U} \qquad ...(4.29)$$

4.3.2 Coupling Coefficient of Resonator

If the cavity is coupled by means of an ideal nil transformer and a series inductance L_s to a generator having external impedance Z_g, then the coupling circuit and its equivalent are shown in Fig. (4.10).

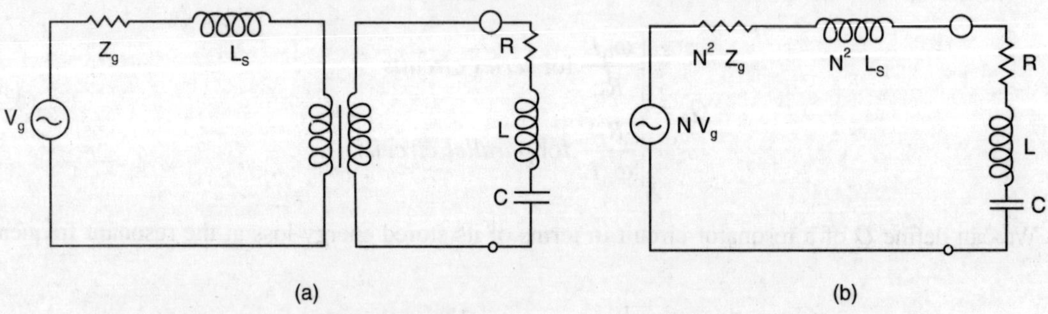

Fig. 4.10 Cavity coupled to a generator *(a)* Coupling circuit *(b)* Equivalent circuit

The loaded Q factor of the system is given by

$$Q_l = \frac{\omega_0 L}{R + N^2 Z_g} \quad \text{for} \quad |N^2 L_s| << |R + N^2 Z_g| \qquad ...(4.30)$$

The coupling coefficient of the system is defined as

$$K \equiv \frac{N^2 Z_g}{R} \qquad ...(4.31)$$

and loaded Q_l would become

$$Q_l = \frac{\omega_0 L}{R(1+K)} = \frac{Q}{1+K} \quad ...(4.32)$$

So,

$$\frac{1}{Q_l} = \frac{1}{Q} + \frac{1}{Q_{ext}} \quad ...(4.33)$$

where $Q_{ext} = \frac{Q}{K} = \omega_0 L/(KR)$ is the external Q.

There are three types of coupling coefficients:

1. **Critical Coupling:** If the resonator is matched to the generator, then

$$K = 1 \quad ...(4.34)$$

The loaded Q_l is given by

$$Q_l = \frac{1}{2} Q_{ext} = \frac{1}{2} Q \quad ...(4.35)$$

2. **Over Coupling:** If $K > 1$, the cavity terminals are at a voltage maximum in the input line at resonance. The normalised impedance at the voltage maximum is the standing wave ratio ρ. That is

$$K = \rho \quad ...(4.36)$$

The loaded Q_l is given by

$$Q_l = \frac{Q}{1+l} \quad ...(4.37)$$

3. **Under Coupling:** If $K < 1$, the cavity terminal are at a voltage minimum and the input terminal impedance is equal to the reciprocal of the standing wave ratio. That is,

$$K = \frac{1}{\rho} \quad ...(4.38)$$

The loaded Q_l is given by

$$Q_l = \frac{\rho}{\rho+1} Q \quad ...(4.39)$$

4.4 EXCITATION OF CAVITIES

In earlier discussion we have dealt with uncoupled cavity resonators. In order to excite a particular mode, the cavity resonator must be properly coupled to an external source. Figure 4.11 shows various methods of excitations of cylindrical cavity. Figure 4.11(*a*) shows cylindrical cavity that is loop coupled to a coaxial line. The conduction current in the loop produces a linking magnetic field as shown in the Fig. 4.11(*a*). Figure 4.11(*b*) shows a probe type coupling between a coaxial line and a cavity. The electric field pattern created by the probe is similar to that of coaxial-to-waveguide transition. Any cavity mode having an electric field parallel to the probe can be excited by this arrangement. For the cylindrical cavity shown in Fig. 4.11(*a, b*) all the TE_{nmp} modes except those with $n = 0$ can be excited.

Part *c* and *d* of this figure, illustrate aperture coupling between a cylindrical cavity and rectangular waveguide operating in the dominant TE_{10} mode. A magnetic field component that is parallel to the long dimension of the slot will be coupled through the aperture. As shown in Fig. 4.11(*c*) the TE_{10} magnetic field couples into the cavity, which allows any of the TM resonant modes to be excited. This coupling arrangement often used to generate the TM_{0lp} modes. In Fig. 4.11(*d*) the rectangular guide and the aperture are rotated by 90°, which allows the TE modes to be excited in the cavity, the TE_{11p} and the TE_{0lp} are the most commonly used modes.

Although many modes can be excited by the coupling mechanism described here, only those modes which have a resonant frequency equal to or nearly equal to the source frequency will be excited. For degenerate modes (that is, those with identical resonant frequency), the coupling should be arranged in such a way that the desired mode gets excited and the unwanted one is not excited. In most cases this is easily achieved since a particular coupling method generally excites either TE or TM modes, but not both.

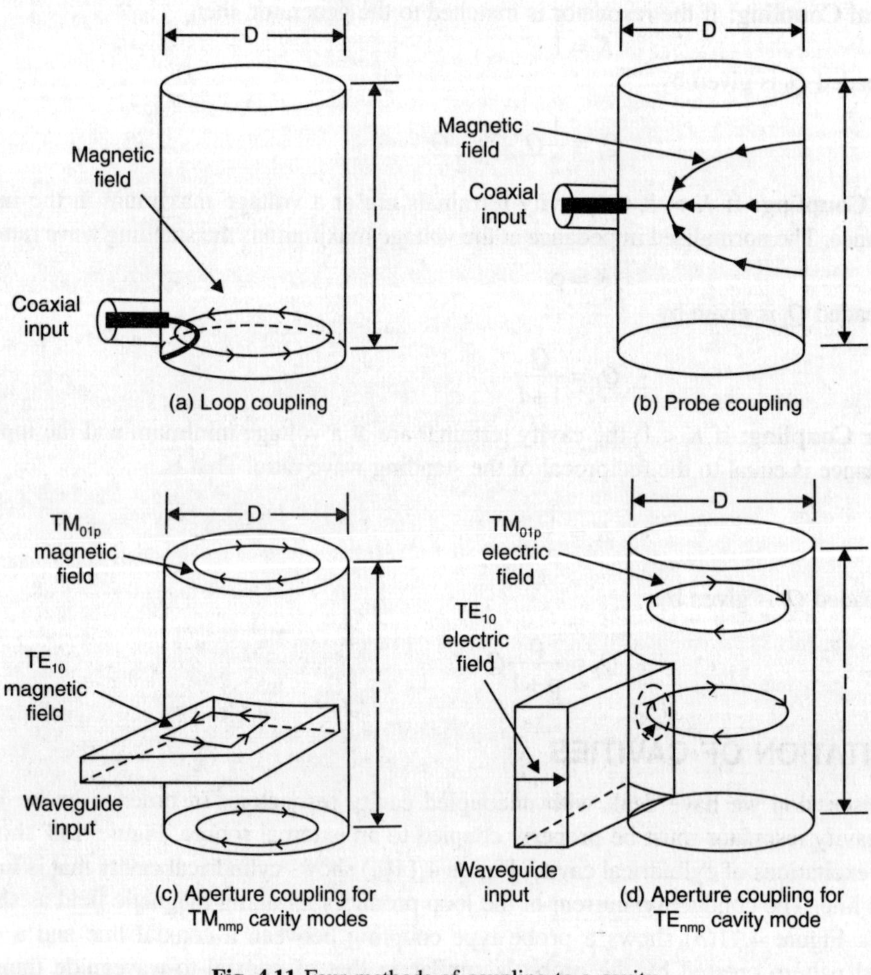

Fig. 4.11 Four methods of coupling to a cavity

These same methods can also be used to couple the cavity to an external load. For example, addition of a coupling loop directly opposite to that shown in Fig. 4.11(*a*) results in a two port cavity with band pass filter characteristics.

Example 4.1. A rectangular—cavity resonator has the following dimension:

$a = 5 \times 10^{-2}$ m, $b = 2 \times 10^{-2}$ m and $d = 15 \times 10^{-2}$ m.

Calculate (*i*) the resonant frequency of the dominant mode TE_{101} for an air filled cavity and (*ii*) dielectric field cavity of $\varepsilon = 4$.

Solution:

(*i*) Resonant frequency of the dominant mode TE_{101} is given by

$$f_{r(\text{air})} = \frac{1}{2\sqrt{\mu\varepsilon}}\sqrt{\left(\frac{m}{a}\right)^2 + \left(\frac{n}{b}\right)^2 + \left(\frac{l}{d}\right)^2} \quad (TE_{mnp})$$

$$f_2 = \frac{3\times 14^8}{2}\sqrt{\left(\frac{1}{5\times 10^{-2}}\right)^2 + \left(\frac{1}{15\times 10^{-2}}\right)^2} = 3.162 \text{ GHz}$$

(*ii*)
$$f_{r(\text{dielectric})} = \frac{f_{np(\text{air})}}{\sqrt{\varepsilon_s}} = \frac{3.162}{\sqrt{4}} = 1.581 \text{ GHz}$$

Example 4.2. An air-filled circular waveguide has a diameter of 6×10^{-2} m and is used as a resonator for TE_{01} mode at 10 GHz by placing two perfectly conducting plates at its two ends. Calculate the minimum distance between the two plates.

Solution:

Resonant frequency f_r is given by $f_r = 2\pi\sqrt{\mu\varepsilon}\sqrt{\left(\frac{P'_{np}}{a}\right)^2 + \left(\frac{r\pi}{d}\right)^2}$ for $r = 1$

Substituting the values we get $\sqrt{\left(\frac{3.832}{3\times 10^{-2}}\right)^2 \times \left(\frac{3.14}{d}\right)^2}$

where d is the minimum distance between the two plates, solving we get

$$\alpha = 1.89 \times 10^{-2} \text{ m}$$

Example 4.3. An air-filled resonant cavity with dimensions $a = 5$ cm, $b = 4$ cm, and $c = 10$ cm is made of copper ($\sigma_c = 5.8 \times 10^7$ mhos/m). It is filled with a lossless material ($\mu_r = 1$), ($\varepsilon_r = 3$). Find the resonant frequency f_r and the quality factor for TE_{101} mode.

Solution:

Resonant frequency $f_r = \frac{1}{2\sqrt{\mu\varepsilon}}\sqrt{\left(\frac{m}{a}\right)^2 + \left(\frac{n}{b}\right)^2 + \left(\frac{l}{c}\right)^2}$

$$\frac{1}{2\sqrt{\mu\varepsilon}} = \frac{c}{2\sqrt{3}} = \frac{1.5\times 10^8}{1.732}$$

$$f_{r_{101}} = \frac{1.5\times10^8}{1.732}\left[\left(\frac{1}{5\times10^{-2}}\right)^2+\left(\frac{1}{10\times10^{-2}}\right)^2\right]^{1/2}$$

$$= 1.936 \text{ GHz}.$$

The quality factor for TE_{101} mode is given as

$$Q_{TE_{101}} = \frac{(a^2+c^2)abc}{\delta\left[2b(a^3+c^3)+ac(a^2+c^2)\right]}$$

$$= \frac{(25+100)200\times10^{-2}}{\delta\left[8(125+1000)+50(25+100)\right]}$$

$$= \frac{1}{61\delta} = \frac{\sqrt{\pi f_{101}\mu_0\sigma_c}}{61}$$

On putting the value of δ

$$= 10.93\times10^3.$$

Example 4.4. A coaxial resonator is constructed of a section of coaxial line and is open-circuited at both ends. The resonator is 5 cm long and filled with dielectric of $\varepsilon_r = 9$. The inner conductor has a radius of 1 cm and the outer conductor has a radius of 2.5 cm.

(*a*) Find the resonant frequency of the rosonator.

(*b*) Determine the resonant frequency of the same resonator with one end open and one end shorted.

Solution: (*a*) A coaxial resonator is open-circuited at both ends

$$Z_{in} = jZ_0 \cot\beta l = \infty$$

$$\beta l = \frac{2\pi}{\lambda}\cdot\frac{m\lambda}{2} = m\pi$$

m = any integer

Then $$\lambda = \frac{2l}{m} = \frac{c}{f\sqrt{\varepsilon_r}}$$

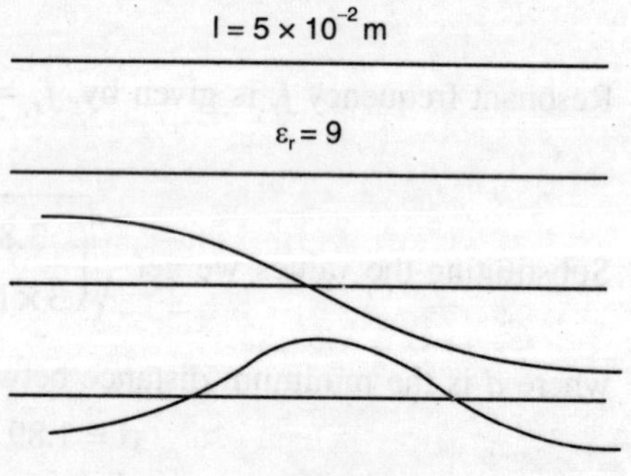

Fig. 4.12

$$f_r = \frac{mc}{2l\sqrt{\varepsilon_r}} = \frac{3\times10^8}{2\times5\times10^{-2}\times3} = 1 \text{ GHz for } m = 1$$

(*b*) A coaxial resonator is open at one end and shorted at the other

$$l = (2m-1)\lambda/4$$

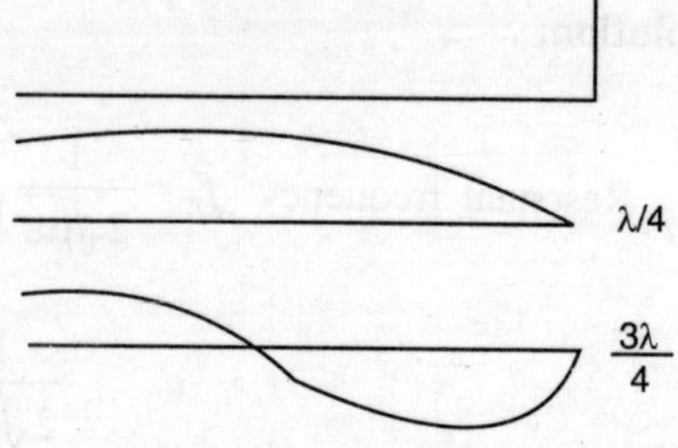

Fig. 4.13

$$f_r = \frac{(2m-1)c}{4l\sqrt{\varepsilon_r}},\ m = 1$$

$$= \frac{3\times10^8}{4\times5\times10^{-2}\times3} = 0.5 \text{ GHz}.$$

Example 4.5. A coaxial resonator is constructed of a section of coaxial line 6 cm long and is short circuited at both ends. The circulator cavity has an inner radius of 1.5 cm and an outer radius of 3.5 cm. The line is dielectric-filled with $\varepsilon_r = 2.25$.

(*a*) Determine the resonant frequency of the cavity for TEM_{001}.

(*b*) Calculate the quality Q of the cavity.

Solution: (*a*) The resonant frequency

$$f_r = \frac{1}{2\sqrt{\mu\varepsilon}}\sqrt{\left(\frac{m}{a}\right)^2 + \left(\frac{n}{b}\right)^2 + \left(\frac{q}{d}\right)^2}$$

Given $a = 1.5$ cm, $b = 3.5$ cm, $d = 6$ cm and $\varepsilon_r = 2.25$

$$f_r = \frac{3\times10^8}{2\sqrt{2.25}}\cdot\frac{1}{6\times10^{-2}}$$

$$= 1.67 \text{ GHz}$$

(*b*) The quality Q of the cavity, assume that high order modes do not exist in the cavity. The fields for the TEM modes are

$$\left.\begin{aligned} E_r &= \frac{A}{r}e^{j\beta z} + \frac{B}{r}e^{-j\beta z} \\ H_Q &= -\frac{A}{rn}e^{j\beta z} + \frac{B}{\ln}e^{-j\beta z} \end{aligned}\right\} \quad ...(1)$$

Let $z = 0$ and $z = l \Rightarrow A = -B$ ...(2)

For $\beta l = n\pi$

$n = \pm1, 2,...$

$\varepsilon_r = 2.25$

$$\eta = \frac{\eta_0}{\sqrt{2.25}} = 251$$

Then $W = \frac{\varepsilon}{2}\iiint\left(\bar{E}\cdot\bar{E}^*\right)dv$

$$= \frac{\mu}{2}\iiint\left(\bar{H}\cdot\bar{H}^*\right)dv \quad ...(3)$$

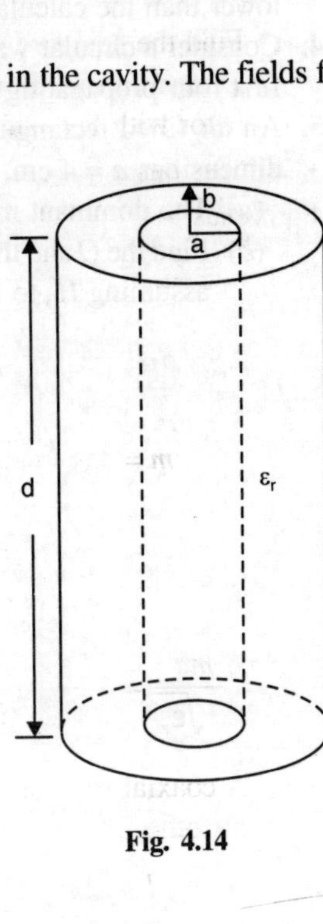

Fig. 4.14

Substitute (1) and (2) in (3), then we have

$$W = \frac{\pi\varepsilon}{2}|A^2|\,l\ln\frac{b}{a}$$

or
$$P = \frac{R_s}{2} \iint_{\text{cavity walls}} \left(\bar{H}_t \cdot \bar{H}_t^* \right) ds \qquad \text{...(4)}$$

where R_s is surface resistance
Substitute (1) and (2) in (4), we have

$$P_1 = \frac{\pi R_s}{2\eta^2} |A^2| l \left(\frac{1}{a} + \frac{1}{b} \right) \text{ and } P_2 = 2\pi R_s \frac{|A^2|}{\eta^2} \ln \frac{b}{a}$$

$$Q = \omega_r \frac{W}{P_1 + P_2} = 2\pi f_r \frac{\varepsilon n^2 d \ln b/a}{R_s \left(\frac{d}{a} + \frac{d}{b} + 4 \ln \frac{b}{a} \right)} = 6880$$

where $R_s = 2.61 \times 10^{-7} \sqrt{f_r}$ for copper surface.

REVIEW QUESTIONS AND PROBLEMS

1. What are cavity resonators? What are their most desirable properties?
2. In terms of field patterns what does TM_{110} and TE_{123} mode signify?
3. Define the quality factor Q of a resonator. Explain why the measured Q of a cavity resonator is lower than the calculated values.
4. Consider a circular waveguide with $a = 0.7$ cm. Determine the cut-off frequencies and identify the first four propagating modes.
5. An air filled rectangular cavity with brass walls ε_0, μ_0, $\sigma = 1.54 \times 10^7$ (S/m) has the following dimensions $a = 4$ cm, $b = 3$ cm and $d = 5$ cm. Determine
   (*a*) The dominant mode and its resonant frequency for this cavity.
   (*b*) Find the Q and the time average stored electric and magnetic energies at the resonant frequency, assuming H_0 to be 0.1 A/m.

CHAPTER 5

Waveguide Component and Application-I

5.1 INTRODUCTION

In previous chapter, we discussed the propagation of microwave power from one point to another point without radiation loss through guiding structure like hollow pipe guides. The rectangular waveguide is widely used guiding structure. For a given operating frequency, waveguide has a greater power handling capacity than coaxial line. Due to absence of center conductor, waveguide components are often less expensive to manufacture than coaxial counterparts, particularly when investment casting and computerised machining techniques are utilised. This chapter describes a variety of reciprocal waveguide component used in microwave systems. Specifically, the material deals with one and two-port networks.

5.2 COUPLING MECHANISMS

The discussion thus far has dealt only with uncoupled cavity resonators. To excite a particular mode, the cavity must be properly coupled to an external source. In Fig. 5.1(*a*) a cylindrical cavity is loop coupled to a coaxial line. The conduction current in the loop produces a linking magnetic field as indicated. Thus the loop is capable of exciting any cavity mode that has a portion of its magnetic field linking the loop. In Fig. 5.1(*b*) illustrate probe-type coupling between a coaxial line and a cavity. The electric field pattern created by the probe is similar to that of the coaxial to waveguide transition. Any cavity mode having an electric field component parallel to the probe can be excited by this coupling arrangement. For the cylindrical cavity shown, this includes all the TE_{nmp} mode except those with $n = 0$. With the probe centrally located along the l-dimension, only TE modes with p an odd integer can be excited since those with p even have an electric field null at the plane of the probe. To produce TE modes with p even, the probe must be offset along the l dimension. When the probe is inserted into one of the end walls of the cavity, a z component of electric field is created, which result in TM modes being excited. The same method can be used to couple into rectangular cavities as well as the coaxial resonator.

In Fig. 5.1(*c*) and (*d*), aperture coupling between a cylindrical cavity and a rectangular waveguide operating in dominant TE_{10} mode. For the slot orientation shown in Fig. 5.1(*c*). The TE_{10} magnetic field couples into the cavity, which allows any of the TM resonant mode to be excited. The coupling arrangement is often used to generate the TM_{01p} modes. When the waveguide and aperture are centrally located along the l-dimension, p must be even since for odd values, the magnetic field is zero at the point in the cavity. The coupling arrangement in Fig. 5.1(*d*), the rectangular guide and aperture have to be

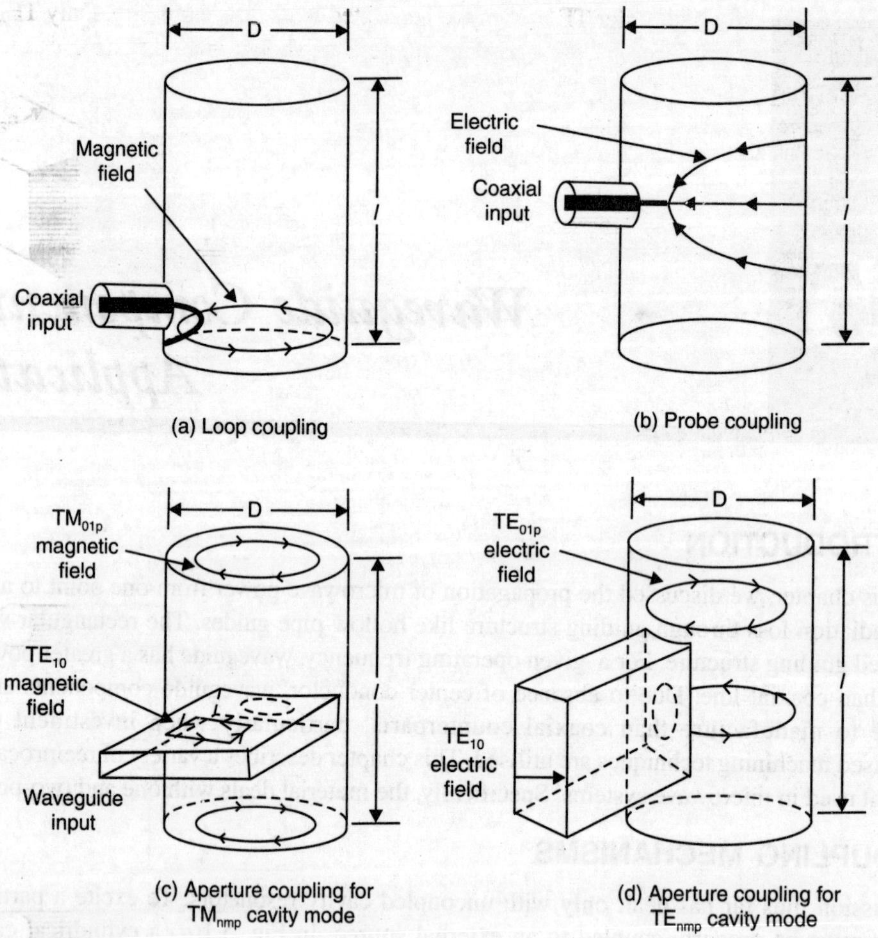

Fig. 5.1 Coupling mechanism

rotated 90°. This allows the TE modes to be excited in the cavity, the TE_{11P} and the TE_{01P} being the most commonly used.

5.3 DISCONTINUITIES – WAVEGUIDE IRISES

Any interruption in the uniformity of a transmission line is defined as a discontinuity. An abrupt change in the *b*-dimensional of the rectangular waveguide or in the *a*-dimensional of waveguide causes a step discontinuity. Other forms of discontinuities in the waveguide circuits are metallic partition extending partially across the guide in a plane perpendicular to the axis. The partition may be of any shape such a partition is called diaphragm or an iris. When a wave is incident, in the neighbourhood of iris higher order mode fields are setup, so that the total field satisfies the proper boundary conditions. The change in guide height as shown in Fig. 5.2(*a*) cause a distortion in the TE_{10} electric field in the vicinity of the step discontinuity. The distorted field can be represented by an infinite set of a propagating TM modes. There presence is required because the distorted pattern contains electric field components parallel to the waveguide axis. It is shown that only TM_{1n} modes with n an even integer are generated. The localized TM modes can be represented by capacitance shunting in the transmission line shown in Fig. 5.3(*a*). Figure 5.2(*b*) shows that change in *a*-dimension of the waveguide can be represented by a shunt inductance. Since only high order TE modes are generated by the discontinuity. Only TE_{mo} modes with m an odd integer are produced by the symmetrical steps as shown in Fig. 5.3(*b*).

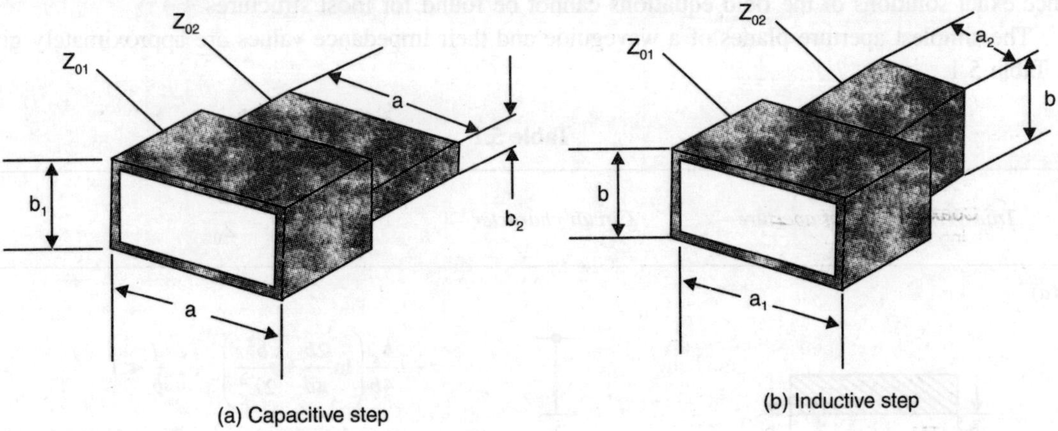

Fig. 5.2 Discontinuities in rectangular waveguide

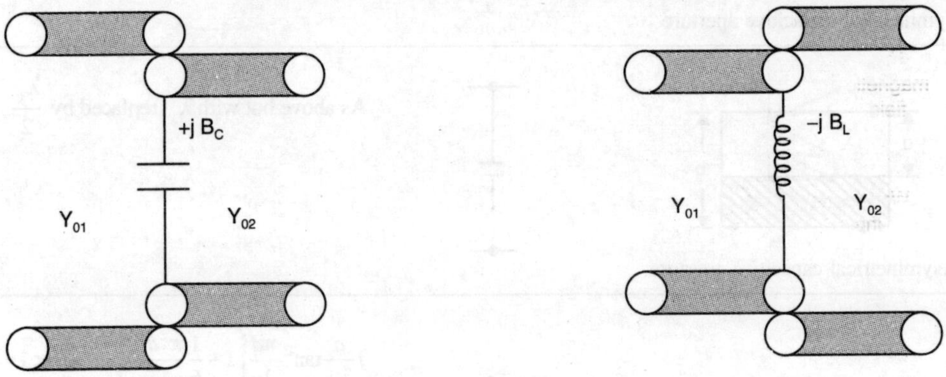

Fig. 5.3 Equivalent circuit

When a dominant mode wave is incident on the iris some of it is reflected from the iris and some of it is transmitted through the opening. An element of this type is essentially a two port device since it includes input and output waveguide connections. This metallic diaphragm along with the reflection less termination presents some impedance or reactance terminating the transmission lines. The metal diaphragm is responsible for the imaginary part of the impedance or the reactance and the magnitude of the imaginary part is proportional to the difference between the stored electric and magnetic energies in the neighbourhood of the iris. In a waveguide far from the iris, the stored electric and magnetic energies are equal.

The irises are classified according to the sign of the imaginary part of the impedance as in low frequency network. Thus an iris that contributes a positive imaginary term to the admittance is called inductive; one that contribute a negative imaginary part to the impedance is called capacitive. Since a metal diaphragm in a waveguide may be considered as a junction between two transmission lines. Its equivalent circuit representation could be that of a two terminal network at low frequencies, for example a T and π-network. Here we assume that the metal diaphragm is sufficiently thin. The impedances can be calculated using variational methods, integral equation methods and method of Green's functions. Most of the solution that have been found using these methods are approximate, since exact solutions of the field equations cannot be found for most structures.

The simplest aperture planes of a waveguide and their impedance values are approximately given in Table 5.1.

Table 5.1

Transverse outline of aperture	*Circuit character*	$\frac{Z}{Z_0}$
(a) 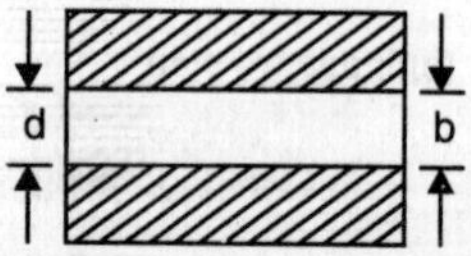 Symmetrical capacitive aperture	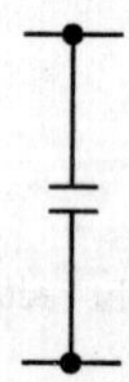	$-j\frac{\lambda_g}{4b}\left(\ln\frac{2b}{\pi d}+\frac{b^2}{2\lambda_g^2}\right)^{-1}$, $\frac{d}{b}\ll 1$ $-j\frac{2b\lambda_g}{\pi^2(b-d)^2}$, $\frac{b-d}{b}\ll 1$
(b) 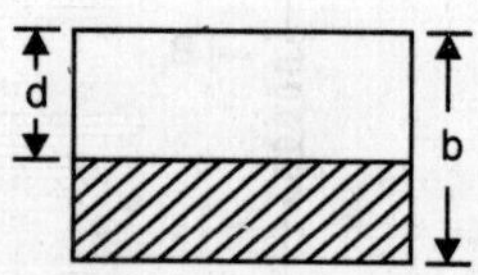 Asymmetrical capacitive aperture	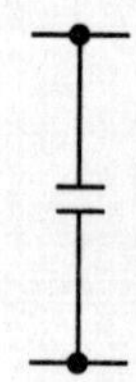	As above but with λ_g replaced by $\frac{\lambda_g}{2}$
(c) 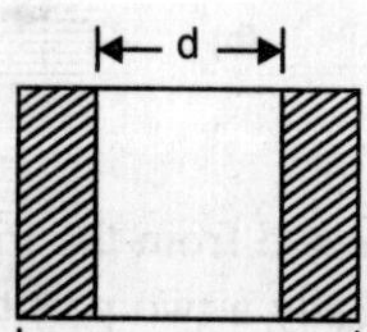 Symmetrical inductive aperture		$j\frac{a}{\lambda_0}\tan^2\frac{\pi d}{2a}\left(1+\frac{1}{6}\frac{\pi^2 d^2}{\lambda^2}\right)$, $\frac{d}{a}\ll 1$ $j\frac{a}{\lambda_g}\cot^2 2\pi\frac{a-d}{a}\left[1+\frac{8\pi^2}{3}\frac{(a-d)^2}{\lambda^2}\right]$, $\frac{a-d}{a}\ll 1$
(d) 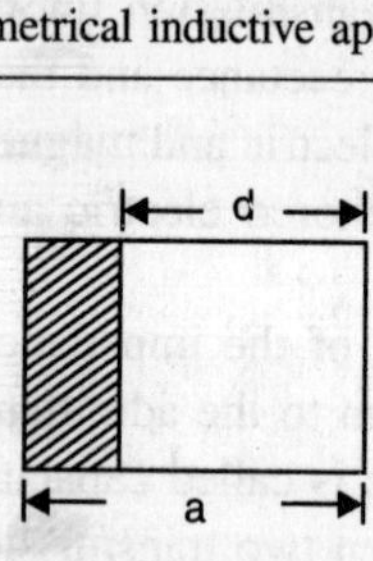 Asymmetrical inductive aperture		$j\frac{2a^3}{\pi^2\lambda_g(a-d)^2}\left[1+\frac{\pi^2}{\lambda^2}(a-d)^2\ln\frac{\pi}{2}\frac{a-d}{a}\right]^{-1}$, $\frac{a-d}{a}\ll 1$

(Contd.)

	Transverse outline of aperture	*Circuit character*	$\frac{Z}{Z_0}$
(*e*)	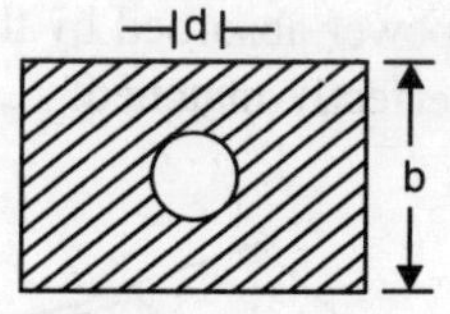 Small circular aperture	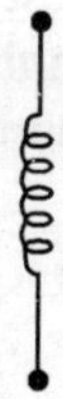	$j\frac{2\pi d^3}{3ab\lambda_g}, \quad d \ll b$
(*f*)	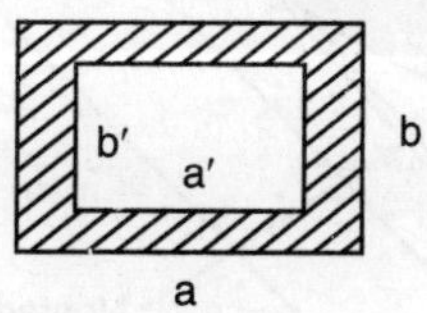 Resonant aperture	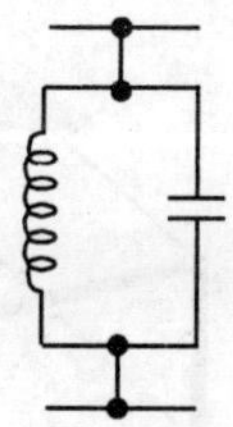	$\frac{Z}{Z_0} \to \infty$ when $\frac{a'}{b'}\sqrt{1-\left(\frac{\lambda}{2a'}\right)^2} = \frac{a}{b}\sqrt{1-\left(\frac{\lambda}{2a}\right)^2}$

5.4 TUNNING SCREW

A tunning screw extends into the waveguide parallel to the *E*-field lines and acts as an impedance element which is much useful, because of its simple construction and ease with which it allows a wide range of admittance to be introduced in parallel with the waveguide as the length of the screw is varied. The tunning screw popularly called a tunning post takes the form of a screw threaded into the broad face of the guide or of a probe with a micrometer, for depth adjustment, and is also provided for change of location along the waveguide axis.

The equivalent circuit behaviour of the post may be represented by a *T*-network shown in Fig. 5.4. The impedance of the shunt arm of the network depends on the length of the post or probe.

At small probe length, the input impedance at the terminals *a a′* of the equivalent circuit is capacitive. As the probe length increases, the capacitive impedance becomes smaller and when the length is approximately $\lambda_0/4$ where λ_0 is a free space wavelength, the tunning screw resonates and presents a vanishing impedance at the location of the screw with further increase in length of the tunning screw, its impedance becomes inductive in character.

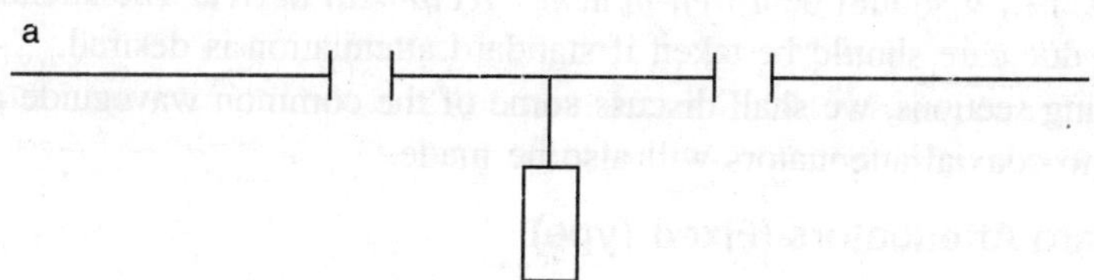

Fig. 5.4 Equivalent circuit of tunning post

5.5 MICROWAVE ATTENUATORS

Attenuation: The term attenuation, usually used in high frequency techniques, is defined as the ratio, expressed in decibels, of the power level or voltage level between two points in a circuit. Clearly, attenuation or (insertion loss) of a component under test will be the ratio, expressed in decibels, of the power absorbed by the load without the component in the line to the power absorbed by the load with the component in the line, when the signal source and the load are perfectly matched.

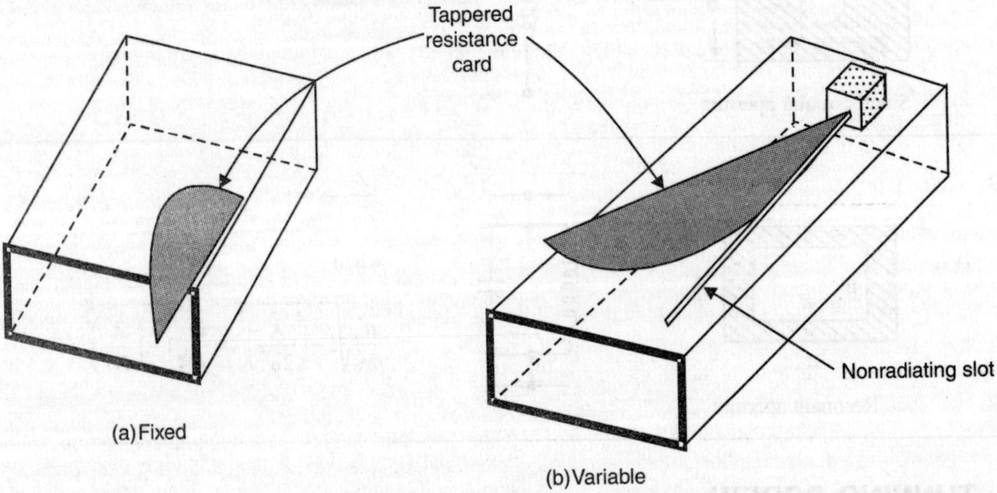

Fig. 5.5 Microwave attenuators

$$\text{Attenuation } A = 10\log_{10}\left(\frac{P_1}{P_2}\right)\text{dB}$$

P_1 = Power absorbed by the load without the given component in the line;
P_2 = Power absorbed by the load with the given component in the line.

Attenuators

The passive elements used to control the amount of microwave power transferred from one point to another on a microwave transmission line are called microwave attenuators. Generally, these elements control the flow of microwave power either by reflecting and/or absorbing it in some dissipating elements. Attenuators may be fixed or variable depending on the requirements.

An ideal attenuator when placed in a transmission line, must present a good impedance match at both the terminals, *i.e.,* it should be a *well-matched reciprocal device.* The attenuation is a function of frequency and so due care should be taken if standard attenuation is desired.

In the following sections, we shall discuss some of the common waveguide attenuators. However, a brief reference to coaxial attenuators will also be made.

5.5.1 Flap or Card Attenuators (Fixed Type)

The fixed type (Fig. 5.6) consists of a dissipative element (pad) placed in a section of the waveguide with its plane parallel to the electric field and at the position where the electric field of the dominant TE_{10} mode is maximum. The dissipating element is generally in the form of card or pad. One class of such pads consists of thin metallic films coated on a glass plate. A baked-on metallic film combining platinum and palladium and also an evaporated film of chromium or nichrome with a protective film of magnesium fluoride have been satisfactorily used. The base material, glass, is chosen because it does not react with the film, its surface is smooth, and it can maintain its original shape at the temperatures we come across. The card is held parallel to the electric field by means of two thin metal rods.

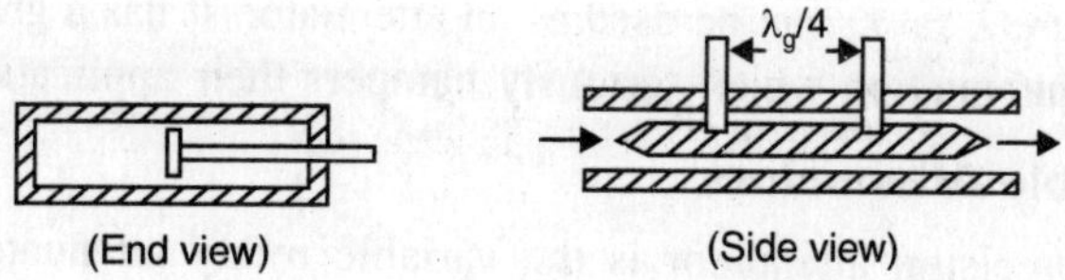

Fig. 5.6 Fixed type flap or card attenuators

To minimize reflections, the rods are held normal to the electric field and spaced quarter-wavelength apart, and the ends of the pad are tapered. When a dominant TE_{10} mode enters the waveguide attenuator, the electric field tangential to the pad causes a dissipative current and is thus absorbed. The amount of microwave power thus dissipated or the attenuation produced depends upon the strength of the electric field, *i.e.*, location of the pad within the waveguide, are of the pad intercepted by the electric field and the frequency.

5.5.2 Variable Attenuators

The waveguide variable attenuators are shown in Fig. 5.7. In Fig. 5.7*a* variable position type attenuator is shown. A nob and gears (not shown in figure) control the movement of the card from the wall to the centre. The amount of attenuation introduced is controlled by changing the position occupied by the absorbing plate inside the waveguide, moving the plate from the narrow wall towards the waveguide axis thus producing an increase in attenuation, which becomes maximum when the plate is moved in the region of maximum electric field intensity, *i.e.*, along the axis.

The second type of waveguide variable attenuator is the variable pad area [Fig. 5.7*b*]. It consists of a slotted waveguide section. The slot is on the centre line of the broad wall and so causes no radiation loss. The attenuation is varied by changing the depth of insertion and hence the intercepted pad area of an absorptive flap into the waveguide through the slot. The attenuation is maximum when the resistive pad or flap extends all the way across the guide and decreases continuously while the card is withdrawn. A suitable deriving mechanism and a dial may be attached to the flap in order to read and calibrate the attenuator.

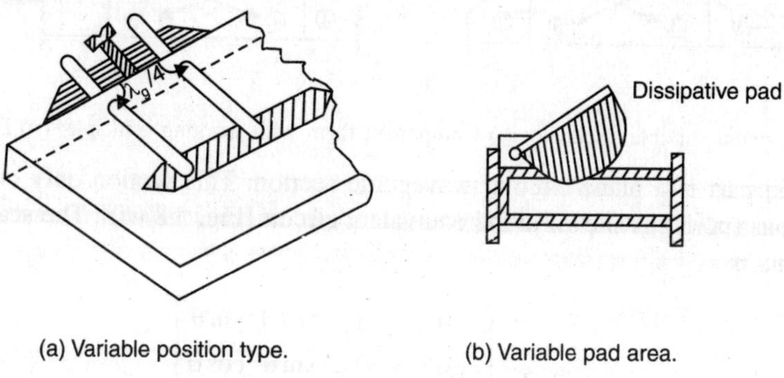

Fig. 5.7 Variable attenuator

5.5.3 Cut-off Attenuators

It was pointed out in chapter 2 that if the dimensions of the waveguide are such that $\lambda >> \lambda_c$ (λ_c is the cut-off wavelength) the wave is exponentially attenuated. Therefore, a waveguide of cross-section sufficiently small to ensure $\lambda >> \lambda_c$ can be used as an attenuator. It has a great operating bandwidth. It has a substantial initial attenuation which seriously hampers their application.

5.5.4 Precision Variable Attenuators

The most satisfactory precision attenuator is the variable rotary attenuator depicted in Fig. 5.8. It comprises the following components:

(*a*) The first part is the tapered rectangular to circular waveguide transition having an attenuating pad placed parallel to the broad wall of the rectangular waveguide.

This component transforms with negligible reflections, the dominant TE_{10} mode in the rectangular waveguide to a TE_{11} mode in a cylindrical waveguide. The orthogonal polarization of the TE_{11} mode, having its electric field parallel to the plane of the attenuating pad is absorbed while the other polarization having electric field perpendicular to the plane of the pad passes unattenuated.

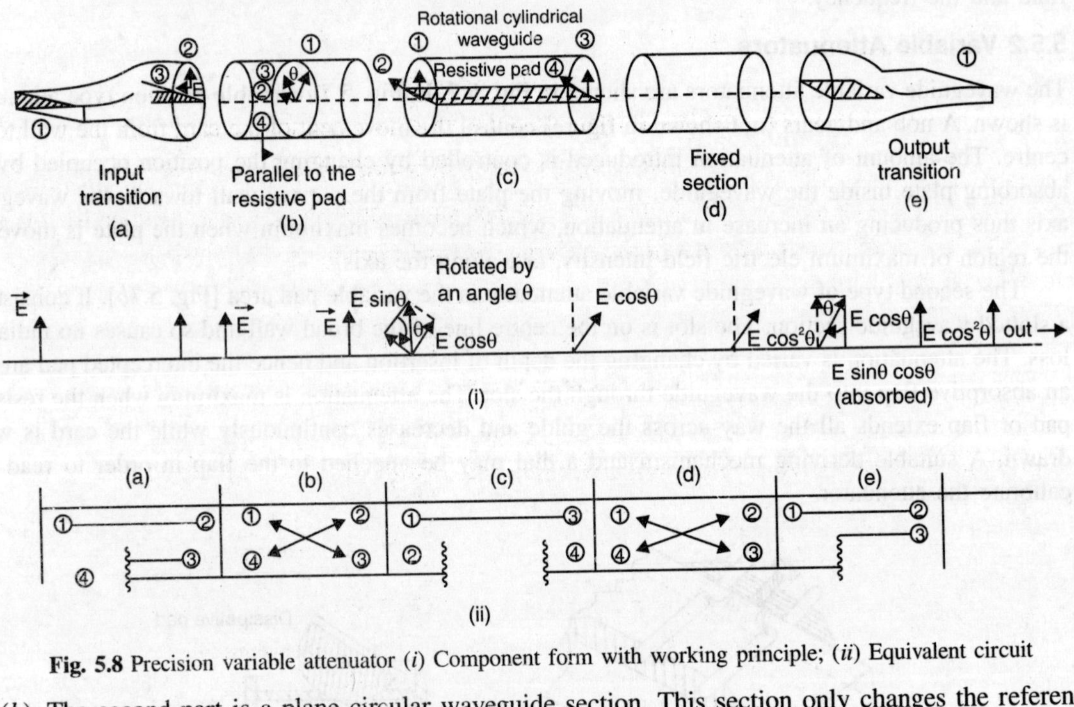

Fig. 5.8 Precision variable attenuator (*i*) Component form with working principle; (*ii*) Equivalent circuit

(*b*) The second part is a plane circular waveguide section. This section only changes the reference polarizations (ports) as shown in the equivalent circuit [Fig. 5.8 (*ii*)]. The scattering matrix of the components is

$$[S] = \begin{bmatrix} 0 & 0 & \cos\theta & \sin\theta \\ 0 & 0 & \sin\theta & \cos\theta \\ \cos\theta & \sin\theta & 0 & 0 \\ -\sin\theta & \cos\theta & 0 & 0 \end{bmatrix} \quad ...(5.1)$$

(*c*) The third and the main component of the instrument is a rotatable circular waveguide section having a tapered resistive pad. At the minimum attenuation condition, the plane of this pad is the same as that of the pad in component (*a*). If the plane of the pad in section (*c*) is rotated through an angle θ, the electric field component $E \sin \theta$, parallel to the resistive pad will be absorbed while the electric field perpendicular to the resistive pad, $E \cos \theta$ passes unaffected.

(*d*) The fourth component is symmetrical with section (*b*).

(*e*) The fifth component is symmetrical with section (*a*).

The electric field $E \cos \theta$ coming out of section (*d*) makes an angle (90 – θ) with the resistive pad of section (*e*); consequently, the normal component of this electric field, *i.e.*, $(E \cos \theta) \cos \theta = E \cos^2 \theta$ passes unattenuated which the tangential component $(E \cos \theta) \sin \theta$ is absorbed. The power transmitted is, therefore, given by

$$P_{\text{trans}} = \frac{1}{R}\left|E \cos^2 \theta\right|^2 = P_{\text{incident}} \cos^2 \theta$$

Or the attenuation is

$$A = -20 \log_{10} \cos^2 \theta$$

$$= -40 \log_{10} \cos \theta \text{ dB} \quad \text{...(5.2)}$$

It is clear from Eqn. (5.2) that the attenuation depends only on the angle of rotation θ. This suggests that the attenuator may be precisely calibrated. Note that the scale will not be linear.

5.5.5 High Power Attenuators

The attenuators discussed so far cannot be used at high powers because of high heat production when power is dissipated in the attenuator pads. A typical high power microwave attenuator consisting of a waveguide hybrid junction is shown in Fig. 5.9. One port of the hybrid is terminated in a variable short-circuit in series with one of the two connecting lines of a cascade arrangement of two hybrids. The high power signal is dissipated in a high power load such as circulating air or water.

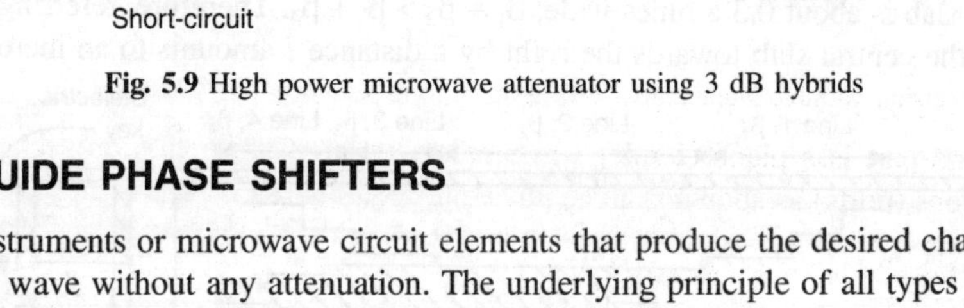

Fig. 5.9 High power microwave attenuator using 3 dB hybrids

5.6 WAVEGUIDE PHASE SHIFTERS

These are the instruments or microwave circuit elements that produce the desired change in the phase of a propagating wave without any attenuation. The underlying principle of all types of phase shifters may be summarized as follows. When a wave propagates on a line, a phase difference prevails between any two arbitrary points along its path. Usually, it is constant and is determined by the velocity of propagation and hence a function of medium parameters. The phase difference between two points, *l* distance apart can be written as

$$\Delta\phi = \phi_2 - \phi_1 = \beta l = \frac{2\pi}{\lambda} l. \quad ...(5.3)$$

Note that $\Delta\phi$ may be varied by varying the velocity of propagation and this is the underlying principle of all phase shifters.

5.6.1 Dielectric Phase Shifter

A simplest waveguide phase shifter consists of a rectangular waveguide loaded with a dielectric slab of thickness t, height h and dielectric constant ε in such a way that dimension h is parallel to the electric field of the dominant TE_{10} mode as shown in Fig. 5.10. The dielectric inserted reduces the velocity of propagation of microwaves which results in an increased electrical path.

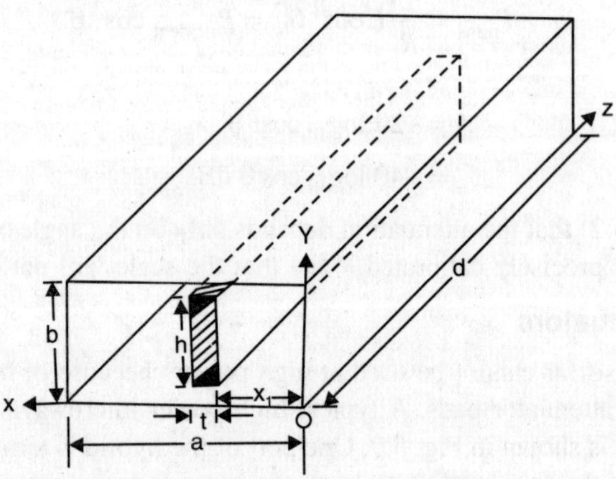

Fig. 5.10 Waveguide loaded with a dielectric slab

5.6.2 Linear Phase Shifter

A typical linear dielectric phase shifter (Fig. 5.11) consists of three dielectric slabs placed in a rectangular waveguide such that the central slab is movable longitudinally by a suitable drive mechanism. A dielectric key protruded through a long centred slot cut in one broad face of the guide is attached to the central slab. The ends of the slabs are so cut that these constitute multisection broad band quarter-wave transformers that provide a match between unloaded and dielectric loaded waveguides. It is found that when the central slab is about 0.3 a times wide, $\beta_1 + \beta_2 > \beta_2 + \beta_4$. Therefore, referring to Fig. 5.11, the displacement of the central slab towards the right by a distance l amounts to an increase in the length

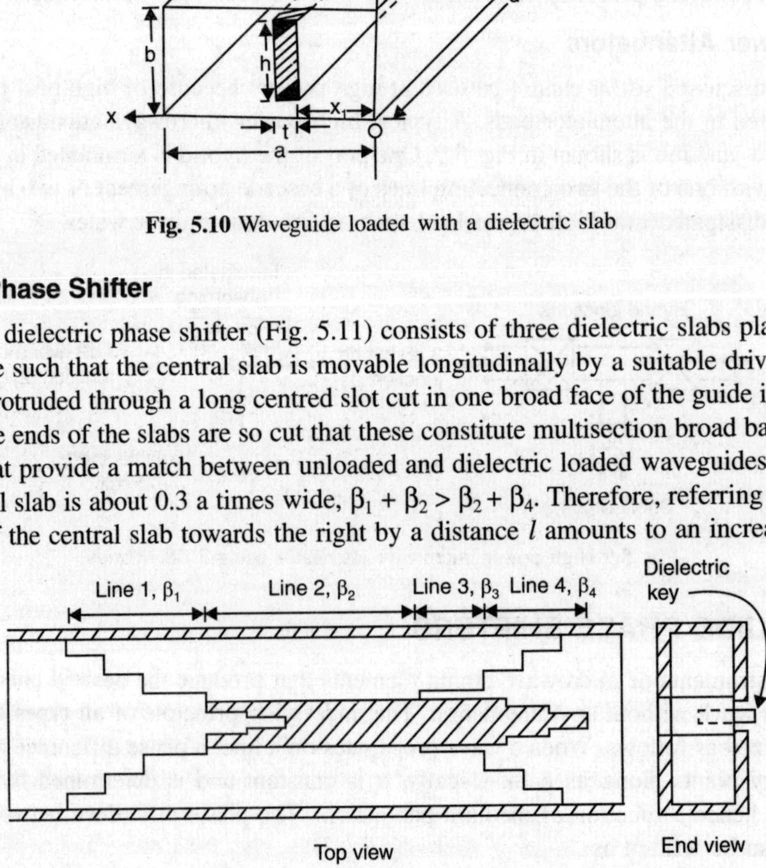

Fig. 5.11 Linear dielectric phase shifter

of the lines 1 and 3 by an amount l and to decrease in the length of the lines 2 and 4 by the same amount. Consequently, the wave propagating through the phase changer undergoes a phase change

$$\Delta\phi = \{(\beta_1 + \beta_3) - (\beta_2 + \beta_4)\}l \qquad ...(5.4)$$

Thus varying l, the desired phase change may be obtained. However, there is an upper limit to the phase change determined by the value of β_1, β_2, β_3 and β_4. The central slab has to be displaced by about 16 cm to obtain a 360° phase shift for a guide 2.28 cm loaded with a dielectric $\varepsilon = 2.56$.

5.6.3 Differential Phase Shifter: Quarter and Half-wave Plates

Consider a circular waveguide diametrically loaded with a thin dielectric slab as shown in Fig. 5.12. Clearly the dominant TE_{11} mode will suffer maximum phase delay when its electric field lies in the principal dielectric plane and minimum (almost zero) when it lies perpendicular to it.

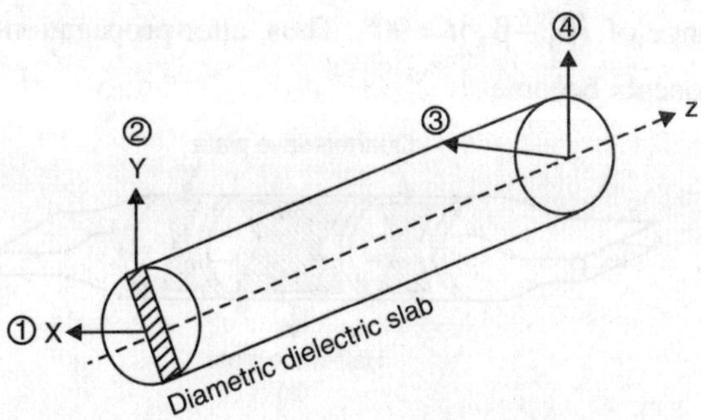

Fig. 5.12 Quarter-half-wave plate using dielectric slab in a circular waveguide

Now a wave approaching the slab, polarized at some intermediate angle may be considered to be made up of two components, one lying in the plane of the slab and another perpendicular to it. Further, if both these degenerate TE_{11} modes are matched (by suitably tapering the slab), then while passing through such a section the mode polarized parallel (||) to the slab will be retarded more than the one perpendicular ($\perp$) to it. After passing through the dielectric region the two degenerate mode combine but the wave suffers a net differential phase shift. Such a device is called a differential phase shifter.

The differential phase shift obtainable from the differential phase shifter depends upon the length, the dielectric constant and the thickness of the slab. If these parameters are suitably adjusted, it is not difficult to obtain a differential phase shift of $\pi/2$ or π and accordingly we can have quarter or half-wave sections. These are respectively designated as $\angle$ 90° and $\angle$ 180°.

5.6.4 Precision Rotary Phase Changer

A precision rotary phase changer is shown in Fig. 5.13 (*a*). The instrument consists of two rectangular to circular waveguide tapered transitions, together with two quarter-wave sections on both the sides of the free rotatable central half-wave section. The quarter-wave sections are oriented at an angle 45° relative to the broad wall of the rectangular waveguide. The incoming linearly polarized TE_{11} mode is decomposed into two modes polarized parallel and perpendicular to the quarter-wave section. When a half-wave section happens to be in its zero set position, the outgoing wave suffers a total phase of 90° + 180° = 270°. Consequently, the wave going out of the second quarter-wave section suffers a total phase 270° + 90° = 360°, *i.e.*, no phase change under ideal conditions. However, when the central half-wave section is rotated by an angle θ, the outgoing wave suffers a phase delay of 2θ.

The basic operation of the device can be understood with reference to Fig. 5.13(*b*) which explicitly shows all the components except transitions.

$E_i < 0$ represents the vertically polarized input wave in the input quarter-wave section. It is decomposed into two transverse components, one E_1, polarised parallel and other, E_2 perpendicular to quarter-wave plate (Fig. 5.13*c* (*i*)). After propagation through the quarter-wave plate these two components are

$$E_1 = E_i \cos 45° e^{-j\beta_1 l} = E_0 e^{-j\beta_1 l}$$

$$E_2 = E_i \sin 45° e^{-j\beta_2 l} = E_0 e^{-j\beta_2 l} \qquad ...(5.5)$$

where $E_0 = \dfrac{E_1}{\sqrt{2}}$. The length l is adjusted such that these two components will have equal magnitude but a differential phase change of $(\beta_1 - \beta_2)l = 90°$. Thus, after propagation through the quarter-wave plate these two field components become

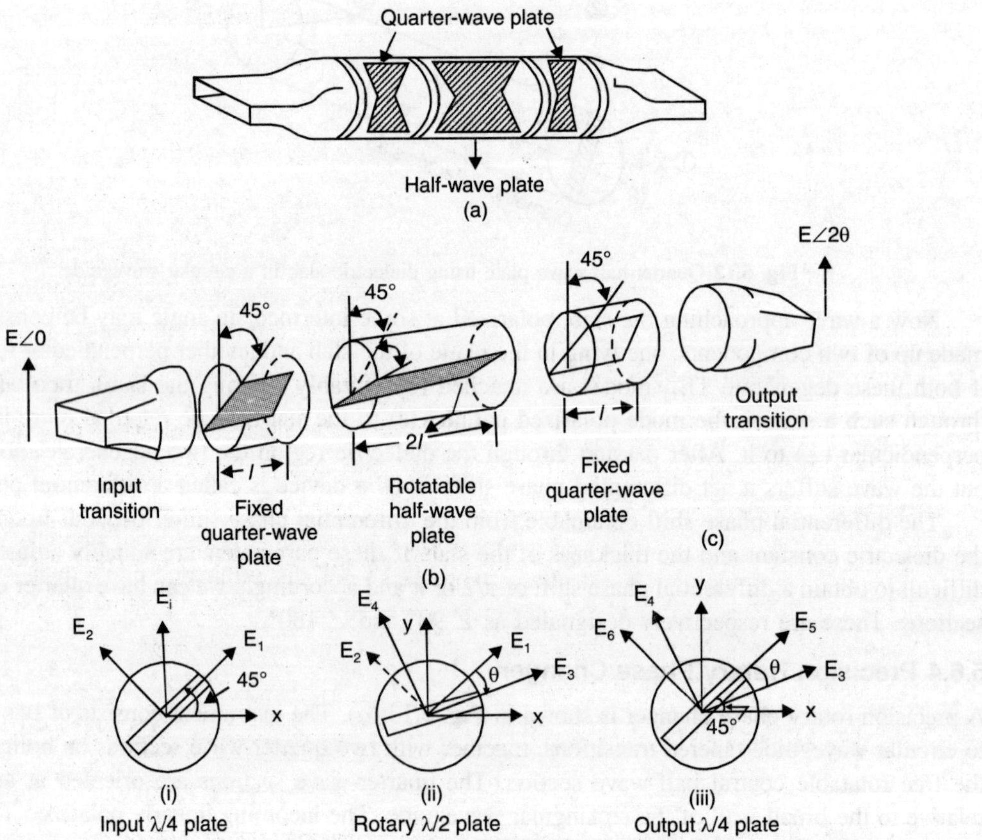

Fig. 5.13 Precision rotary phase shifter: (*a*) Compact form (*b*) Component form (*c*) Vector-phasor

$$E_1 = E_0 e^{-j\beta_2 l}$$

$$E_2 = jE_0 e^{-j\beta_2 l} = -jE_1 = E_1 e^{j\pi/2} \qquad ...(5.6)$$

We note that the quarter-wave section converts a linearly polarized TE_{11} wave to a circularly polarised wave and vice versa,

After propagation through half-wave plate the field components parallel and perpendicular to the half-wave plate are as (Fig. 5.13*c*(*ii*)).

$$E_3 = (E_1 \cos\theta - E_2 \sin\theta)e^{-j2\beta_1 l} = E_0 e^{-j\theta} e^{-j3\beta_1 l} \quad ...(5.7)$$

and

$$E_4 = (E_2 \cos\theta + E_1 \sin\theta)e^{-j2\beta_2 l} = E_0 e^{-j\theta} e^{-3\beta_1 l} e^{-j\pi/2}$$

Since

$$2(\beta_1 - \beta_2)l = \pi \text{ or } -2\beta_2 l = (\pi - 2\beta_1 l)$$

These two field components (E_3 and E_4) are again decomposed after propagation through half-wave plate into two TE_{11} modes, polarised parallel and perpendicular to the output quarter-wave plate (Fig. 5.13*c*(*iii*)). At the output end of this quarter-wave plate, the decompose components can be written as

$$E_5 = (E_3 \cos\theta + E_4 \sin\theta)e^{-j\beta_1 l} = E_0 e^{-j2\theta} e^{-j4\beta_1 l} \quad ...(5.8)$$

$$E_6 = (E_4 \cos\theta - E_3 \sin\theta)e^{-j\beta_2 l} = E_0 e^{-j2\theta} e^{-j4\beta_1 l}$$

We find from Eqn. (5.8) that parallel component E_5 = perpendicular component E_6 at the output end in magnitude as well as in phase. Therefore, these two will produce a resultant field which in linearly polarised TE_{11} wave

$$E_{out} = \sqrt{2}\, E_0 e^{-j2\theta} e^{-j4\beta_1 l}$$

$$= E_i e^{-j2\theta} e^{-j4\beta_1 l} \quad ...(5.9)$$

Which has the same direction of polarisation as the incident field E_1 with a phase charge of $(2\theta + 4\beta_1 l)$. At a given frequency $4\beta_1 l$ is fixed and angle θ can be varied hence a phase shift of 2θ can be obtained by rotating the half-wave plate precisely through an angle θ with respect to the quarter-wave plate.

The simple dependence of the phase change on a mechanical rotation may be used to calibrate the phase changer precisely and is the chief advantage of the rotary phase changer.

5.7 DIRECTIONAL COUPLER

A directional coupler is a useful hybrid waveguide joint which couples power in an auxiliary waveguide arm in one direction. It is a four-port device but one of the ports is terminated into a matched load. To understand its operation we consider the block diagram in Fig. 5.14.

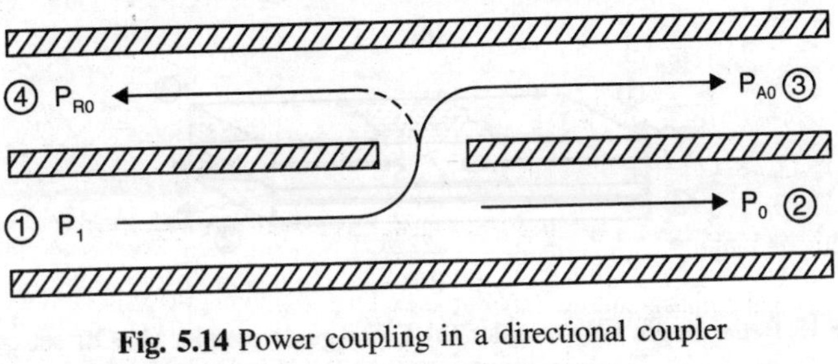

Fig. 5.14 Power coupling in a directional coupler

(*i*) If power is fed into port (1), the power is coupled in ports (2) and (3), *i.e.*, power flows in the forward direction of the auxiliary arm port (3) but no power couples in port (4), *i.e.*, in backward direction. Similarly power fed in (2) couples into ports (1) and (4) and not in (3).

(*ii*) All the four ports are matched, *i.e.,* if three of them are terminated in matched loads, the fourth is automatically terminated in a matched load.

(*iii*) If power couples in reverse direction, *i.e.,* power fed in (1) appears in ports (2) and (4) and nothing in (3), then such type of coupler is known as backward *directional coupler.* The conclusion is that in the auxiliary section the power is coupled in only one direction.

5.7.1 Directivity and Coupling of a Directional Coupler

The characteristics of directional couplers are expressed in two parameters namely *directivity* and *coupling* defined as follows:

(*a*) **Directivity** (*D*) **of a directional coupler** is defined as the ratio, expressed in decibels, of the power output in the coupled auxiliary arm to the power flowing in uncoupled auxiliary arm. Theoretically, *D* should be infinite because no power should couple in arm 3.

(*b*) **Coupling factor or coupling** (*C*) is defined as the ratio, expressed in decibels, of the power (entering) at input of the main line to the power coupled at the output in the auxiliary arm.

The high value of *C* is desired. Both the parameters are frequency-dependent. Hence, to have a wide-band directional coupler, ideal conditions cannot be achieved. However, total directivity of the signal, in addition to frequency, depend upon reflections from internal termination, intrinsic directivity and reflections from the output flange.

5.7.2 Types of Couplers

There are a variety of ways in which auxiliary arm may be coupled to the main guide to have a directional coupler. The most common is the aperture-coupled based on Bethe hole theory.

(a) Bethe-hole coupler (Single hole)

This type of directional coupler consists of two rectangular waveguides coupled by means of a circular aperture located at the centre of the common broad wall. To achieve directional coupling, the axes of the two waveguides are inclined at an angle θ° (Fig. 5.15*a*), determined by frequency and the dimension of the hole. If θ is zero then the hole is to be taken offset and not at the centre. This construction may be understood by considering TE_{10} mode incident on port (1).

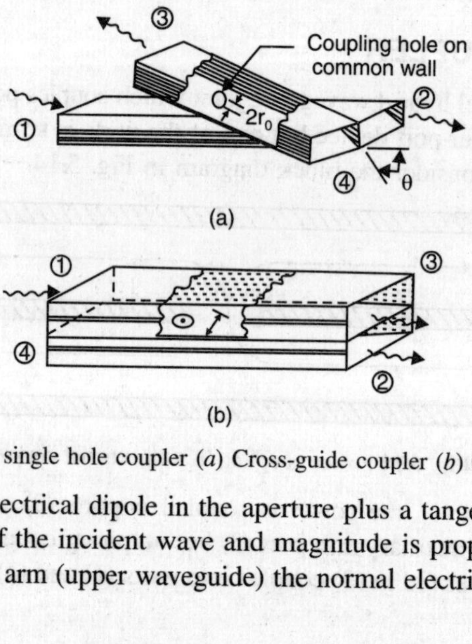

Fig. 5.15 Bethe single hole coupler (*a*) Cross-guide coupler (*b*) Off set coupler

This produces a normal electrical dipole in the aperture plus a tangential magnetic dipole, whose direction is the same as that of the incident wave and magnitude is proportional to the incident wave strength. Now, in the auxiliary arm (upper waveguide) the normal electric dipole and axial component of magnetic dipole radiate symmetrically in both the directions but the transverse component of magnetic dipole radiates antisymmetrically, so proper adjustment of the axes of two waveguides at an angle θ causes the complete cancellation of backward power in the auxiliary arm. This cancellation may also be achieved by offsetting the hole as shown in Fig. 5.15*b*.

By varying the angle θ between the two waveguides or by changing the distance *d*, powers in port 3 and port 4 can be controlled. When the transmission take place from port 1 to port 2, the uncoupled port is called *isolated port* (power coupled = 0), however practically it is not zero. Coupling and directivity of this Bethe-hole coupler (centre hole) are given by

$$C = -20\log \frac{4}{3}\frac{\beta r_0^3}{ab}\left(\cos\theta + \lambda_g^2/2\lambda^2\right)\text{dB} \qquad ...(5.10)$$

$$D = 20\log \frac{2\beta^2\cos\theta + k_0^2}{2\beta^2\cos\theta - k_0^2}\left(\cos\theta - \lambda_g^2/2\lambda^2\right)dy \qquad ...(5.11)$$

Where r_0 = Radius of hole

$a \times b$ = Waveguide cross-sectional dimension

$$\beta = 2\pi/\lambda_g$$

$$\lambda_g = \frac{\lambda_0}{\left[\sqrt{1-(\lambda_0/2a)^2}\right]}$$

and

$$k_0 = 2\pi/\lambda_0$$

The optimum angle θ (for $P_4 = 0$) is given by the relation $\cos\theta = \lambda_g^2/2\lambda_0^2$.

For the off-set hole, the optimum forward coupling ($P_4 = 0$) is obtained for the off-set value '*d*' given by

$$\sin(\pi d/a) = \frac{\lambda_g}{\sqrt{6}\,a}$$

The maximum coupling and directivity are:

$$C = 20\log\left[(1+x^2)/x\right] \qquad ...(5.12)$$

$$D = 20\log 1/x \qquad ...(5.13)$$

where

$$x = \left(16\pi r_0^3/3ab\lambda_g\right)\sin^2(\pi d/a)$$

In these relations the guide wall thickness is assumed negligibly small. For the derivation readers are referred to J.L. Altman, op. cit.

(b) Double-hole or multi-hole couplers

Two-hole couplers are constructed by coupling two identical rectangular waveguides by means of two identical apertures spaced by a distance of $\lambda_g/4$, *i.e.*, quarter guide wavelength apart. In this type of coupler, as shown in Fig. 5.16*a*, the signal which travels back to the first hole from the second hole is 180° out of phase so the two signals tend to cancel in the backward direction. The signals in the forward direction are in phase hence they add together. If the distance between the holes is other than $\lambda_g/4$, the said characteristics of coupling and directivity cannot be achieved so this type of coupler is a narrow-band coupler. It is to be noted that directivity and coupling parameters are the sums of inherent property of the single Bethe-hole and the one associated with array (two-hole array in this case).

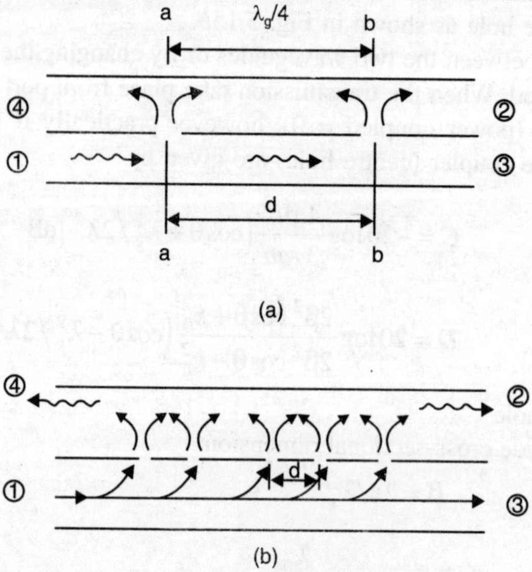

Fig. 5.16 Bethe hole coupler (*a*) Two-hole type (*b*) Multi-hole type

Multi-hole couplers (Fig. 5.16*b*) operate on the same basic principles as two-hole coupler. The coupling array holes are each separated by a distance of $\lambda_g/4$. The hole dimensions are so varied that a wide frequency band is covered. Here coupling versus frequency variations are ± 0.5 dB over the entire waveguide range. The holes are drilled in the broad side of the waveguides for high power operation.

(c) Schwinger reversed phase coupler

This coupler is designed to make directivity less dependent on frequency. To achieve this, one aperture is made to radiate a field which is in negative direction to the one radiated by another. This acts as a reverse coupler because fields in the auxiliary arm cancel in forward direction but are added together in the reverse direction (Fig. 5.17). Considering TE_{10} mode incident on port (1), the operation of this coupler may be explained as before.

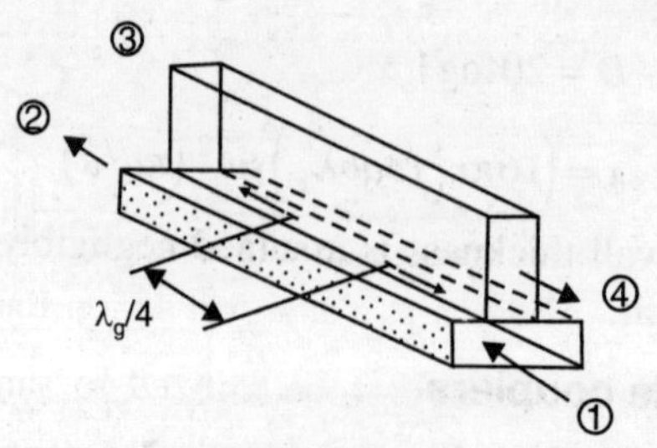

Fig. 5.17 Schwinger reversed phase coupler

In applications where multi-hole coupler is not needed, we employ less expensive and more compact couplers as cross-guide coupler and *T*-slot coupler (Figs. 5.18 and 5.19). Although directional coupler is a four-port device, but as frequently stated earlier, one of its uncoupled arms is terminated and it is thus a three-port device as shown in Fig. 5.20.

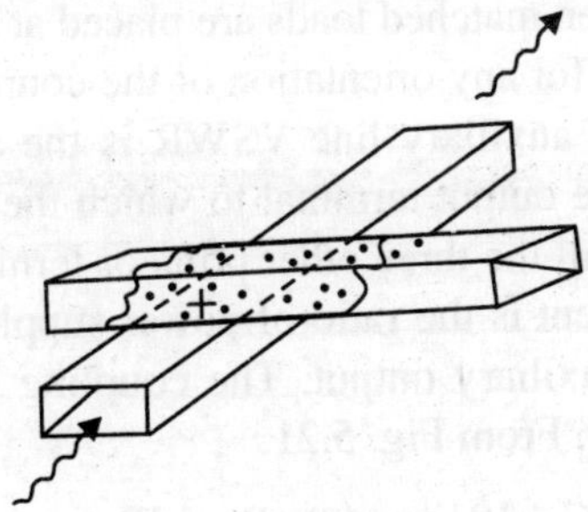

Fig. 5.18 Double aperture cross-guide coupler

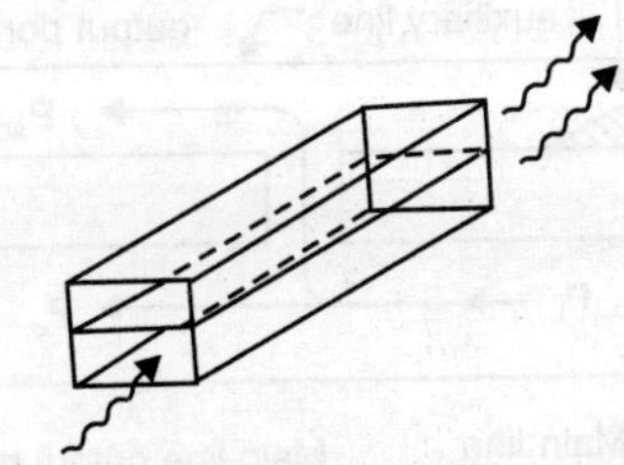

Fig. 5.19 Waveguide *T*-slot coupler

Directional couplers are used in bench test set-ups in microwave systems. They are ideal for power monitoring, frequency, reflection coefficient, noise and pulse shape detection. They are also used for injecting local oscillator and coding signals into a transmission line as well as for the connection of subsidiary equipment to the line. Broad-band applications include measuring low level VSWR, tuning waveguide systems and measuring VSWR by refractometer set-ups. The cross-guide couplers find special applications in monitoring power in waveguide transmission lines, in measurement of reflections, in attenuation of power, in decoupling microwave instruments such as frequency meters, in mixing microwave signals and in measuring receiver sensitivity.

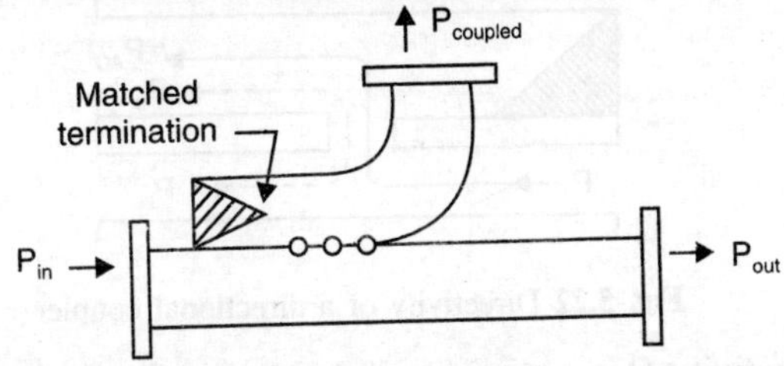

Fig. 5.20 Directional coupler as a three-port device: Unidirectional coupler

The unterminated directional couplers can be utilized to sample incident and reflected power simultaneously.

5.7.3 Directional Coupler Parameters

We now study the characteristics of a directional coupler. We will measure (*i*) main line and auxiliary line VSWR as a function of frequency, (*ii*) coupling coefficient as a function of frequency, (*iii*) directivity as a function of frequency, and (*iv*) the main line insertion loss as a function of frequency. These parameters are defined as follows:

(*i*) **Main line VSWR:** The main line VSWR is the standing wave ratio measured looking into the main line input terminals when matched loads are placed at all the three other terminals. The main line VSWR will be the same for any orientation of the coupler in the transmission line.

(*ii*) **Auxiliary line VSWR:** The auxiliary line VSWR is the standing wave ratio measured in the auxiliary line looking into the output terminal to which the detector is normally connected when matched loads are placed at all the three other ports or terminals.

(*iii*) **Coupling:** Coupling coefficient is the ratio of power supplied to the main line input (P_i) to the power output (P_{AO}) at the auxiliary output. The coupling coefficient C is usually expressed in decibels as a positive number. From Fig. 5.21.

$$C = 10\log_{10}\left(P_i/P_{AO}\right)\text{dB} \qquad ...(5.14)$$

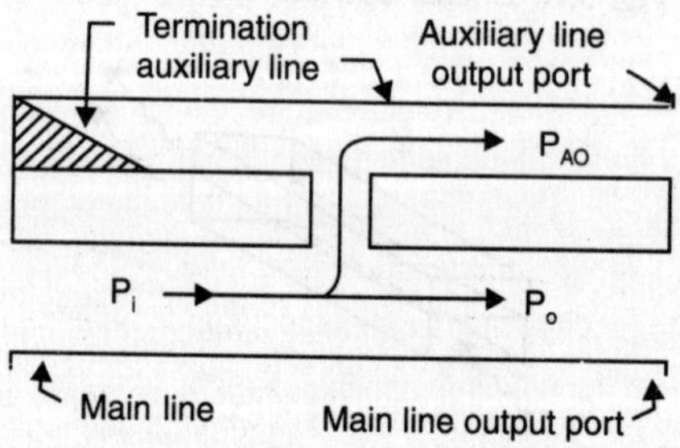

Fig. 5.21 Power coupling in a directional coupler

(*iv*) **Main Line Insertion Loss:** The main line insertion loss is the attenuation introduced in a transmission line by the insertion of the directional coupler. The auxiliary line of the coupler is assumed to be properly terminated. Therefore, referring to Fig. 5.21, the main line insertion loss, L is given by

$$L = 10\log_{10}\left(\frac{P_i}{P_O}\right)\text{dB} \qquad ...(5.15)$$

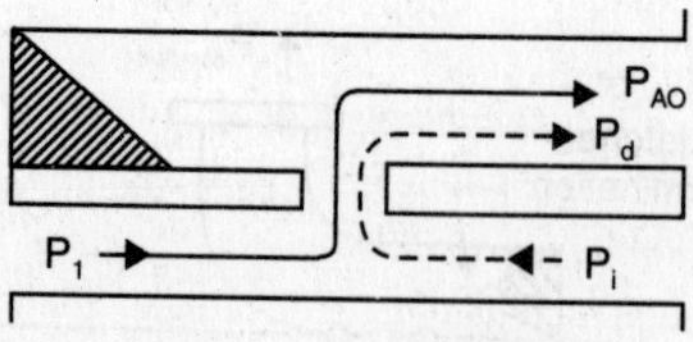

Fig. 5.22 Directivity of a directional coupler

(*v*) **Directivity:** The directivity D is a measure of the discrimination property of a directional coupler between the waves travelling in the two directions in the main line. It is measured as the ratio of the two power outputs from the auxiliary line when a given amount of power is successively applied to each terminal of the main line. The other terminals or ports of the coupler not in use in the particular measurement are assumed to be terminated in matched loads. Referring to Fig. 5.22.

$$D = 10\log_{10}\left(P_{AO}/P_d\right) \qquad ...(5.16)$$

5.8 WAVEGUIDE MULTIPORT JUNCTIONS

Waveguide junctions play an important role in microwave techniques. Most commonly used are *E*-plane or series tee, *H*-plane or shunt tee, and combination of the two, is a hybrid or magic tee. These junctions are used to split the line power into two or combine the power from the two lines with proper consideration of the phase. In this section *E* and *H*-plane tees are described and their properties discussed.

5.8.1 Waveguide *E*-plane Tee

Figures 5.23*a* and 5.23*b* respectively show the sketch of the *E*-phane tee and its equivalent circuit. It is clear from the sketch of the *E*-plane tee that an auxiliary waveguide arm (*E*-arm or arm 3) is fastened to the broader wall of the main guide (arms 1 and 2). Thus it is also a three-port device in which the auxiliary arm axis is parallel to the plane of the electric fields of the main guide and the coupling from the main guide to the auxiliary guide (*E*-arm) is by means of electric fields. Therefore, it is also known as *E*-plane tee. From Fig. 5.23(*a*) and 5.23(*b*), it is clear that it causes load connected to its branches to appear in series. So it is often referred to as a series tee.

As indicated in Fig. 5.23*a*, and 5.23*b*, the two main guide arms are symmetrical with respect to the auxiliary guide arm. As such if power is fed from the auxiliary arm (*E*-arm), it is equally distributed in the two arms 1 and 2 (Fig. 5.23*c*), when they are terminated in equal loads. However, as depicted in the field configuration (Fig. 5.23*c*), the power flowing out in arm 1 is 180° out of phase to the one in arm 2. As such this tee is known as 'subtracter' or 'differencer'.

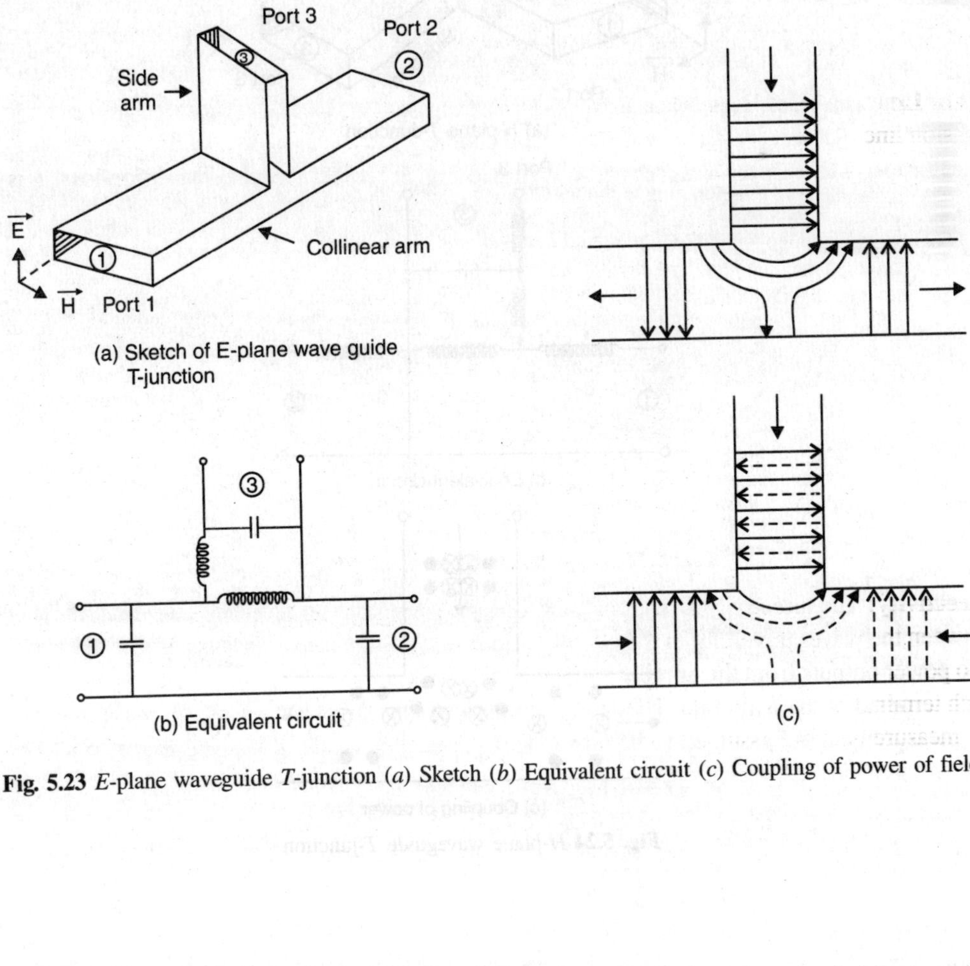

Fig. 5.23 *E*-plane waveguide *T*-junction (*a*) Sketch (*b*) Equivalent circuit (*c*) Coupling of power of fields

5.8.2 Waveguide *H*-plane Tee

Figure 5.24(*a*) shows the sketch of an *H*-plane tee. It is clear from the sketch that an auxiliary waveguide arm (*H*-arm or arm 3) is fastened perpendicular to the narrow wall of a main guide (arms 1 and 2), thus it is a three-port device in which axis of the auxiliary or side arm is parallel to the planes of the magnetic field of the main guide and the coupling from the main guide to the branch guide (arm 3) is by means of magnetic fields. Therefore, it is also known as *H*-plane tee.

The perpendicular arm (*H*-arm) is generally taken as input and other two arms are in shunt to the input and hence it is also called as shunt tee. Because of symmetry of the tee [see Fig. 5.24(*b*); equivalent circuit of *H*-plane tee], when power enters the auxiliary arm (*H*-arm), and the two main arms 1 and 2 are terminated in identical loads, the power supplied to each load is equal and in phase with one another.

If two signals of equal amplitude and in same phase are fed into two main arms (ports) 1 and 2, they will be added together in the side arm (*H*-arm). Thus *H*-plane tee is an 'adder'. This is clear from the equivalent circuit given in Fig. 5.24(*b*).

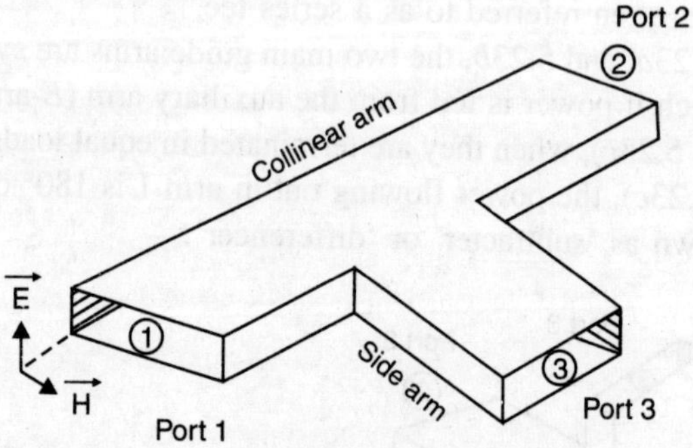

(a) *H*-plane *T*-Junction

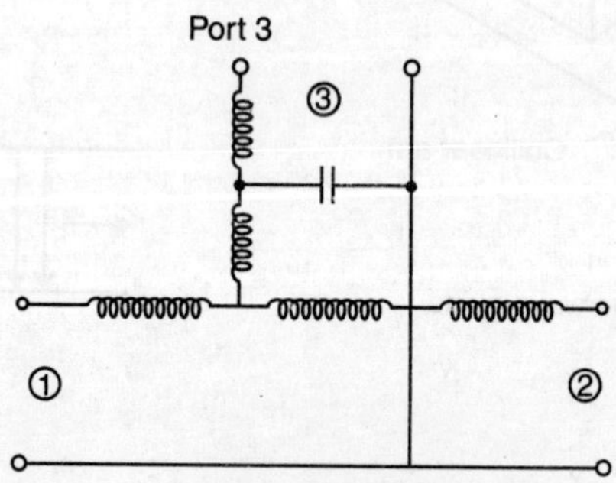

(b) Equivalent circuit

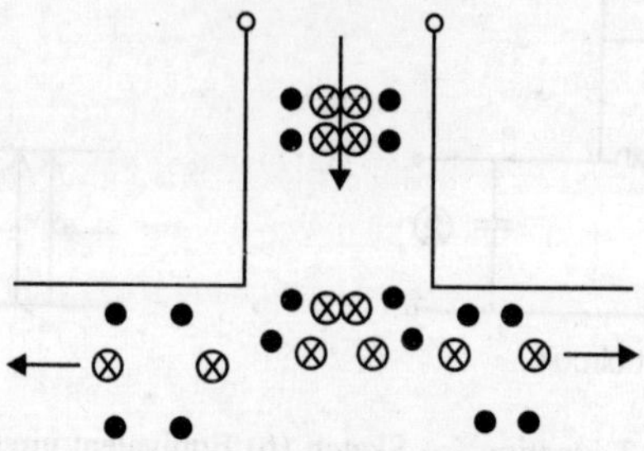

(c) Coupling of power

Fig. 5.24 *H*-plane waveguide *T*-junction

5.8.3 Characteristic Parameters

(a) Isolation

The isolation of a *T*-junction is the ratio of power supplied from a matched generator to one of the arms, to the power coupled to a matched detector in any other arm when the remaining arm is terminated in a matched load. From Fig. 5.25(*a*)

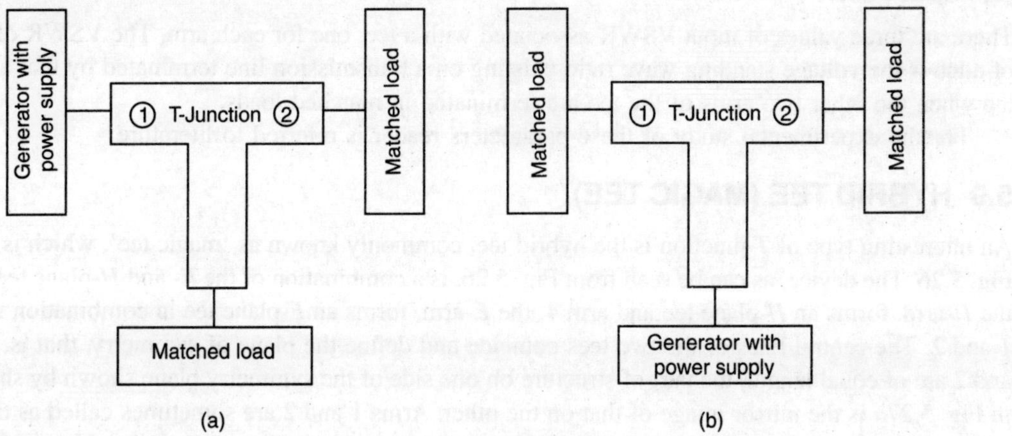

Fig. 5.25 Terminated *T*-junction (*a*) Generator at the input port 1 and matched detector at the port 2; (*b*) Generator at the coupled port 3 and matched detector at port 1.

Isolation between port 1 and 2 is

$$I_{12} = 10\log_{10}\frac{P_1}{P_2}\,\text{dB} \quad \text{...(5.17a)}$$

and when matched load and detector are interchanged,

$$I_{13} = 10\log_{10}\frac{P_1}{P_3}\,\text{dB} \quad \text{...(5.17b)}$$

Similarly from Fig. 5.25(*b*)

$$I_{31} = 10\log_{10}\frac{P_3}{P_1}\,\text{dB} \quad \text{...(5.17c)}$$

and when matched load and detector are interchanged,

$$I_{32} = 10\log_{10}\frac{P_3}{P_2}\,\text{dB} \quad \text{...(5.17d)}$$

When arm 2 becomes the input, we will have other two values of isolation, *i.e.*, I_{21} and I_{23}. Due to reciprocity property, I_{21} will be the same as I_{12}. Therefore, we shall measure only the first four isolation coefficients.

(b) Coupling coefficient

Corresponding to the values of isolation, we can compute the coupling coefficients by the formula

$$C = 10^{-\alpha/20} \quad \text{...(5.18a)}$$

where α is the attenuation in dB between the input and detector arm when the third arm is terminated in a matched load. For example, the attenuation measured between arms 1 and 2 is 3 dB when arm 3 is terminated in matched load, that is, the coupling coefficient between arms 1 and 2,

$$C_{12} = 10^{-\alpha/20} = 10^{-3/20} = 0.708 \text{ dB} \qquad ...(5.18b)$$

(c) Input VSWR

There are three values of input VSWR associated with a tee, one for each arm. The VSWR of any arm of a tee is the voltage standing wave ratio existing on a transmission line terminated by that arm of the tee when the other two arms of the tee are terminated in matched loads.

For the experimental study of these parameters reader is referred to literature.

5.9 HYBRID TEE (MAGIC TEE)

An interesting type of *T*-junction is the hybrid tee, commonly known as 'magic tee', which is shown in Fig. 5.26. The device, as can be seen from Fig. 5.26, is a combination of the *E*- and *H*-plane tees. Arm 3, the *H*-arm, forms an *H*-plane tee and arm 4, the *E*-arm, forms an *E*-plane tee in combination with arms 1 and 2. The central lines of the two tees coincide and define the plane of symmetry, that is, if arms 1 and 2 are of equal length, the part of structure on one side of the symmetry plane shown by shaded area in Fig. 5.27*a* is the mirror image of that on the other. Arms 1 and 2 are sometimes called as the side or collinear arms.

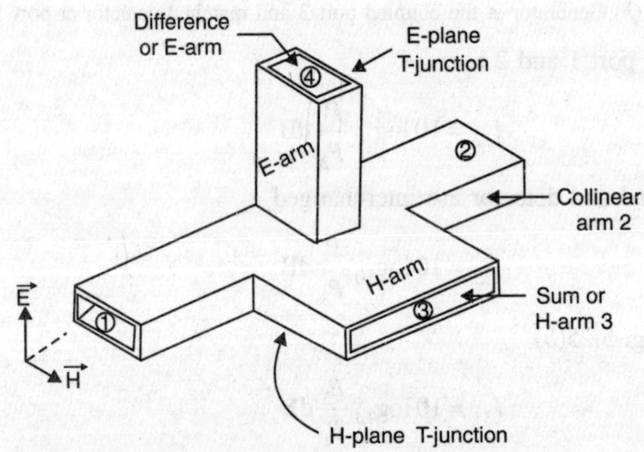

Fig. 5.26 *E-H* tee or magic tee

5.9.1 Magic of the Magic Tee

The name 'magic tee' is derived from the manner in which power divides among various arms. If power is fed into arm 3 (*H*-arm), the electric field divides equally between arms 1 and 2 as shown in Fig. 5.27*b* and the fields are in phase. Because of symmetry of the *T*-junction, no net electric field parallel to the narrow dimension of the waveguide is excited in arm 3. (Required to propagate power in the TE_{10} dominant mode). Thus no power is coupled in port 3. Reciprocity demands no coupling in port 3 (*H*-arm) if power is fed in 4 (*E*-arm).

Another property that results from the symmetry of the junction is, if power is fed in *E*- or *H*-arm, it is equally divided between arms 1 and 2.

Further, magic tee being combination of *E*- and *H*-plane tees, if power is fed from arms 1 and 2, it is added in *H*-arm (3) while it is subtracted in *E*-arm (4).

A simple *E-H* tee has disadvantage of not being matched when seen from *E*- and *H*-arms when side arms are terminated in matched loads. The VSWR being > 2 (generally), the most commonly used method to reduce VSWR is to introduce discontinuity such as port iris in or near *T*-junction to cancel out reflections occurring therein.

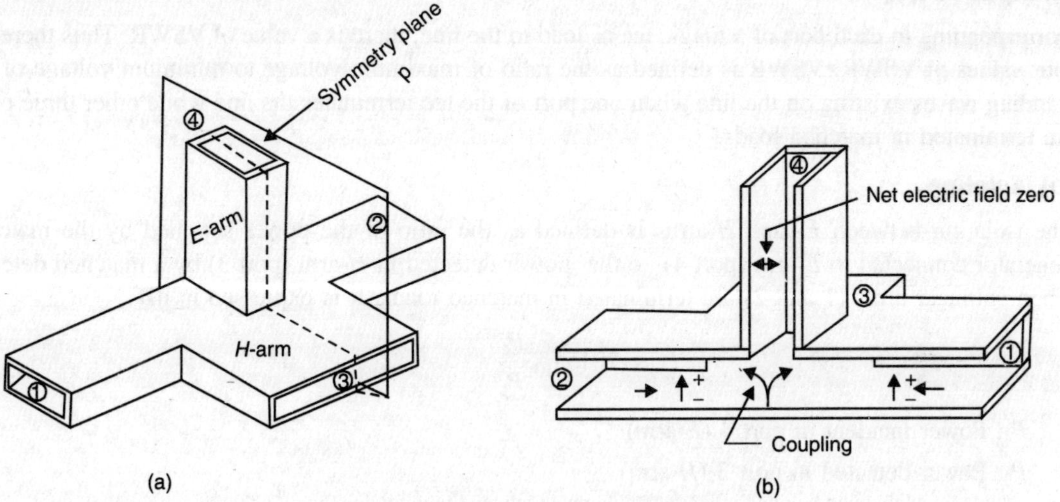

Fig. 5.27 (*a*) Symmetry plane shown by shaded area, (*b*) Cutway rear view of a waveguide *E-H* tee

5.9.2 Rat Race

Another form of hybrid junction is a ring-type structure known as 'rat race' and is depicted in Fig. 5.28. As is clear, it has four ports, each at a circumdistance of $\lambda_g/4$ and hence power incident in port 1 is divided equally in two directions and adds in phase in ports 2 and 4, while it cancels in port 3. Thus port

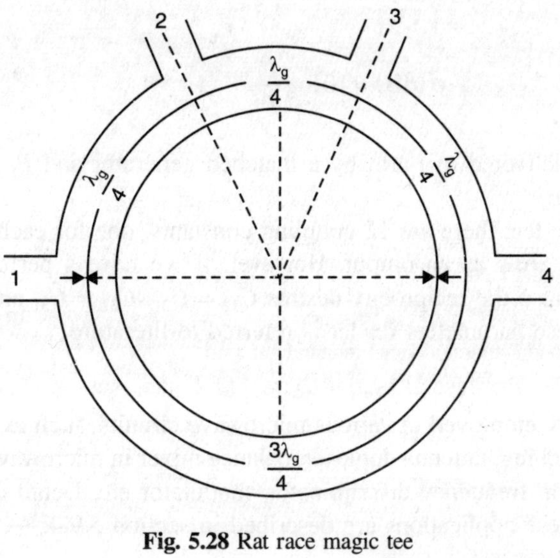

Fig. 5.28 Rat race magic tee

3 is uncoupled to port 1. Similarly, ports 2 and 4 are uncoupled. Sometimes, magic tee is used in mixer, frequency discriminator circuits, etc.

5.9.3 Magic Tee Parameters

The basic properties of a magic tee are defined as follows:

(a) Input VSWR

Corresponding to each port of a magic tee as load to the line, there is a value of VSWR. Thus there are four values of VSWR. VSWR is defined as the ratio of maximum voltage to minimum voltage of the standing waves existing on the line when one port of the tee terminates the line while other three ports are terminated in matched loads.

(b) Isolation

The isolation between *E*- and *H*-arms is defined as the ratio of the power supplied by the matched generator connected to *E*-arm (port 4), to the power detected in *H*-arm (port 3) by a matched detector when collinear arms (1 and 2) are terminated in matched loads. It is expressed in dB.

$$I_{34} = 10\log_{10}\frac{P_4}{P_3} \qquad \text{...(5.19)}$$

P_4: Power incident in port 4 (*E*-arm)

P_3: Power detected in port 3 (*H*-arm)

Similarly isolation between other ports may also be defined and measured.

(c) Coupling coefficient

The voltage coupling coefficient from *i* to arm *j* is defined as

$$C_{ij} = 10^{-\alpha/20} \qquad \text{...(5.20)}$$

where α is the attenuation in dB when *i* is the input and *j* the output arm.

Thus

$$\alpha(\text{dB}) = 10\log\frac{P_i}{P_j} \qquad \text{...(5.21)}$$

where P_i is the power delivered to *i*-arm by a matched generator and P_j is the power detected by a matched detector in arm *j*.

In the case of magic tee, there are 12 coupling constants, one for each of the arms as an input to each of the other three arms as an output. However, if we have a perfectly matched detector and generator, $C_{ij} = C_{ji}$ and also the reciprocity desires $C_{12} = C_{21}$, $C_{32} = C_{31}$ and $C_{41} = C_{42}$.

For the study of these parameters reader is referred to literature.

Applications

A magic tee is frequently employed in various microwave circuits, such as impedance bridge, the *E-H* tuner for impedance matching, antenna duplexer, balance mixer in microwave super-heterodyne receiver to balance phase detector, frequency discriminator, modulator etc. Detail discussion may be found in the literature. Few of these applications are described in section 5.9.4.

5.9.4 Applications of the Magic Tee

Impedance bridges

A magic tee is frequently employed in microwave impedance measuring bridges. These bridges are similar to low frequency wheatstone bridges. Figure 5.29 depicts a typical microwave impedance bridge designed by utilizing a magic tee. Power from a matched source is fed in the *H*-arm (3) of the magic tee. A standard variable impedance is connected to port 1 as reference impedance and arm 2 is terminated in the impedance to be measured. A matched detector is connected to port 4 (*E*-arm) to receive power reflected from arms 1 and 2. These powers will be out of 180° phase.

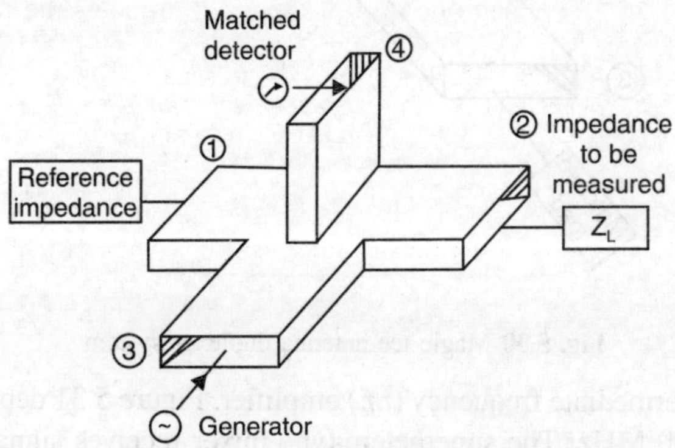

Fig. 5.29 Impedance bridge using magic tee

The reference impedance is adjusted so as to have no signal in the detector. Under this condition the power reflected from the reference impedance and reaching the detector (half of the reflected power) equals the power reflected from the unknown impedance reaching the detector. Because the two powers are out of 180° phase and the lengths of the two ports are equal (or in other words, when the bridge is balanced), then

$$\Gamma_P = \Gamma_\Lambda$$

or

$$\frac{Z_R - 1}{Z_R + 1} = \frac{Z_L - 1}{Z_L + 1} \quad \text{...(5.22)}$$

Γ_P and Z_R are respectively the reflection coefficient and normalized impedance for the reference impedance whereas Γ_Λ and Z_L are the respective quantities for the given unknown impedance. The above equation may be used to determine Z_L in terms of Z_R.

Antenna duplexer

A duplexer system couples two circuits to the same load but avoids mutual coupling. A typical duplexing system is depicted in Fig. 5.30. The same antenna is used for the transmitter as well as receiver but the transmitter power is kept out of the receiver and vice-versa. In this system the matched generator and matched detector are connected to arms 1 and 2 respectively. *H*-arm (3) is terminated in a matched load and *E*-arm (4) is connected to the matched antenna. In this case, power received by the antenna is coupled to the detector due to the coupling properties of the magic tee.

In this *tee,* since half of the power is transmitted, it is useful at low power levels.

Balanced mixer

A balanced microwave mixer is used in superheterodyne receivers to balance out the local oscillator

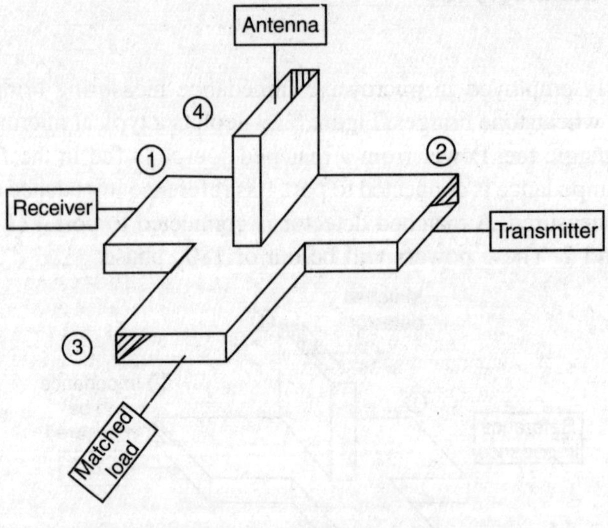

Fig. 5.30 Magic tee antenna duplexer system

noise at the input of the intermediate frequency (*i.f.*) amplifier. Figure 5.31 depicts a typical arrangement for *i.f.* frequencies 1 to 100 MHz. The superheterodyne mixer receives signals from both the antenna and the local oscillator of the receiver but the local oscillator power is prevented from reaching the antenna by the isolation properties of E and H-arms.

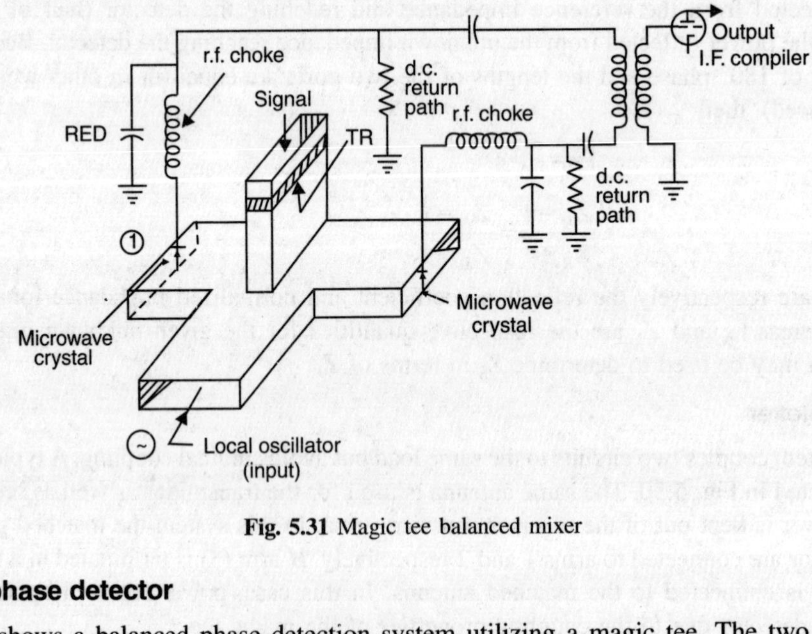

Fig. 5.31 Magic tee balanced mixer

Balanced phase detector

Figure 5.32 shows a balanced phase detection system utilizing a magic tee. The two signals whose relative phase has to be detected, are fed from the E- and H-arms. The powers detected by matched microwave crystals at ports 1 and 2 are fed to the differential amplifier. If the power fed at ports 1 and 2 is equal, *i.e.*, $a_3 = a_4$, then the output power recorded by the differential amplifier will be

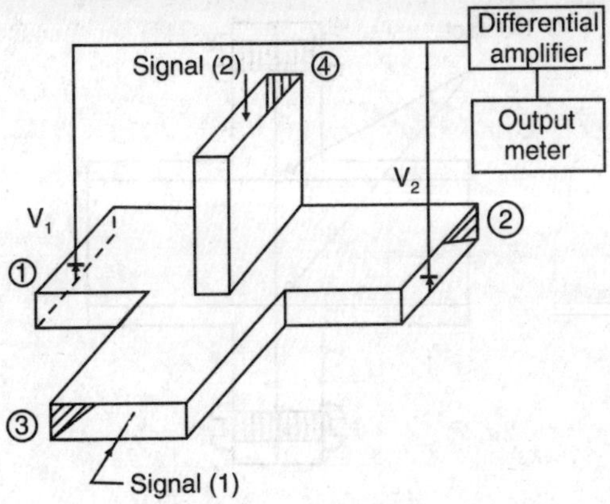

Fig. 5.32 Magic tee balance phase-detector

$$P_{\text{output}} = G\left[\left\{\frac{1}{\sqrt{2}}(a_3 + a_4)\right\}^2 - \left\{\frac{1}{\sqrt{2}}(a_3 - a_4)\right\}^2\right] \quad ...(5.23)$$

G being the amplifier gain factor, now, if there is a phase relation $a_4 = a_3 e^{j\theta}$,
Then

$$P_{\text{output}} \propto G\left(\cos^2\frac{\theta}{2} - \sin^2\frac{\theta}{2}\right)$$

$$\propto G\cos\theta \quad ...(5.24)$$

Thus, phase can be detected and a meter can be calibrated in cos θ to indicate the phase difference directly on the scale.

5.10 WAVEGUIDE CONNECTORS AND FITTINGS

Regardless of the purpose for which waveguide may be used, some means must be available for connecting adjacent sections. The basic requirement is that joint should offer no appreciable discontinuity to flow of microwave power. Various joints are available. In this section we will describe few of these, such as waveguide flanges, bends, twists and corners, transitions, rotary joints etc.

5.10.1 Waveguide Flanges

Rectangular waveguide sections are joined by means of flanges which may be of either the cover type or the choke type.

(*i*) **Cover-flange coupling:** Two waveguide sections with cover-flanges can be bolted together to form a contacting joint as shown in Fig. 5.33.

A cover flange connector has flat, smooth, clean and square contacting surfaces. It avoids reflections and resistive loss at this joint.

In high power applications voltage breakdown may occur at this joint, however, these are widely used because of simplicity of the cover-to-cover connection. In general, the VSWR from such a joint is of the order of 1.03 or less.

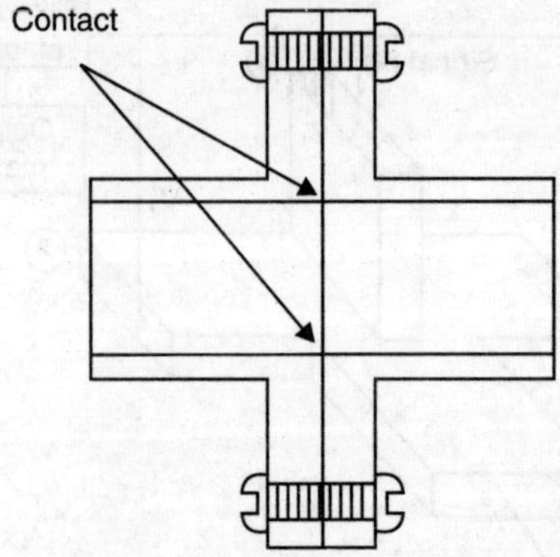

Fig. 5.33 Cover to cover coupling

(*ii*) **Choke flange:** Figure 5.34 depicts the choke-type waveguide connectors. It utilizes the transmission line theory in a way that largely obviates the discontinuity due to a break in the conductivity of a waveguide wall. Two quarter-wave sections present a short circuit at the joint and electrical continuity is thus preserved.

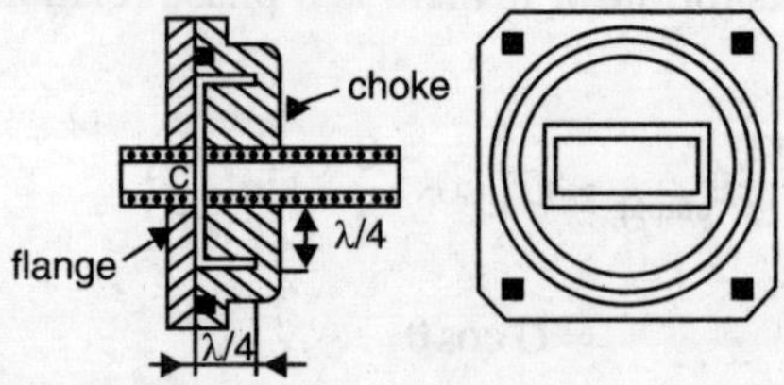

Fig. 5.34 Waveguide choke-flange coupling

In this there is negligible voltage drop across the ohmic contact between the flanges as such voltage breakdown is avoided. Therefore cover-to-choke connections can be used for high power applications. In general, the VSWR for this joint is of the order of 1.05 or less but it is more frequency dependent than the cover to cover joint.

5.10.2 Waveguide Corners, Bends and Twists

Waveguide corners (*E* and *H* Plane) and bends, which are shown in Fig. 5.35, are normally used to change the direction of the guide through a desired angle. Corners are also used to reduce the overall system of dimensions. These produce discontinuities and in order to minimize reflections from the discontinuities, the mean length (Fig. 5.35*a*, *b*), *L* between the junction of the corners is kept equal to **an odd number of quarter-wavelengths**. That is

$$L = (2n+1)\frac{\lambda_\gamma}{4} \qquad \text{...(5.25)}$$

where $n = 0, 1, 2, 3,...$ and λ_γ is the guide wavelength. In this situation the reflected waves from both ends of waveguide sections are completely cancelled.

In case of bends the radius of curvature for minimum reflection is given by Southworth [2 opt. cit.] as

$$R = 1.5b \text{ for an } E\text{-bend [Fig. 5.35}d]$$

and

$$R = 1.5a \text{ for a } H\text{-bend [Fig. 5.35}c]$$

For a good quality of match, long radius bends should preferably be used or the length of a bend be made a multiple of a half of a guide wavelength.

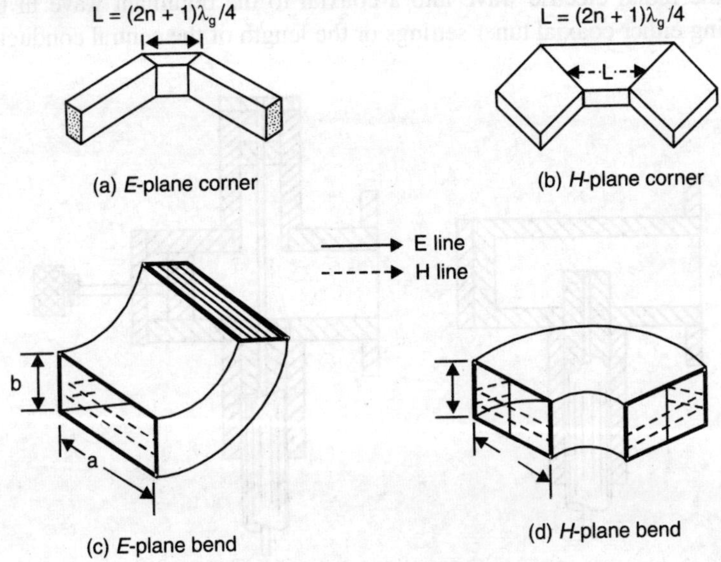

Fig. 5.35 Waveguide corners and bends

Twists

Waveguide twists are used to rotate the plane of polarization in a waveguide. In case of rectangular waveguides this may be done by filling the interior with a low melting point material and twisting the whole waveguide to the desired angle. The low melting point material is then removed. The circular waveguide rotator is made by placing a diametral septum with a progressive twist in the guide. The twists are shown in Fig. 5.36. These may also be electroformed.

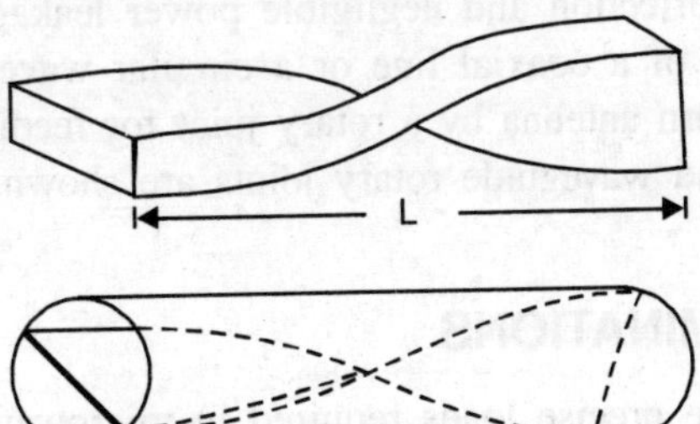

Fig. 5.36 Waveguide twists or rotators

The length of the twist is kept equal to an odd multiple of the quarter guide wavelength ($\lambda_g/4$).

Bends and other more complicated waveguide structures may be electroformed, *i.e.*, constructed by depositing electrolytically the required thickness of metal on a carefully made matrix.

5.10.3 Transitions

In a microwave system, it is often necessary to change from one form of waveguide to another or from waveguide to coaxial line or back. This is achieved by transitions, also known as adopters.

It is mostly desired to have a transition from a coaxial line to a waveguide and vice versa. Figure 5.37 illustrates such transitions. Discontinuities resulting from sudden termination of the dielectric or transformation of the radial electric wave into a coaxial to the dominant wave in the waveguide are turned out by varying either coaxial tuner settings or the length of the central conductor protruding into the guide.

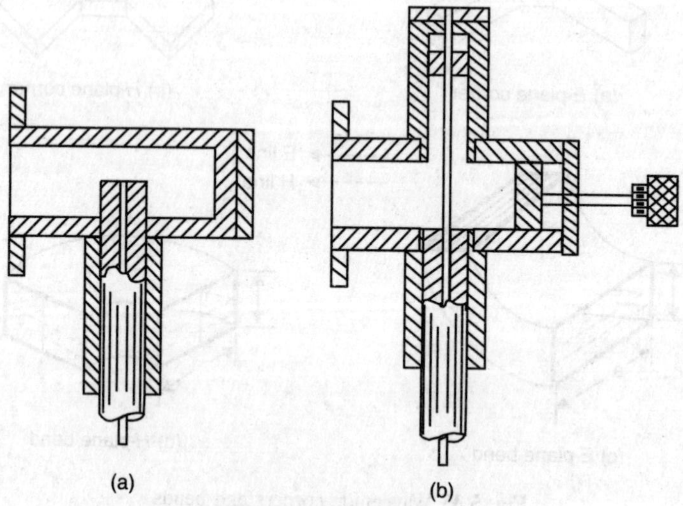

Fig. 5.37 (*a*) Transition from coaxial line to waveguide
(*b*) Transition from one form of waveguide to another

The transition from one form or shape of a waveguide to another may be achieved by waveguide transformers shown in Fig. 5.37*b*. The reader could find more information in literature [2, 3, opt. cit.]

5.10.4 Rotary Joints

A rotary joint allows one section of a waveguide to rotate with respect to the other, while passing microwave energy, with low reflection and negligible power leakage through it. Rotary joints are assembled from short sections of a coaxial line or a circular waveguide. In case of radar a fixed waveguide is connected to a horn antenna by a rotary joint for feeding a paraboloid reflector which rotates for tracking. Coaxial and waveguide rotary joints are shown in Fig. 5.38 *a*, *b*. These joints produce a VSWR < 1.1.

5.11 MICROWAVE TERMINATIONS

Microwave terminations are the precise loads required in microwave measurement and calibration systems. Most common among them are matched terminations and variable precession short-circuits. A matched load or a matched termination absorbs all the power incident on it without any reflection. In this respect it is equivalent to an absolutely black body in microwave range. On the other hand, a microwave line terminated in matched load does not have any reflection and hence is equivalent to terminating the line with its characteristic impedance. The short circuit termination produces an adjustable reactive load at the desired point on a microwave line.

5.11.1 Matched Terminations

Matched waveguide terminations are constructed by mounting a power-absorbing card or pad in the

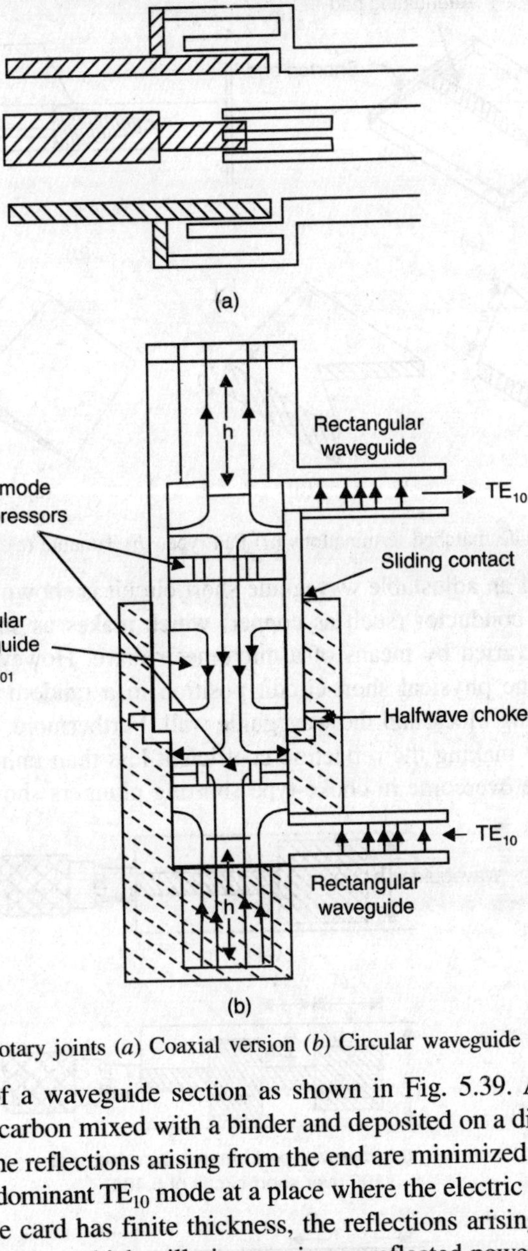

Fig. 5.38 Rotary joints (*a*) Coaxial version (*b*) Circular waveguide version

space near the closed end of a waveguide section as shown in Fig. 5.39. A typical attenuator card consists of powdered iron or carbon mixed with a binder and deposited on a dielectric strip or porcelain containing silicon carbide. The reflections arising from the end are minimized by tapering the card. The card is placed parallel to the dominant TE_{10} mode at a place where the electric field is maximum to have maximum attenuation. As the card has finite thickness, the reflections arising from it cannot be ruled out. Moreover, it is located at E_{max} which will give maximum reflected power. To avoid this the pad is kept closer to the side wall and its length is increased. Tunable matched termination is shown in Fig. 5.39*b*.

Matched loads may also be designed by loading the whole of the waveguide with lossy materials such as powdered metal, lossy dielectric, wood, flowing water, sand, etc. as shown in Fig. 5.39 *c*, *d* and *e*.

5.11.2 Variable Waveguide Short-Circuit Termination

Waveguide short-circuit terminations provide standard reflection at desired, precisely measurable position. The basic idea behind it is to provide short-circuit by changing reactance of the termination.

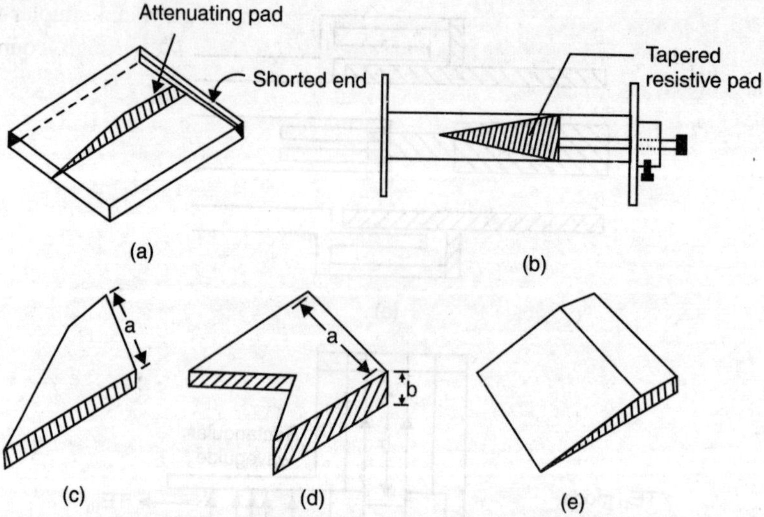

Fig. 5.39 Waveguide matched terminations *(a)* Pad type, *(b)* Tunable *(c)*, *(d)* and *(e)* Lossy type

The simplest form of an adjustable waveguide short-circuit is shown in Fig. 5.40*a*. It consists of a sliding block of a good conductor (such as copper) which makes as snug fit in the waveguide. The position of the block is varied by means of a micrometer drive. However, the electrical short-circuit position deviates from the physical short-circuit position in a random manner owing to the erratic contact between the sliding block and the waveguide wall. Furthermore, some power leakage past the block may occur thereby making the reflection coefficient less than unity. This too is highly undesirable. These problems are overcome in choke-type shorting plungers shown in Figs. 5.40*b* and *c*.

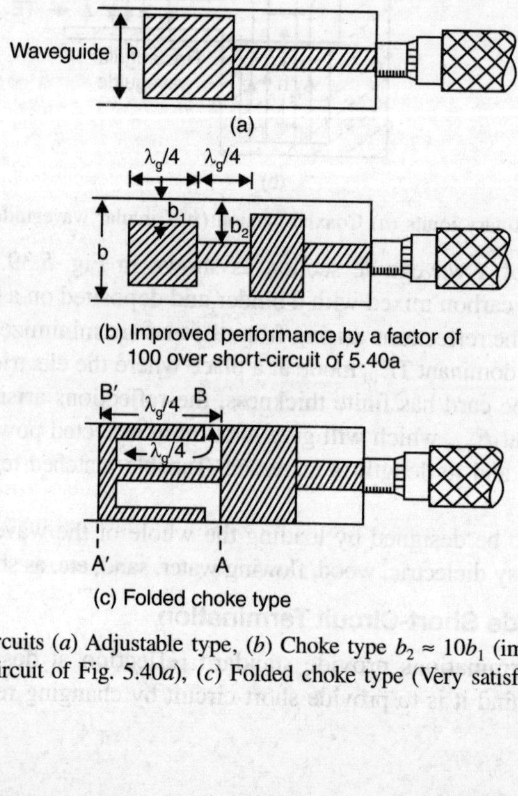

Fig. 5.40 Waveguide short-circuits *(a)* Adjustable type, *(b)* Choke type $b_2 \approx 10b_1$ (improved performance by a factor of 100 over short-circuit of Fig. 5.40*a*), *(c)* Folded choke type (Very satisfactory performance)

Example 5.1. A 90 W power source is connected to the input of a directional coupler with $C = 20$ dB, $D = 35$ dB and an insertion loss of 0.5 dB. Find the output powers at the through, coupled and isolated ports. Assume all ports to be matched.

Solution: Given $C = 20$ dB, $D = 35$ dB and $P_i = 90$ W

We know,

$$C = 10\log\frac{P_i}{P_f} = 20$$

$$\therefore \quad \frac{P_i}{P_f} = 10^2 = 100$$

$$\therefore \quad P_f = \frac{P_i}{100}$$

$$P_f = 0.9 \text{ W}$$

Also we know

$$D = 10\log\frac{P_f}{P_b} = 35$$

$$\frac{P_f}{P_b} = 10^{3.5}$$

$$\therefore \quad P_b = 284.6 \ \mu\text{W}$$

So Received Power

$$P_r = P_i - P_f - P_b$$

$$= 90 - 0.9 - 284.6 \times 10^{-6}$$

$$P_r = 89.09 \text{ W}$$

So

$$P_r (\text{in dB}) = 10\log\frac{P_i}{P_r} = 10\log\left(\frac{90}{89.09}\right)$$

$$= 0.044 \text{ dB}$$

$\therefore$ Effective Received Power

$$P_r' = P_r - \text{insertion loss}$$

$$= 0.044 - 0.5 = -0.455 \text{ dB}.$$

Example 5.2. A under coupled resonant cavity is connected to a lossless transmission line as shown in Fig. 5.41. The directional coupler is assumed to be ideal and matched on all arms. The unloaded Q of the cavity is 1000 and the VSWR at resonance is 2.5.

(*a*) Calculate the loaded Q_l of the cavity.

(*b*) Find the reading of bolometer 2 if bolometer 1 reads 4 mW.

(*c*) Compute the power dissipated in the cavity.

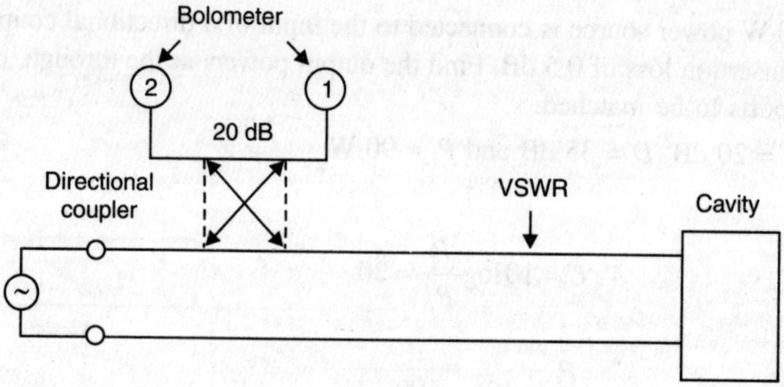

Fig. 5.41

Solution: Given Q = 1000, VSWR (ρ) = 2.5

(*a*) We know

$$Q_l = \frac{\rho Q}{1+\rho}$$

$$= \frac{2.5 \times 1000}{1+2.5} = 714.286$$

(*b*) $\Gamma = \frac{\rho - 1}{\rho + 1} = \frac{2.5-1}{2.5+1} = 0.43$ (Reflection coefficient)

Given, power at #1 bolometer is 4 mW,

and 20 dB power = 100,

So power from the generator is

P_{in} = 4 mW × 100 = 400 mW

or Power reflected from the cavity is

$P_{ref.}$ = 400 mW × $(0.43)^2$ = 73.96 mW

∴ Power at #2 bolometer is $P_2 = \frac{73.43}{100} = 0.73$ mW

(*c*) Power in the cavity = $P_{in} - P_{ref.}$ = 400 – 73.4 = 326.6 mW.

Example 5.3. A microwave transmission system consists of a generator, an over coupled cavity, two ideal but not identical dual directional couplers with matched bolometers, and a load Z_l. The lossless transmission line has a characteristic impedance Z_0. The reading of the four bolometer (1, 2, 3, and 4) are 2 mW, 4 mW; 0 mW and 1 mW, respectively. The system is shown in Fig. 5.42.

(*a*) Find the load impedance Z_l in terms of Z_0.

(*b*) Calculate the power dissipated by Z_l.

(*c*) Compute the power dissipated in the cavity.

(*d*) Determine the VSWR on the input transmission line.

(*e*) Find the ratio of Q_l/Q_o for the cavity.

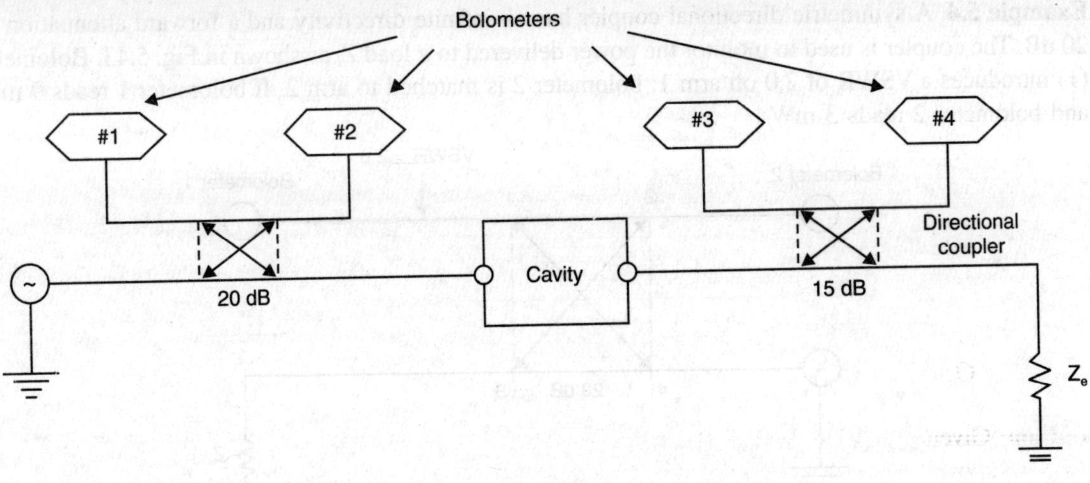

Fig. 5.42

Solution: Given $P_1 = 2$ mW, $P_2 = 4$ mW, $P_3 = 0$ and $P_4 = 1$ mW

Power 20 dB = 100 and 15 dB = 32

(*a*) Given $P_3 = 0$ so, $\Gamma_l = 0$

So, $$Z_l = Z_0$$

(*b*) Power into the cavity is,

$$P_{in} = 4 \times 100 = 400 \text{ mW}$$

or Power reflected from the cavity is,

$$P_{ref.} = 2 \times 100 = 200 \text{ mW}$$

or Power dissipated by $$Z_l = P_4 \times \text{Power}$$

$$= 1 \times 32 = 32 \text{ mW}$$

(*c*) Power dissipated in the cavity is

$$P_{ca} = P_{in} - P_{ref.} - P_{zl}$$

$$= 400 - 200 - 32 = 168 \text{ mW}.$$

(*d*) VSWR on the input transmission line

$$\Gamma_{in} = \sqrt{\frac{P_{ref.}}{P_{in}}} = \sqrt{\frac{200}{400}} = 0.707$$

So, $$VSWR = \rho = \frac{1+\Gamma}{1-\Gamma} = \frac{1+0.707}{1-0.707}$$

$$= 5.8$$

(*e*) Ratio of Q_l/Q_o,

$$\frac{Q_l}{Q_o} = \frac{1}{1+\rho} = \frac{1}{1+5.8} = 0.15$$

Example 5.4. A symmetric directional coupler has an infinite directivity and a forward attenuation of 20 dB. The coupler is used to monitor the power delivered to a load Z_l as shown in Fig. 5.43. Bolometer (1) introduces a VSWR of 2.0 on arm 1; bolometer 2 is matched to arm 2. If bolometer 1 reads 9 mW and bolometer 2 reads 3 mW.

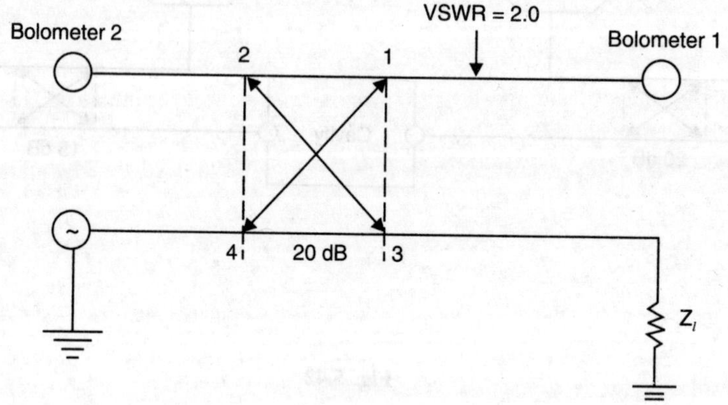

Fig. 5.43

(*a*) Find the amount of power dissipated in the load Z_l.
(*b*) Determine the VSWR on arm 3.

Solution:

Given $\rho_1 = 2.0$, $P_1 = 9$ mW and $P_2 = 3$ mW
Power 20 dB = 100

(*a*) We know

$$\Gamma_1 = \frac{\rho_1 - 1}{\rho + 1} = \frac{2-1}{2+1} = 0.33$$

$$\Gamma_1^2 = \frac{P_{\text{ref.}}}{P_{\text{in}}} = \frac{1}{9} = \frac{P_{\text{ref.}}}{P_{\text{ref.}} + 9}$$

$$\therefore \quad P_{\text{ref.}} = 1.125 \text{ mW}$$

So power into port 1 is $P_1 = 9 + 1.125 = 10.13$ mW.

or Power into the line from the generator is

$$P_4 = P_1 \times 100 = 1013 \text{ mW}$$

Since port 2 is matched and Bolometer (2) reads 3 mW, so power reflected from Z_l is

$$P_{3\text{ref.}} = (3 - 1.13) \times 100 = 1.87 \times 100 = 187 \text{ mW}$$

$$\therefore P_l = 1013 - 187 = 826 \text{ mW}$$

(*b*) We know $\Gamma_3 = \sqrt{\dfrac{P_{3\text{ref}}}{P_4}} = \sqrt{\dfrac{187}{1013}} = 0.43$

$$\therefore \quad SWR = \rho = \frac{1+0.43}{1-0.43} = 2.5.$$

REVIEW QUESTIONS AND PROBLEMS

1. Distinguish between *E*-plane and *H*-plane tees and hence discuss the construction and working of a magic-tee.
2. Explain the construction and working of waveguide phase shifter.
3. Explain the construction and working of precision variable attenuator. Discuss its frequency response and reflection losses.
4. Describe Bethe-hole coupler and explain its working. Obtain scattering matrix of a directional coupler.
5. Discuss various parameters of a directional coupler. Mention any two uses of directional coupler.
6. Write short notes on:
   (*i*) Rotary attenuator. (*ii*) Rotary phase shifter.
   (*iii*) Multi-hole directional coupler
7. Write short note on "Cavity resonators and its applications".
8. Distinguish between slotted line and direction coupler methods of VSWR measurements.
9. (*a*) Draw a neat sketch of magic *T*-junction. Image that a source is connected to arm '*P*' and arm '*S*' is match terminated. Arm 1 and 2 are terminated in reflection coefficients of 0.2 and 0.3 respectively. What is the VSWR seen by the source?
   (*b*) Draw *H*-plane tee junction and explain its properties.
10. (*a*) Describe the expression for coupling factor and directivity of a two hole direction coupler.
    (*b*) What are the different types of matching elements normally used in waveguide system. Distinguish between magic tee and rat race hybrid.
11. (*a*) Explain the working of two hole direction coupler with a neat diagram.
    (*b*) Explain about *E*-plane tee junction with a neat sketch. Why it is called a series tee?
12. (*a*) Explain the difference between
    (*i*) *E*-plane tee (*ii*) *H*-plane tee.
    Explain clearly why do you call them series and parallel tee respectively.
    (*b*) Describe with a neat sketch a precision Attenuator, and explain its operation.
13. Explain the construction, operation and application of the following microwave components.
    (*a*) Directional coupler (*b*) Waveguide tees.
14. (*a*) Derive the expression for coupling factor and directivity of a two hole directional coupler.
    (*b*) What are the different types of matching elements normally used in waveguide system? Distinguish between magic tee and rate race hybrid.
15. Write short notes on the following:
    (*a*) Directional coupler (*b*) Waveguide windows (*c*) Flap attenuator.

CHAPTER 6

Waveguide Components and Application-II

6.1 INTRODUCTION

'Ferrite' is a generic name given to ferromagnetic (or ferrimagnetic) insulators with specific resistivity as high as 10^{14} times that of metals and with dielectric constants around 10 to 15 times greater and relative permeabilities of several thousand. Ferrites are made by sintering a mixture of metallic oxides and have the general chemical composition M-O-Fe_2O_3, where M is a divalent metal such as Zn, Ni, Co, Mn, etc. or a mixture of these. Ferrites are transparent to electromagnetic waves. An electromagnetic wave propagating through the ferrite encounters strong interaction with spinning electrons and gives rise to desirable magnetic properties in ferrites. The paired electrons in certain materials cancel their magnetisation effects due to electrons rotating in opposite directions. However, if unparied electrons are present, they give rise to magnetic moments and hence magnetic properties. Thus a spinning electron may be taken as a small magnetic dipole. When a magnetic field is applied, it tries to align (in addition

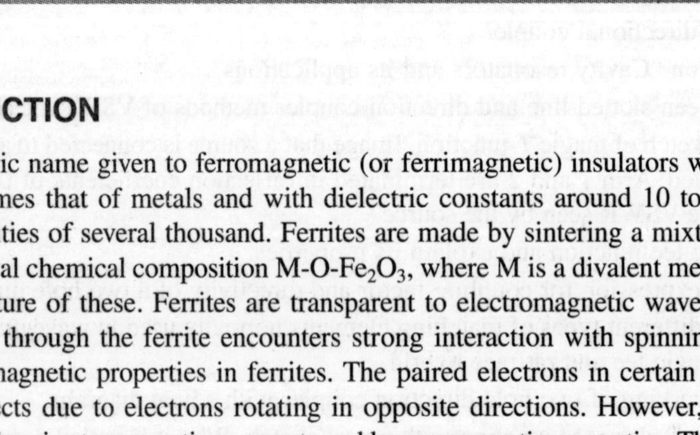

(a) (b)

Fig. 6.1 (*a*) and (*b*) Illustration of spinning electron as gyroscope top

to its rotation) in the direction of the magnetic field. As the magnetic movement of spinning electron has a fixed value, it cannot align along the magnetic field. This places strain over it and in order to get ride of this strain, it starts precessing (Fig. 6.1*b*), forming a 'classical spinning gyroscope top'. In ferrites the unpaired electron spins are mutually parallel throughout domains of uniform magnetisation. Thus macroscopic magnetisation vector of the domain behave as one unit under the influence of field, as such in investigating the response of ferrites towards *r.f.* fields, classical mechanics may be applied to the system of electron spins.

6.2 BASIC PROPERTIES OF FERRITES

The use of ferrites in a number of microwave devices is based on the fact that propagation constants for different modes of polarisation (*i.e.*, the left- and right-handed circular polarisations) of an electromagnetic ray are different. This gives rise to nonreciprocal Faraday rotation. The desirable magnetic properties of ferrites can be obtained by considering the gyroscopic motion of the electron.

6.2.1 Permeability Tensor

If τ be the torque exerted by the steady magnetic field H_0 when an electron is placed in it, then

$$\tau = \frac{d\mathbf{J}}{dt} \qquad \text{...(6.1)}$$

where **J** is spin angular momentum vector. This torque acts on the magnetic dipole (electron) having magnetic moment **M** tending to rotate it into the alignment of the field. Due to angular momentum, spin does not change its angle with respect to the applied magnetic field but precesse around H_0. It is this tendency precesse in the presence of magnetic field that endows the magnetic properties to such materials. The magnetic moment per unit volume, **M** is related to the angular momentum **J** by a constant, the gyromagnetic ratio, *i.e.*,

$$\mathbf{M} = \gamma \mathbf{J} \qquad \text{...(6.2)}$$

Again

$$\tau = \mu_0 (\mathbf{M} \times \mathbf{H}) \qquad \text{...(6.3)}$$

Substituting Eqns. 6.2 and 6.3 in Eqn. 6.1, we find

$$\frac{1}{\gamma}\frac{d\mathbf{M}}{dt} = \mu_0 (\mathbf{M} \times \mathbf{H}) \qquad \text{...(6.4)}$$

To consider propagation of microwave in ferrites, we assume harmonic variation of **M** with time because *r.f.* field is harmonic ($\alpha e^{j\omega t}$). Thus

$$\mathbf{M} = \frac{\tau\mu_0}{j\omega}(\mathbf{M} \times \mathbf{H}) \qquad \text{...(6.5)}$$

Equation (6.5) connects **M** and **H**, and by solving it, magnetic susceptibility constants $\left(\chi = \frac{\mathbf{M}}{\mathbf{H}}\right)$ may be obtained, which is related to an assembly of unpaired spins. To do so, we assume the magnetic field **H** as consisting of a small high frequency (*r.f.*) magnetic field H_1, superimposed on a constant polarising field H_z in *z*-direction. The response of the medium (ferrite) to this H_z is assumed to be a large constant magnetisation M_z in the *z*-direction and time varying components of magnetization

$\hat{x}M_x$ and $\hat{y}M_y$ in the x- and y-direction respectively. Thus we have

$$H = \left(\hat{x}H_x + \hat{y}H_y\right)e^{j\omega t} + \hat{z}H_z \quad ...(6.6)$$

and

$$M = \left(\hat{x}M_x + \hat{y}M_y\right)e^{j\omega t} + \hat{z}M_z \quad ...(6.7)$$

We have omitted time-varying components in z-direction because these are small as compared to high steady field H_z in this direction. Substitution of Eqns. 6.6 and 6.7 in Eqn. 6.5 gives components of magnetisation in terms of the *r.f.* field components, as

$$M_x = \frac{\omega_0\omega_m}{\omega_0^2 - \omega^2}H_x + j\frac{\omega\omega_m}{\omega_0^2 - \omega^2}H_y \quad ...(6.8)$$

$$M_y = -j\frac{\omega_0\omega_m}{\omega_0^2 - \omega^2}H_x + \frac{\omega\omega_m}{\omega_0^2 - \omega^2}H_y \quad ...(6.9)$$

$$M_z = 0 \quad ...(6.10)$$

where

$$\omega_0 = -\mu_0\gamma H_z \quad ...(6.11)$$

and

$$\omega_m = -\mu_0\gamma M_z \quad ...(6.12)$$

Equation 6.8 through 6.12 establishes relations between components of magnetization and the magnetic field. In the scalar case the ratio **M/H** expresses magnetic susceptibility χ of the medium. However, in our present case magnetization is not the same in all the directions so χ can be said to be a tensor. The components of χ can be obtained by transforming the above equations in the following matrix form:

$$\begin{bmatrix} M_x \\ M_y \\ M_z \end{bmatrix} = \begin{bmatrix} \chi_{xy} & \chi_{xy} & 0 \\ \chi_{yz} & \chi_{yy} & 0 \\ 0 & 0 & 0 \end{bmatrix} \begin{bmatrix} H_x \\ H_y \\ H_z \end{bmatrix} \quad ...(6.13)$$

where

$$\chi_{xx} = \chi_{yy} = \frac{\omega_0\omega_m}{\omega_0^2 - \omega^2}$$

$$\chi_{xy} = -\chi_{yx} = \frac{j\omega\omega_m}{\omega_0^2 - \omega^2}$$

Again magnetic induction **B** is related to the magnetization **M** as

$$\mathbf{B} = \mu\mathbf{H} = \mu_0(\mathbf{H} + \mathbf{M}) \quad ...(6.14)$$

$$\begin{bmatrix} B_x \\ B_y \\ B_z \end{bmatrix} = \mu_0 \begin{bmatrix} 1+\chi_{xx} & \chi_{yy} & 0 \\ -\chi_{yx} & 1+\chi_{yy} & 0 \\ 0 & 0 & 1 \end{bmatrix} \begin{bmatrix} H_x \\ H_y \\ H_z \end{bmatrix} \quad ...(6.15)$$

or in matrix notation

$$[B]=[\mu][H] \qquad ...(6.16)$$

where μ is the **Permeability Tensor** for the ferrite and can be written as

$$[\mu]=\begin{bmatrix} \mu & -jk & 0 \\ jk & \mu & 0 \\ 0 & 0 & \mu_0 \end{bmatrix} \qquad ...(6.17)$$

where
$$\mu=\mu_0\left(1+\chi_{xx}\right)=\mu_0\left(1+\frac{\omega_0\,\omega_m}{\omega_0^2-\omega^2}\right)$$

$$k=\mu_0\chi_{yx}=-\mu_0\left(\frac{\omega\,\omega_m}{\omega_0^2-\omega^2}\right)$$

where ω_0 is generally known as the **Gyromagnetic resonance frequency** because at $\omega = \omega_0$, $\mu = \mu_0$, and permeability is no more a tensor but becomes a scalar. ω_m is a constant of a material and depends upon the strength of the saturation magnetization M_0 (see Eqn. 6.12) in the ferrimagnetic domains and has values of the order of $2\pi \times 400$ to $2\pi \times 1000$ MHz.

If losses are present in ferrites, then a new term of damping is to be introduced in Eqn. 6.13, which tends to reduce the precession angle ϕ. Equation 6.13 may be solved in a similar way to give permeability and susceptibility constants to Microwave Magnetics by R.F. Soohoo.

6.3 PLANE WAVE PROPAGATION IN AN INFINITE FERRITE MEDIUM IN THE DIRECTION OF BIAS (FARADAY ROTATION)

We consider an infinite ferrite-filled region with a d.c. bias field given by $H_0 = z\hat{H}_0$, and a tensor permittivity $[\mu]$ given by (6.14). Maxwell's equation are:

$$\nabla \times E = -j\omega[\mu]H \qquad ...(6.14a)$$

$$\nabla \times H = j\omega\varepsilon E \qquad ...(6.14b)$$

$$\nabla \cdot D = 0 \qquad ...(6.14c)$$

$$\nabla \cdot B = 0 \qquad ...(6.14d)$$

Suppose plane wave propagation is in the z-direction, with $\dfrac{\partial}{\partial x}=\dfrac{\partial}{\partial y}=0$. Then the electric and magnetic fields are given in the form.

$$E = E_0 e^{-j\beta z} \qquad ...(6.15a)$$

$$H = H_0 e^{-j\beta z} \qquad ...(6.15b)$$

Using Eqn. 6.13, the two curl equations 6.14*a*, *b* then reduce to the following:

$$j\beta E_y = -j\omega\left(\mu H_x + jkH_y\right) \qquad ...(6.16a)$$

$$-j\beta E_x = -j\omega\left(-jkH_x + \mu H_y\right) \qquad ...(6.16b)$$

$$0 = -j\omega\mu_0 H_z \qquad \text{...(6.16c)}$$

$$j\beta H_y = j\omega\varepsilon E_y \qquad \text{...(6.16d)}$$

$$-j\beta H_y = -j\omega\varepsilon E_y \qquad \text{...(6.16e)}$$

$$0 = j\omega\varepsilon E_y \qquad \text{...(6.16f)}$$

From Eqns 6.16*c* and 6.16*f*, $E_z = H_z = 0$, as expected for TEM planes. We also have $\nabla \cdot D = \nabla \cdot B = 0$, since wave is propagating in source free medium.

The transverse field components are given by the Eqns. 6.16*d*, *e* as

$$Y = \frac{H_y}{E_x} = \frac{-H_x}{E_y} = \frac{\omega\varepsilon}{\beta} \qquad \text{...(6.17)}$$

where Y is the wave admittance.

Eliminating H_x and H_y in Eqns. (6.16*a*) and (6.16*b*) using Eqn. (6.17) gives

$$j\omega^2\varepsilon k E_x + \left(\beta^2 - \omega^2\mu\varepsilon\right)E_y = 0 \qquad \text{...(6.18a)}$$

$$\left(\beta^2 - \omega^2\mu\varepsilon\right)E_x - j\omega^2\varepsilon k E_y = 0 \qquad \text{...(6.18b)}$$

The determinant of set of Eqns. 6.18*a*, *b* must vanish for a nontrivial solution,

$$\omega^4\varepsilon^2k^2 - \left(\beta^2 - \omega^2\mu\varepsilon\right)^2 = 0$$

$$\beta_{\pm} = \omega\sqrt{\varepsilon(\mu \pm k)} \qquad \text{...(6.19)}$$

Thus there are two possible propagation constants β_+ and β_-.

6.3.1 Fields Associated with β_+ and β_-

Substituting β_+ in the Eqn. (6.18*a* or *b*), we get

$$j\omega^2\varepsilon k E_x + \omega^2\varepsilon k E_y = 0$$

or

$$E_y = -jE_x$$

Then the electric field of Eqn. 6.15*a* will have the following form:

$$E_+ = E_0\left(\hat{x} - j\hat{y}\right)e^{-j\beta_+ z} \qquad \text{...(6.20a)}$$

It is a right hand circularly polarised plane wave (RHCP).

The associated magnetic field is obtained by using Eqn. 6.17 as

$$H_+ = E_0 Y_+\left(j\hat{x} + \hat{y}\right)e^{-j\beta_+ z} \qquad \text{...(6.20b)}$$

where Y_+ is the wave admittance for this wave

$$Y_+ = \frac{\omega\varepsilon}{\beta_+} = \sqrt{\frac{\varepsilon}{\mu + k}} \qquad \text{...(6.20c)}$$

Similarly, the fields associated with β_- are left-hand circularly polarized (LHCP).

$$E_- = E_0(\hat{x} - j\hat{y})e^{-j\beta_- z} \quad ...(6.21a)$$

$$H_- = E_0 Y_-(-j\hat{x} + \hat{y})e^{-j\beta_- z} \quad ...(6.21b)$$

where $$Y_- = \frac{\omega\varepsilon}{\beta_-} = \sqrt{\frac{\varepsilon}{\mu - k}} \quad ...(6.21c)$$

We note that for an RHCP wave, the ferrite material can be represented with an effective permeability of ($\mu + k$) and for LHCP it is equal to ($\mu - k$).

6.3.2 Faraday Rotation

The sum of electric field of an RHCP and an LHCP represent a linearly polarised electric field at $z = 0$

$$E|_{z=0} = \hat{x}E_0 = \frac{E_0}{2}(\hat{x} - j\hat{y}) + \frac{E_0}{2}(\hat{x} + j\hat{y}) \quad ...(6.22)$$

From Eqns. (6.21*a*, *b*) the total field of Eqn. 6.22 will propagate as

$$E = \frac{E_0}{2}(\hat{x} - j\hat{y})e^{-j\beta_+ z} + \frac{E_0}{2}(\hat{x} - j\hat{y})e^{-j\beta_- z}$$

$$= \frac{E_0}{2}\hat{x}\left(e^{-j\beta_+ z} + e^{-j\beta_- z}\right) - j\frac{E_0}{2}\hat{y}\left(e^{-j\beta_+ z} - e^{-j\beta_- z}\right)$$

$$= E_0\left[\hat{x}\cos\left(\frac{\beta_+ - \beta_-}{2}\right)z - \hat{y}\sin\left(\frac{\beta_+ - \beta_-}{2}\right)z\right]e^{-j(\beta_+ + \beta_-)z/2} \quad ...(6.23)$$

This resultant wave is linearly polarised wave that has undergone a phase delay of $(\beta_- + \beta_+)z/2$. The new plane of polarisation makes an angle θ with the old (*i.e.*, $\hat{x}$)

$$\theta = \tan^{-1}(E_y/E_x) = \tan^{-1}\{-\tan(\beta_+ - \beta_-)z/2\}$$

$$= -(\beta_+ - \beta)z/2 \quad ...(6.24)$$

Thus plane of polarisation is rotated at the rate of $(\beta_+ - \beta_-)/2$ rad/metre.

If the direction of propagation is reversed, the plane of polarisation continues to be rotated by an angle θ in the same direction because β_+ and β_- are functions of magnetic field and not of the direction of motion. This property is known as **non-reciprocity** and used in designing non-reciprocal devices such as isolators circulators.

The preceding discussion is valid so far as $\omega < \omega_0$ (ω'*s* are functions of applied magnetic field). At $\omega = \omega_0$, the rotating magnetic field (*r.f.* circularly polarised waves) assumes the direction of precession and have the same frequency. Thus resonance occurs and travelling *r.f.* field aids electron precession thereby setting it to violent oscillations and absorbing the energy from the *r.f.* field. This is called **resonant absorption.**

Further, in this discussion we have assumed lossless ferrites. As a matter of fact when wave passes through ferrites, it will be attenuated in addition to phase delay. The attenuation is maximum at resonance.

6.4 MICROWAVE FERRITE DEVICES

6.4.1 Isolator

This microwave device permits unattenuated transmission in one direction (forward direction) but provides very high attenuation in the reverse direction (backward direction). This is generally used between the source and rest of the set-up to avoid overloading of the source due to deflected power. Two types of ferrite isolators are possible:

(*i*) **Faraday-rotation isolator:** As shown in Fig. 6.2, it consists of a 45° waveguide twist connected with a circular guide, which in turn is connected to another rectangular guide at the other end. The circular part has a known tapered length of ferrite placed under the biasing of the magnetic field in the forward direction. Two resistive cards are placed perpendicular to the electric field at the two ends.

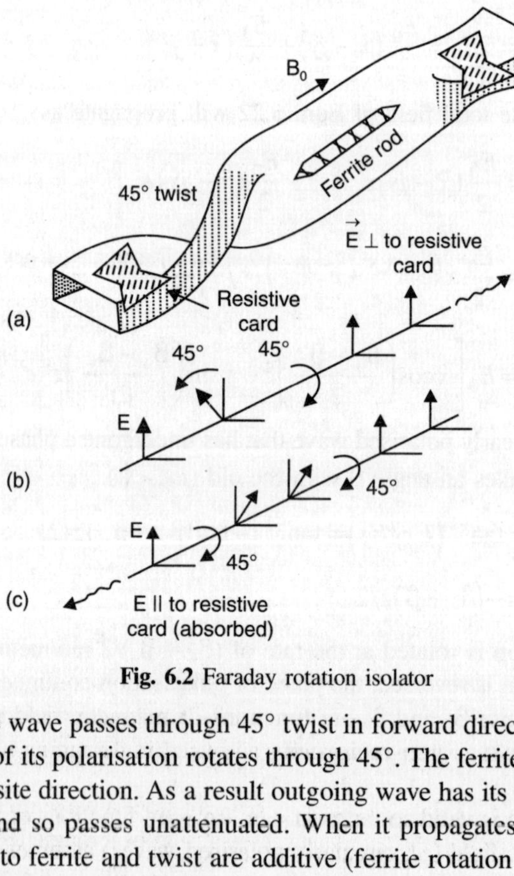

Fig. 6.2 Faraday rotation isolator

When TE_{10} mode wave passes through 45° twist in forward direction (port 1 to 2) as shown in Fig. 6.2*b* the plane of its polarisation rotates through 45°. The ferrite rod causes an additional 45° rotation in the opposite direction. As a result outgoing wave has its electric field perpendicular to the resistive pads and so passes unattenuated. When it propagates in the reverse direction, the phase rotations due to ferrite and twist are additive (ferrite rotation is still in the same direction) and electric vector is rotated by 90° making it parallel to the resistive pad and hence absorbed.

(*ii*) **Resonance isolator:** In the vicinity of frequency $\omega_0 \approx \omega$, the attenuation constant for negative (left-handed) circular polarisation is much smaller as compared to positive (right-handed) polarisation. This property is put to use in a resonance isolator. Resonance isolator is designed by placing ferrite slab (or two slabs at a distance) such that *r.f.* magnetic field of TE_{10} mode is circularly polarised. The ferrite is magnetised by a static field applied in *y*-direction as shown in Fig. 6.3*a*, *b* and *c*.

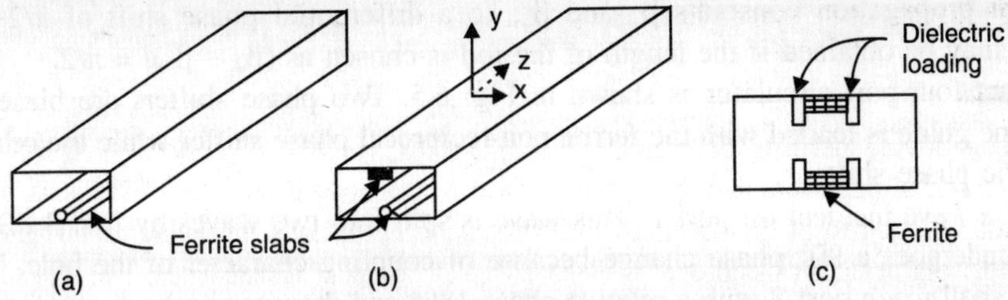

Fig. 6.3 Typical resonance isolator geometries

When wave travels from port 1 to 2 (in forward direction, *i.e.* +*z*-direction), magnetic field is negative circularly polarised and wave passes unattenuated. But when it passes in backward direction it shall have positive circularly polarised magnetic field thereby experiencing heavy attenuation and hence absorbed.

6.4.2 Circulator

Circulator is a multiport device such that wave incident on n^{th} port is coupled to $(n + 1)^{th}$ port only, *i.e.*, if power is incident in port 1, port 2 is coupled but not ports 3 and 4. Microwave circulators with four ports (Fig. 6.4) are most common. A four-port circulator may be constructed from two magic tees or hybrid junctions and a gyrator which produces an additional phase shift of 180° for propagation in the direction from *a* to *b*. The electric lengths of propagation from *b* to *a, c* to *d* and *d* to *c* are equal. Since tees and gyrator ports are all matched, the circulator is thus a matched device.

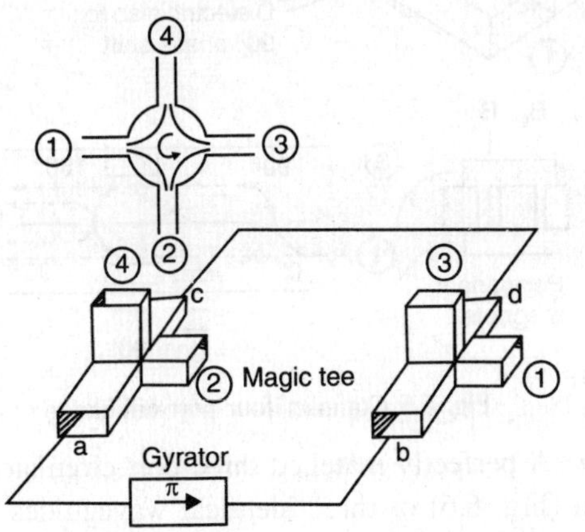

Fig. 6.4 A four-port microwave circulator utilizing magic tee

To understand the basic operation through Fig. 6.4, consider a wave incident on port 1, it splits into two equal in-phase waves propagating inside arms (see action of a magic tee), reach at *a* and *c* in phase thereby emerging in the 'adder' arm of another tee, *i.e.*, port 2. The incident wave in port 2 will split in two waves which arrive at *b* and *d* with 180° phase shift and as a result, shall be added in subtractor arm of the second magic tee, *i.e.*, port 3 and so on.

(*i*) *Compact four-port circulator:* A compact four-port circulator may be constructed using 3 dB side-hole directional couplers and two non-reciprocal 90° ferrite phase shifters. The latter consists of thin ferrite rod of length *l* placed in a rectangular waveguide where the magnetic field of TE_{10} mode is circularly polarised. A biasing magnetic field B_0 is applied in *y*-direction. Since two circular polarisations have different propagation constants β_+ and β_-; so a differential phase shift of $\pi/2$ between two polarisations may be obtained if the length of the rod is chosen as $(\beta_+ - \beta_-)l = \pi/2$.

A compact four-port circulator is shown in Fig. 6.5. Two phase shifters are biased in opposite directions. One guide is loaded with the ferrite non-reciprocal phase shifter while the other by a reciprocal dielectric phase shifter.

Consider a wave incident on port 1. This wave is split into two waves by first 3 dB coupler. The upper wave undergoes a 90° phase change because of coupling character of the hole. This wave (in upper guide) shall reach port 2 with a relative phase 180° and the wave in lower waveguide at port 4, with a phase 90°. The second coupler again splits the waves such that resultant waves go out of phase and hence cancel out in port 4, while they go in phase and hence add in port 2. A similar analysis may be carried out to show that ports are coupled as 1 → 2 → 3 → 4 → 1 and not in the reverse direction.

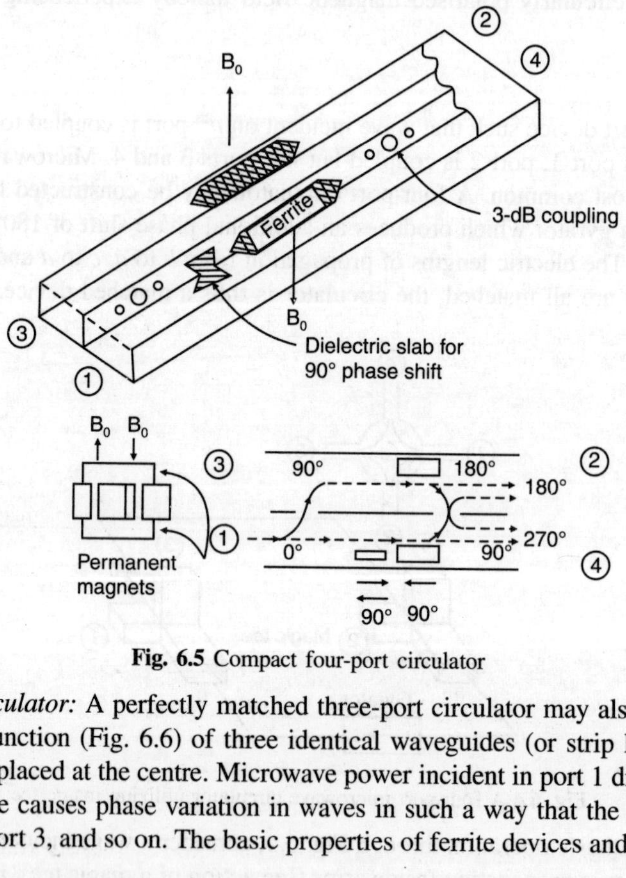

Fig. 6.5 Compact four-port circulator

(ii) Three-port circulator: A perfectly matched three-port circulator may also be designed. It consists of symmetric *y*-junction (Fig. 6.6) of three identical waveguides (or strip lines), with an axially magnetised ferrite rod placed at the centre. Microwave power incident in port 1 divides in the other two ports equally but ferrite causes phase variation in waves in such a way that the two are added in port 2 while subtracted in port 3, and so on. The basic properties of ferrite devices and various quantities are defined as follows.

6.4.3 Gyrator

A gyrator is defined as a two-port device that has a relative difference in phase shift of 180° for transmission from port 1 to port 2 as compared with the phase shift for transmission from port 2 to port 1. A gyrator may be obtained by employing the non-reciprocal property of Faraday rotation. In Fig. 6.7, it consists of a rectangular guide with a 90° twist connected to a circular guide, which in turn is connected to another rectangular guide at the other end. The two rectangular guides have the same

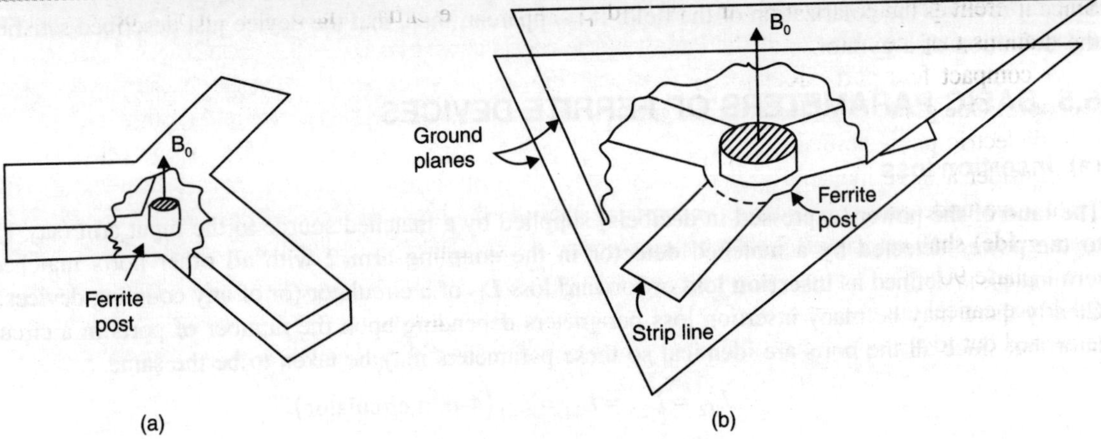

Fig. 6.6 Three-port circulator

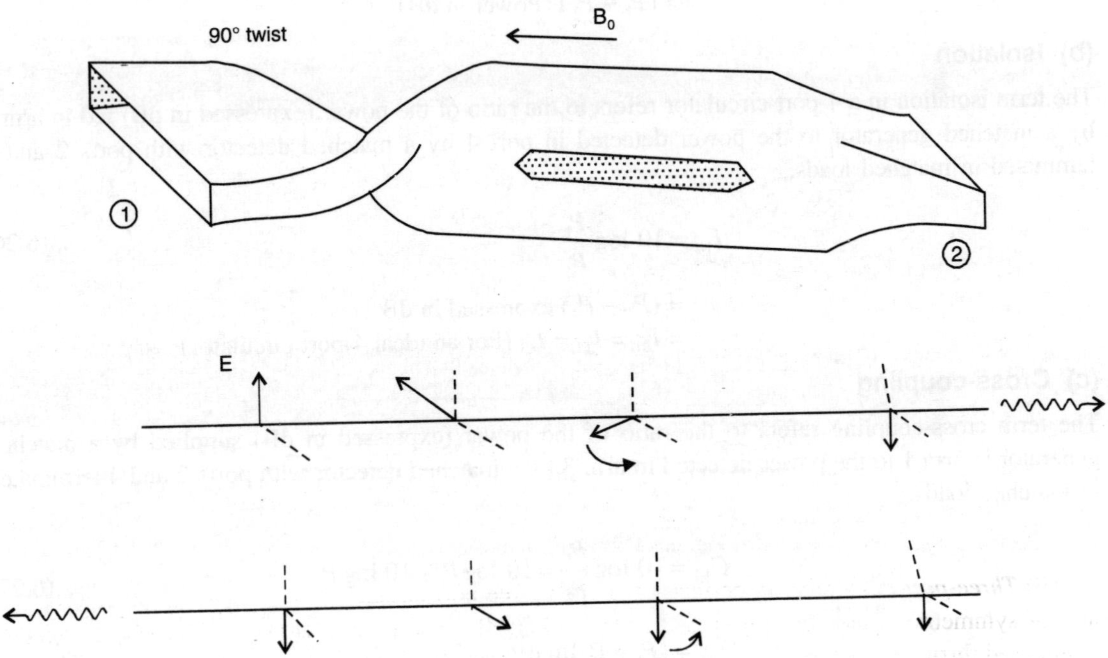

Fig. 6.7 A Microwave gyrator

orientation at the input ports. The circular guide contains a thin cylindrical rod of ferrite with the ends tapered to reduce reflections. A static axial magnetic field is applied so as to produce 90° Faraday rotation of the TE_{11} dominant mode in the circular guide. Consider a wave propagating from left to right. In passing through the twist the plane of polarization is rotated by 90° in counter clockwise direction. If the ferrite produces an additional 90° of rotation, the total angle of rotation will be 180°. For a wave propagating from right to left, the Faraday rotation is still 90° in the same sense. However, in passing through the twist, the next 90° of rotation is in a direction to cancel the Faraday rotation. Thus the transmission from port 2 to port 1, there is no net rotation of the plane of polarization. The 180° rotation for transmission from port 1 to port 2 is equivalent to an additional 180° of phase shift.

Since it reverses the polarization of the field. It is apparent, then, that the device just described satisfies the definition of a gyrator.

6.5 BASIC PARAMETERS OF FERRITE DEVICES

(a) Insertion loss

The ratio of the power (expressed in decibels) supplied by a matched source to the input arm (say 1), to the power detected by a matched detector in the coupling arm 2 with all other ports matched terminated, is defined as **insertion loss** or forward loss L_{12} of a circulator (or of any coupling devices). Clearly there may be many insertion loss parameters depending upon the number of ports in a circulator, but since all the ports are identical so these parameters may be taken to be the same.

$$L_{12} = L_{23} = L_{34} = L_{41} \left(\text{4-port circulator}\right)$$

$$L_{12} = 10 \log P_1/P_2 = 10 \log P_1 - 10 \log P_2 \qquad \text{...(6.25)}$$

$$= \left(P_1 - P_2\right) (\text{Power in dB})$$

(b) Isolation

The term isolation in a 4-port circulator refers to the ratio of the power (expressed in dB) fed in arm 1 by a matched generator to the power detected in port 4 by a matched detector with ports 2 and 3 teminated in matched loads.

$$I_{14} = 10 \log \frac{P_1}{P_4} \qquad \text{...(6.26)}$$

$$= (P_1 - P_4) \text{ expressed in dB}$$

$$= I_{21} = I_{32} = I_{43} \text{ (For an ideal 4-port circulator)}$$

(c) Cross-coupling

The term cross-coupling refers to the ratio of the power (expressed in dB) supplied by a matched generator in arm 1 to the power detected in arm 3 by a matched detector with ports 2 and 4 terminated in matched loads.

$$C_{13} = 10 \log \frac{P_1}{P_3} = 10 . \log P_1 - 10 \log P_3 \qquad \text{...(6.27)}$$

$$= \left(P_1 - P_3\right) \text{in dB}$$

Clearly $\qquad C_{13} = C_{24} \qquad C_{31} = C_{42}$ for an ideal 4-port circulator.

(d) Input VSWR

Input VSWR of a 4-port circulator is the ratio of the maximum voltage to the minimum voltage of the standing waves existing on the line (slotted) when one-port of the calculator (all ports being identical) terminates the line and all others have matched terminations.

Computer-Aided Design (CAD) for microwave circuits

CAD programs for microwave circuit analysis and optimization are fast, powerful and accurate. These have the capability of analysing microwave circuits consisting of several lumped, passive and active components. There are several commercially available microwave CAD software packages such as SUPER COMPACT* and TOUCHSTONE* which are very useful tools for microwave engineer. With the use of CAD the microwave engineer could optimize the design by adjusting some of the circuit parameters to achieve the best performance. A paper model is developed which meets the specification. Then an engineering product can be built and tested. Thus, it saves the time and reduces cost of a design, while enhancing the quality.

6.6 SCATTERING MATRIX REPRESENTATION OF MICROWAVE NETWORK

A waveguide section is a two-port network *i.e.,* it has one input and one output. The output port is terminated by some means it is one-port network. Similarly, tees (Fig. 6.8) and coupler (Fig. 6.9) form three-port networks. The relationship between incoming and outgoing waves at various ports of a network describe the system or network completely. The input-output relationship (the wave approach) is defined by scattering matrix which is widely used to characterize multiports. However, there are other two types of matrices, namely, impedance matrix and admittances matrix, these two are based on Kirchhoff's laws for electrical networks. These two are not convenient since voltage currents and impedances cannot be measured precisely and directly at microwaves. On the other hand transmission coefficient, reflection coefficient, and standing wave ratio at microwaves are easily and precisely measurable. These form the basis of the scattering matrix formulation. **Scattering matrix is a linear relationship between input and output that involves precisely measurable parameters.**

6.6.1 Scattering Matrix Formulation

Consider the n-port network as shown in Fig. 6.8.

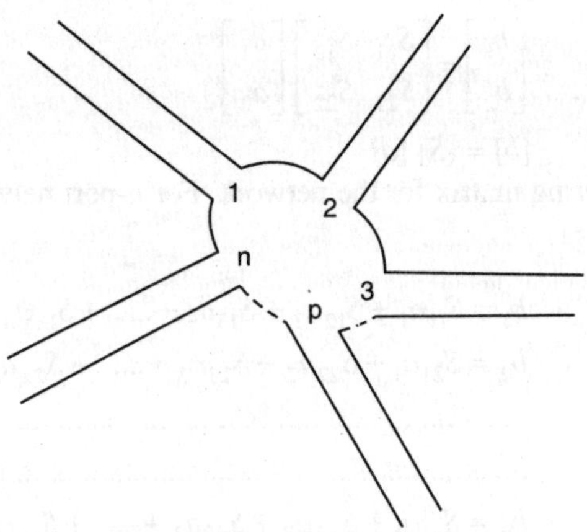

Fig. 6.8 n-port network

Let us define a parameter a_n proportional to the incident wave at the nth port and parameter b_n proportional to the outgoing wave at n^{th} port in the following manner,

$$a_n = V_n^+ / \sqrt{Z_c} \qquad \text{...(6.28)}$$

$$b_n = V_n^- / \sqrt{Z_c} \qquad \text{...(6.29)}$$

where V_n^+ represents incident wave along the line connected to the nth port.

and V_n^- represents outgoing wave along the line connected to the nth port.

and Z_c is the characteristic impedance of the line. For a two-port network the relationship between b_n's and a_n's may be written as

$$\begin{aligned} b_1 &= S_{11}a_1 + S_{12}a_2 \\ b_2 &= S_{21}a_2 + S_{22}a_2 \end{aligned} \qquad \text{...(6.30)}$$

where S_{11} is a reflection coefficient at port 1

S_{22} is a reflection coefficient at port 2

S_{12} is a transmission coefficient from port 2 to port 1

S_{21} is a transmission coefficient from port 1 to port 2.

Figure 6.9*a* shows the *S*-parameter of a two-port network.

Equation 6.30 can be put in the matrix form.

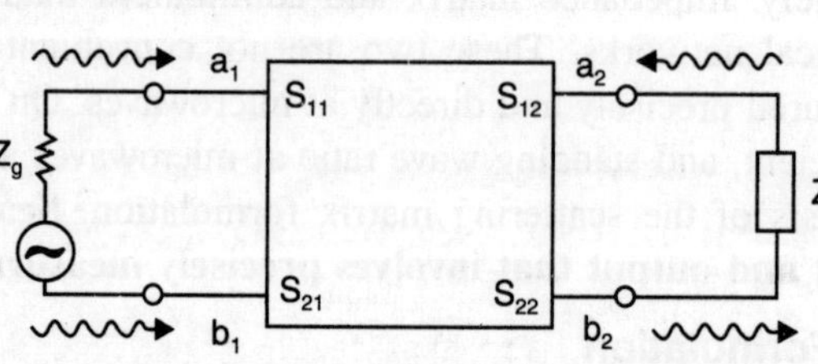

Fig. 6.9 (*a*) *S*-parameters of two-port network

$$\begin{bmatrix} b_1 \\ b_2 \end{bmatrix} = \begin{bmatrix} S_{11} & S_{12} \\ S_{21} & S_{22} \end{bmatrix} \begin{bmatrix} a_1 \\ a_2 \end{bmatrix} \qquad \text{...(6.31)}$$

or $\qquad [b] = [S]\,[a]$

where *S* is known as the scattering matrix for the network. For *n*-port network the relationship between b_n's and a_n's may be written

$$\begin{aligned} b_1 &= S_{11}a_1 + S_{12}a_2 + S_{13}a_3 + \ldots\ldots + S_{1n}a_n \\ b_2 &= S_{21}a_1 + S_{22}a_2 + S_{23}a_3 + \ldots\ldots + S_{2n}a_n \\ &\ldots\ldots\ldots\ldots\ldots\ldots\ldots\ldots\ldots\ldots \\ &\ldots\ldots\ldots\ldots\ldots\ldots\ldots\ldots\ldots\ldots \\ b_n &= S_{n1}a_1 + S_{n2}a_2 + S_{n3}a_3 + \ldots\ldots + S_{nn}a_n \end{aligned} \qquad \text{...(6.32)}$$

as each incident wave will make a contribution to the total resultant wave from n^{th} port. Equation 6.32 can be put in the following matrix form:

$$\begin{bmatrix} b_1 \\ b_2 \\ \vdots \\ b_n \end{bmatrix} = \begin{bmatrix} S_{11} & S_{12} & \cdots & S_{1n} \\ S_{21} & S_{22} & \cdots & S_{2n} \\ \vdots & \vdots & & \vdots \\ S_{n1} & S_{n2} & \cdots & S_{nn} \end{bmatrix} \begin{bmatrix} a_1 \\ a_2 \\ \vdots \\ a_n \end{bmatrix} \qquad \text{...(6.33)}$$

Equation 6.33 can be rewritten in matrix notations

$$[b] = [S]\,[a] \qquad ...(6.34)$$

where $[a]$ and $[b]$ are the respective matrices of the incident (input) and reflected (output) waves and $[S]$ by definition the **scattering matrix.**

Equation 6.34 resembles to relation $E_{ref.} = \rho E_{inc.}$

where ρ is the reflection coefficient. The scattering coefficients S_{11}, S_{22}, S_{33} carry the simple physical meaning of reflection coefficients ρ_i, when the multiport device is energized from the ith port provided all the other ports terminate in matched loads.

6.6.2 Average Power Flowing into the Port *n*

Equations 6.28 and 6.29 can be used to evaluate the average power flowing into the port n. From transmission line theory we can write

$$V_n = V_n^+ + V_n^- = \sqrt{Z_c}\,(a_n + b_n) \qquad ...(6.35)$$

and

$$I_n = \frac{1}{\sqrt{Z_c}}\left(V_n^+ - V_n^-\right) = \frac{1}{\sqrt{Z_c}}(a_n - b_n) \qquad ...(6.36)$$

The average power flow into the port n is given by

$$W_n = \frac{1}{2}\mathrm{Re}\left(V_n I_n^*\right)$$

$$= \frac{1}{2}\mathrm{Re}\left\{\left(a_n a_n^* - b_n b_n^*\right) + \left(b_n a_n^* - b_n a_n^*\right)\right\}$$

$$= \frac{1}{2}\left(a_n a_n^* - b_n b_n^*\right) \qquad ...(6.37)$$

Equation 6.37 shows that W_n is equal to the power carried into the port n by the incident wave (which is given by $\frac{1}{2}a_n a_n^*$) less the power reflected back (given by $\frac{1}{2}b_n b_n^*$).

6.6.3 Insertion Loss, Transmission Loss, Reflection Loss and Return Loss in Terms of *S*-parameters when the Ports are Matched Terminated

Let us consider two-port network in which

P_i is power fed at port 1,

P_r is power reflected from port 1

P_o is output power at port 2 then the following losses are defined in terms of S-parameters.

(*i*) Insertion loss (dB) $= 10\log\dfrac{P_i}{P_o} = 10\log\dfrac{|a_1|^2}{|a_2|^2}$

$$= 20\log\frac{1}{|S_{21}|}$$

$$= 20 \log \frac{1}{|S_{12}|} \qquad \text{...(6.38)}$$

(*ii*) Transmission loss or attenuation (dB) $= 10 \log \dfrac{P_i - P_r}{P_o}$

$$= 10 \log \frac{1 - |S_{11}|^2}{|S_{12}|^2} \qquad \text{...(6.39)}$$

(*iii*) Reflection loss (dB) $= 10 \log \dfrac{P_i}{P_i - P_r}$

$$= 10 \log \frac{1}{1 - |S_{11}|^2} \qquad \text{...(6.40)}$$

(*iv*) Return loss (dB) $= 10 \log \dfrac{P_i}{P_r}$

$$= 20 \log \frac{1}{|P|}$$

$$= 20 \log \frac{1}{|S_{11}|} \qquad \text{...(6.41)}$$

6.6.4 Properties of *S*-parameters

S-parameters are, in general, complex quantities having following properties:

(*i*) **Zero property of [*S*] matrix:** The sum of the products of each term of any column (or row) multiplied by the complex conjugate of the corresponding terms of any other column (or row) is zero *i.e.,*

$$S_{11} S_{12}^* + S_{21} S_{22}^* + \ldots\ldots + S_{N1} S_{N2}^* = 0$$

or

$$\sum_{n=1}^{N} S_{ns} S_{nr}^* = 0 \ (s \neq r) \qquad \text{...(6.42)}$$

(*ii*) **Unity property of [*S*] matrix:** The sum of the products of each term of any one row (or column) multiplied by its complex conjugate is unity, that is

$$\sum_{n=1}^{N} S_{ni} S_{ni}^* = 1 \qquad \text{...(6.43)}$$

(*iii*) **Symmetric property of [*S*] matrix:** For any lossless linear and reciprocal network (same transmission characteristics in either direction of a pair of ports) the scattering matrix is symmetrical, *i.e.,* $S_{ij} = S_{ji}$ $(i \neq j)$.

(*iv*) **Phase shift property:** Suppose the reference plane of 1 and 2 of two-port network are shifted to 1′ and 2′ (shown in Fig. 6.9*b*) by electrical phase shift $\phi_1 = \beta_1 l_1$ and $\phi_2 = \beta_2 l_2$ respectively, the new wave variables are $a_1 e^{j\phi_1}$, $a_2 e^{-j\phi_1}$, $b_1 e^{j\phi_2}$, $b_2 e^{-j\phi_2}$.

The new *S*-matrix *S′* is then given by

$$[S'] = \begin{bmatrix} e^{j\phi_1} & 0 \\ 0 & e^{j\phi_2} \end{bmatrix} \begin{bmatrix} S_{11} & S_{12} \\ S_{21} & S_{22} \end{bmatrix} \begin{bmatrix} e^{-j\phi_1} & 0 \\ 0 & e^{-j\phi_2} \end{bmatrix}$$

This property is valid for any number of ports and is called phase shift property when reference planes are shifted.

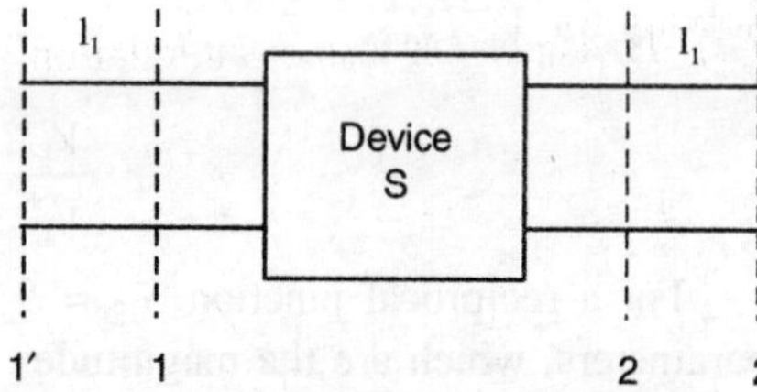

Fig. 6.9 (*b*) Reference plane shifted

6.7 S-MATRIX CALCULATIONS

The directly measurable quantities are the amplitudes and phase angles of the waves reflected, or scattered, from a junction relative to the incident-wave amplitudes and phase angles. In view of the linearity of the field equations and most microwave devices, the scattered-wave amplitudes are linearly related to the incident wave amplitudes. The matrix describing their linear relationship is called the scattering matrix.

6.7.1 Scattering Matrix for a Two-port Junction

In Fig. 6.10, let the scattering matrix parameters of the junction be S_{11}, S_{21}, S_{12} and S_{22}. The incident and scattered wave amplitudes are related by

$$[V^-] = [S][V^+] \qquad \text{...(6.44}a\text{)}$$

$$[V_1^-] = S_{11}V_1^+ + S_{12}V_2^+ \qquad \text{...(6.44}b\text{)}$$

$$[V_2^-] = S_{21}V_1^+ + S_{22}V_2^+ \qquad \text{...(6.44}c\text{)}$$

If the output guide is terminated in a matched load, $V_2^+ = 0$. From Eqn. (6.44*b*) it is seen that S_{11} is reflection coefficient in the input guide 1, with guide 2 terminated in a matched load. Also S_{21} is a transmission coefficient into guide 2 from 1. Similarly apply to the parameters S_{22} and S_{12}.

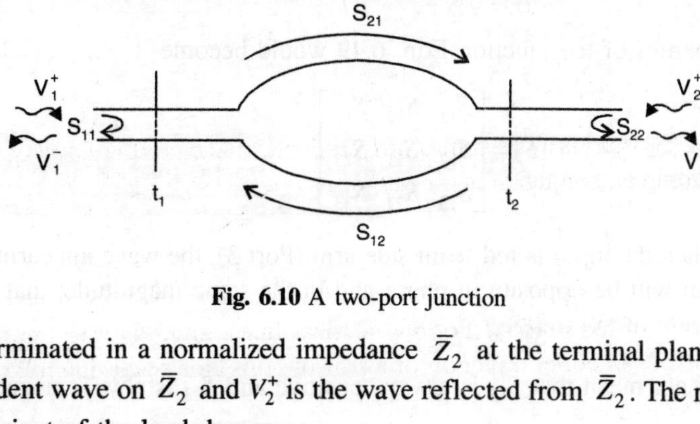

Fig. 6.10 A two-port junction

If guide 2 is terminated in a normalized impedance $\bar{Z}_2$ at the terminal plane t_2, then V_2^- may be regarded as the incident wave on $\bar{Z}_2$ and V_2^+ is the wave reflected from $\bar{Z}_2$. The ratio must be equal to the reflection coefficient of the load; hence

$$\frac{V_2^+}{V_2^-} = \frac{\bar{Z}_2 - 1}{\bar{Z}_2 + 1} = \Gamma_L \qquad ...(6.45)$$

Solving for V_1^- / V_1^+,

$$\frac{V_1^-}{V_1^+} = S_{11} - \frac{S_{12} S_{21} \Gamma_L}{S_{22} \Gamma_L - 1} \qquad ...(6.46)$$

For a reciprocal junction $S_{12} = S_{21}$, and the scattering matrix contains, at most, six independent parameters, which are the magnitudes and phase angles of S_{11}, S_{21} and S_{22}. If the junction is lossless, the scattering matrix contains only three parameters,

$$S_{11} S_{11}^* + S_{12} S_{12}^* = 1 \qquad ...(6.47a)$$

$$S_{22} S_{22}^* + S_{12} S_{12}^* = 1 \qquad ...(6.47b)$$

$$S_{11} S_{12}^* + S_{12} S_{22}^* = 0 \qquad ...(6.47c)$$

hence the reflection coefficients in the input and output guides are equal in magnitude for a lossless junction. So

$$|S_{12}| = \sqrt{1 - |S_{11}|^2} \qquad ...(6.48)$$

6.7.2 Scattering Matrix for Tee Junction

Tee is a three-port device, so the scattering matrix, in general, be written as

$$[S] = \begin{bmatrix} S_{11} & S_{12} & S_{13} \\ S_{21} & S_{22} & S_{23} \\ S_{31} & S_{32} & S_{33} \end{bmatrix} \qquad ...(6.49)$$

Since $[S]$ is symmetric, we have

$$S_{12} = S_{21},\ S_{13} = S_{31},\ S_{23} = S_{32}$$

and more ever tee is symmetric about port 1 and 2 (symmetric arms)

$$S_{11} = S_{22}$$

Using these properties of tee junction Eqn. 6.49 would become

$$[S] = \begin{bmatrix} S_{11} & S_{12} & S_{13} \\ S_{12} & S_{11} & S_{23} \\ S_{13} & S_{23} & S_{33} \end{bmatrix} \qquad ...(6.50)$$

***E*-plane tee**—When the input is fed from side arm (Port 3), the wave appearing at the two-ports 1 and 2 of collinear arm will be opposite in phase and in the same magnitude, that is

$$S_{13} = -S_{23} \qquad ...(6.51)$$

However, it does not mean that S_{13} is always positive and S_{23} is always negative.

S-matrix will be

$$[S]=\begin{bmatrix} S_{11} & S_{12} & S_{13} \\ S_{12} & S_{11} & -S_{13} \\ S_{13} & -S_{13} & S_{33} \end{bmatrix} \qquad ...(6.52)$$

***H*-plane tee**—When the input is fed from side arm (port 3), the wave appearing at the two ports 1 and 2 of collinear arm will be in phase and in the same magnitude, that

$$S_{13} = S_{23}$$

Thus its [S] marix is written as

$$[S]=\begin{bmatrix} S_{11} & S_{12} & S_{13} \\ S_{12} & S_{11} & S_{13} \\ S_{13} & S_{13} & S_{33} \end{bmatrix} \qquad ...(6.53)$$

We note that (for both E- and H-plane tees) from purely isotropy and symmetry properties, the number of unknown scattering coefficients reduced to four.

If we assume

$$S_{11} = \alpha,\ S_{13} = \gamma,\ S_{12} = \delta,\ S_{33} = \beta$$

Hence scattering matrix for E-plane tee (Eqn. 6.52) can be written as

$$[S]=\begin{bmatrix} \alpha & \delta & \gamma \\ \delta & \alpha & -\gamma \\ \gamma & -\gamma & \beta \end{bmatrix} \qquad ...(6.54)$$

and for H-plane tee (Eqn. 6.53) can be written as

$$[S]=\begin{bmatrix} \alpha & \delta & \gamma \\ \delta & \alpha & \gamma \\ \gamma & \gamma & \beta \end{bmatrix} \qquad ...(6.55)$$

Further these devices are lossless hence [S] must be **unitary**.

6.7.3 Scattering Matrix for Hybrid Tee

A hybrid tee is a four-port device, so in general [S] matrix is

$$[S]=\begin{bmatrix} S_{11} & S_{12} & S_{13} & S_{14} \\ S_{21} & S_{22} & S_{23} & S_{24} \\ S_{31} & S_{32} & S_{33} & S_{34} \\ S_{41} & S_{42} & S_{43} & S_{44} \end{bmatrix} \qquad ...(6.56)$$

Assuming the tee to be lossless and medium inside to be isotropic, the symmetry of the tee requires

$$S_{12} = S_{21},\ S_{13} = S_{31},\ S_{14} = S_{41},$$

$$S_{23} = S_{32},\ S_{24} = S_{42},\ S_{34} = S_{43} \,..........$$

Thus,

$$[S]=\begin{bmatrix} S_{11} & S_{12} & S_{13} & S_{14} \\ S_{12} & S_{22} & S_{23} & S_{24} \\ S_{13} & S_{23} & S_{33} & S_{34} \\ S_{14} & S_{24} & S_{34} & S_{44} \end{bmatrix} \quad ...(6.57)$$

Since the magic tee is symmetrical about plane P as shown in Fig. 6.27*a*, arms 1 and 2 are mirror images of one another with respect to this plane. Moreover, the boundary conditions which are to be satisfied by the electric field are also the same. It, therefore, follows from the uniqueness theorem that the electric field distribution in arms 1 and 2 must be identical. Hence, the electric coupling from arms 3 to arm 1 and from arm 3 to arm 2 must be identical.

$$S_{13} = S_{23} \quad ...(6.58)$$

If power is fed in port 4, *i.e.*, *E*-arm, it divides anti-symmetrically in arms 1 and 2, *i.e.*, the electric field in these arms is equal in magnitude but π radians or 180° is out of phase as illustrated in Fig. 6.11. It, therefore, follows that coupling coefficients from 1 to 4 and 2 to 4 are equal in magnitude and opposite in phase, that is

$$S_{14} = -S_{24} \quad ...(6.59a)$$

The symmetry of the magic tee, looking into port 1, with all other ports matched terminated, is equivalent to looking into port 2, with all other ports matched terminated. Hence

$$S_{11} = S_{22} \quad ...(6.59b)$$

Moreover, ports 3 and 4 are uncoupled because of cross-polarizations of the electric field.

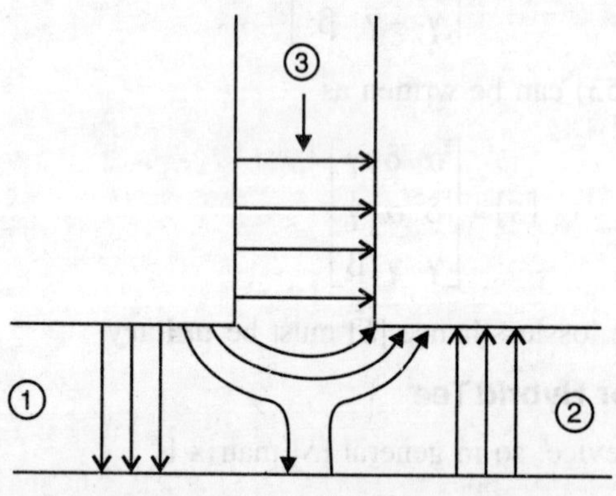

Fig. 6.11 Power incident in port 4 of magic tee is divided into equal and out of π phase powers in ports 1 and 2

$$S_{34} = 0 \quad ...(6.60)$$

Using Eqns. 6.58 through 6.60, Eqn. 6.57 can be reduced to

$$[S]=\begin{bmatrix} S_{11} & S_{12} & S_{13} & S_{14} \\ S_{12} & S_{11} & S_{13} & -S_{14} \\ S_{13} & S_{13} & S_{33} & 0 \\ S_{14} & -S_{14} & 0 & S_{44} \end{bmatrix} \quad ...(6.61)$$

Equation 6.61 is the scattering matrix of any symmetrical, 4-port tee of the form shown in Fig. 5.27*a*. It does not matter what is inside the junction provided it is symmetrical and propagates the TE_{10} mode. However, in practice, the device is made more useful by making ports 3 and 4 well-matched, *i.e.*,

$$S_{33} = S_{44} = 0 \qquad \text{...(6.62)}$$

This can be done by having a proper choice of the reference planes on ports 3 and 4. Using Eqn. 6.62 in Eqn. 6.61, we obtain

$$[S] = \begin{bmatrix} S_{11} & S_{12} & S_{13} & S_{14} \\ S_{12} & S_{11} & S_{13} & -S_{14} \\ S_{13} & S_{13} & 0 & 0 \\ S_{14} & -S_{14} & 0 & 0 \end{bmatrix} \qquad \text{...(6.63)}$$

Since the device is assumed to be lossless,

$$[S]\,[S]^* = [1].$$

whose

$$(1, 1) \text{ term} = |S_{11}|^2 + |S_{12}|^2 + |S_{13}|^2 + |S_{14}|^2 = 1 \qquad \text{...(6.64}a\text{)}$$

$$(3, 3) \text{ term} = 2|S_{13}|^2 = 1 \qquad \text{...(6.64}b\text{)}$$

$$(4, 4) \text{ term} = 2|S_{14}|^2 = 1. \qquad \text{...(6.64}c\text{)}$$

From Eqns. (6.64*b* and 6.64*c*), we have respectively,

$$S_{13} = \frac{1}{\sqrt{2}} e^{j\phi} \text{ and } S_{14} = \frac{1}{\sqrt{2}} e^{j\psi} \qquad \text{...(6.65)}$$

Phases ϕ and ψ in Eqn. (6.65) are the choices of the reference planes and can be made each equal to zero, *i.e.*, $\phi = \psi = 0$. It then follows from Eqn. (6.64a) and (6.65), that

$$S_{11} = S_{12} = 0 \qquad \text{...(6.66)}$$

because each of $|S_{11}|^2$ and $|S_{12}|^2$ is positive or zero.

Therefore, using Eqns. (6.65) and Eqn. (6.66), Eqn. (6.63) can be reduced to

$$[S] = \frac{1}{\sqrt{2}} \begin{bmatrix} 0 & 0 & 1 & 1 \\ 0 & 0 & 1 & -1 \\ 1 & 1 & 0 & 0 \\ 1 & -1 & 0 & 0 \end{bmatrix} \qquad \text{...(6.67)}$$

This equation is used to describe hybrid tee with various terminal conditions such as the ports of the tee are well matched, the ports of the tee are not necessarily matched etc.

6.7.4 *S*-matrix of a Directional Coupler

Directional coupler is a four-port matched device so in general [*S*] matrix is a four by four matrix with diagonal element

$$S_{11} = S_{22} = S_{33} = S_{44} = 0$$

Thus
$$[S] = \begin{bmatrix} 0 & S_{12} & S_{13} & S_{14} \\ S_{21} & 0 & S_{23} & S_{24} \\ S_{31} & S_{32} & 0 & S_{34} \\ S_{41} & S_{42} & S_{43} & 0 \end{bmatrix} \quad ...(6.68)$$

Moreover, the device is symmetric, hence

$$S_{ij} = S_{ji}; \; i \text{ and } j = 1, 2, 3, 4 \quad ...(6.69)$$

Using Eqn. 6.69 in Eqn. 6.68.

$$[S] = \begin{bmatrix} 0 & S_{12} & S_{13} & S_{14} \\ S_{12} & 0 & S_{23} & S_{24} \\ S_{13} & S_{23} & 0 & S_{34} \\ S_{14} & S_{24} & S_{34} & 0 \end{bmatrix} \quad ...(6.70)$$

As noted, there is no coupling between port 1 and port 4 and between port 2 and port 3 *i.e.*, referring to Fig. 5.14, from the property of the directional coupler.

$$S_{13} = S_{31} = S_{24} = S_{42} = 0$$

Consequently Eqn. (6.70) will be,

$$[S] = \begin{bmatrix} 0 & S_{12} & 0 & S_{14} \\ S_{12} & 0 & S_{23} & 0 \\ 0 & S_{23} & 0 & S_{34} \\ S_{14} & 0 & S_{34} & 0 \end{bmatrix} \quad ...(6.71)$$

For a lossless network the *S*-matrix is unitary, hence

$$S_{12}S_{12}^* + S_{14}S_{14}^* = 1 \quad ...(6.72)$$

$$S_{12}S_{12}^* + S_{23}S_{23}^* = 1 \quad ...(6.73)$$

$$S_{23}S_{23}^* + S_{34}S_{34}^* = 1 \quad ...(6.74)$$

$$S_{14}S_{14}^* + S_{34}S_{34}^* = 1 \quad ...(6.75)$$

$$S_{12}S_{23}^* + S_{14}S_{34}^* = 0 \quad ...(6.76)$$

From the Eqns. 6.72 and 6.73, we get

$$|S_{14}| = |S_{23}| \quad ...(6.77)$$

and from Eqns. 6.73 and 6.74, we get

$$|S_{12}| = |S_{34}| \quad ...(6.78)$$

By choosing reference plane of ports 1 with respect to that of port 2 and reference plane of port 3 with respect to that of 4, we can make *S*-parameters real,

$$S_{12} = S_{34} = \alpha \qquad ...(6.79)$$

where α is positive real number. Then from Eqn. 6.76

$$\alpha\left(S_{23}^{*} + S_{14}\right) = 0$$

Since $$\alpha \neq 0,\ S_{23}^{*} + S_{14} = 0 \qquad ...(6.80)$$

Now again selecting the reference plane of port 4 with respect to port 1, we can make S_{14} real, so that

$$S_{23} = -S_{23}^{*} = S_{14} = \beta \qquad ...(6.81)$$

Thus [S] matrix may be written as

$$[S] = \begin{bmatrix} 0 & \alpha & 0 & \beta \\ \alpha & 0 & \beta & 0 \\ 0 & \beta & 0 & \alpha \\ \beta & 0 & \alpha & 0 \end{bmatrix} \qquad ...(6.82)$$

where $\alpha^2 + \beta^2 = 1$ for conservation of energy, α is called the transmission factor and β is called the coupling factor.

Coupling of such an ideal coupler is given by

$$C = -20 \log \beta \, \text{dB} \qquad ...(6.83)$$

If the coupler is not perfect the directivity is given to

$$D = 20 \log \left|\beta / S_{13}\right| \text{dB} \qquad ...(6.84)$$

Note that $S_{13} = 0$ for an ideal coupler.

The amount of coupling and the directivity depend upon the size and location of the hole in the common wall.

6.7.5 *S*-matrix of a Circulator

A lossless, matched, nonreciprocal three-port microwave junction is a perfect three-port circulator. A perfect matched three-port junction has a scattering matrix of the form

$$[S] = \begin{bmatrix} 0 & S_{12} & S_{13} \\ S_{21} & 0 & S_{23} \\ S_{31} & S_{32} & 0 \end{bmatrix} \qquad ...(6.85)$$

For a nonreciprocal junction the scattering matrix is no longer symmetrical that is, $S_{ij} \neq S_{ji}$. Applying the unitary condition to Eqn. (6.85) gives,

$$S_{12}S_{12}^{*} + S_{13}S_{13}^{*} = 1 \qquad ...(6.86a)$$

$$S_{21}S_{21}^{*} + S_{23}S_{23}^{*} = 1 \qquad ...(6.86b)$$

$$S_{31}S_{31}^{*} + S_{32}S_{32}^{*} = 1 \qquad ...(6.86c)$$

$$S_{13}S_{23}^{*} = S_{12}S_{32}^{*} = S_{21}S_{31}^{*} = 0 \qquad ...(6.86d)$$

Let us assume that $S_{21} \neq 0$. Then in Eqn. (6.86*d*) gives $S_{31} = 0$. The Eqn. (6.86*c*) now requires $|S_{32}| = 1$, and thus $S_{12} = 0$ from Eqn. (6.86*d*), $|S_{13}| = 1$ and $S_{23} = 0$.

So

$$|S_{21}| = |S_{32}| = |S_{13}| = 1$$

$$S_{12} = S_{23} = S_{31} = 0 \qquad ...(6.87)$$

There is perfect transmission from port 1 into port 2, from port 2 into port 3, and from port 3 into port 1. There is zero transmission in any other direction. The resultant scattering matrix of any matched, lossless, non-reciprocal three-port junction must then have the form

$$[S] = \begin{bmatrix} 0 & 0 & S_{13} \\ S_{21} & 0 & 0 \\ 0 & S_{23} & 0 \end{bmatrix} \qquad ...(6.88)$$

6.7.6 S-matrix of Isolator

A Faraday rotation isolator is a circular waveguide section shown in Fig. 6.2. For an ideal lossless, matched isolator,

$$|S_{21}| = 1, \ |S_{12}| = |S_{11}| = |S_{22}| = 0 \qquad ...(6.89)$$

i.e.,

$$[S] = \begin{bmatrix} 0 & 0 \\ 1 & 0 \end{bmatrix} \qquad ...(6.90)$$

Example 6.1. Determine the scattering parameters for a 10 dB direction coupler. The Directivity D = 30 dB. Assume that it is lossless and the VSWR at each port is 1.0 under matched conditions. Designate the port in the main guide as 1 or 2 and the ports in the auxiliary guide as 3 and 4.

Solution. *S*-parameters for a 10 dB direction coupler.

Given D = 30 dB and VSWR = 1.0

The coupling factor(C),

$$C = -10 \log \frac{P_1}{P_4}$$

$$\therefore \qquad -10 = 10 \log \frac{P_1}{P_4} = |S_{41}^2|$$

or

$$S_{41} = \sqrt{0.1} = 0.3162$$

or

$$S_{41} = S_{14} = 0.3162 \text{ (as matched and lossless)} \qquad ...(1)$$

We know the directivity (D),

$$D = 10 \log \frac{P_4}{P_3}$$

$$30 = 10 \log \frac{P_4}{P_3}$$

$$10^3 = \frac{|S_{41}^2|}{|S_{31}^2|}$$

$\therefore$ $$S_{31} = 0.01 = S_{13} \qquad ...(2)$$

We know,

$$S_{11} = \frac{VSWR - 1}{VSWR + 1}$$

$$VSWR = 1.0 \text{ given}$$

$$S_{11} = S_{22} = S_{33} = S_{44} = 0 \qquad ...(3)$$

So S-matrix becomes
$$\begin{bmatrix} 0 & S_{12} & 0.01 & 0.3162 \\ S_{12} & 0 & S_{23} & S_{24} \\ 0.01 & S_{23} & 0 & S_{34} \\ 0.3162 & S_{24} & S_{34} & 0 \end{bmatrix} \qquad ...(4)$$

$$[\because \; S_{13} = S_{31}; S_{12} = S_{21}; \; S_{42} = S_{24}; S_{43} = S_{34}]$$

Given that the main guide 1 and 2 ports and auxiliary guide as 3 and 4 ports. Let input power is given at port 1,

$$P_1 = P_2 + P_3 + P_4$$

$\therefore$ $$1 = \frac{P_2}{P_1} + \frac{P_3}{P_1} + \frac{P_4}{P_1}$$

$$1 = |S_{21}|^2 + |S_{31}|^2 + |S_{41}|^2$$

From Eqns. (1) and (2),

$$1 = |S_{21}|^2 + 10^{-4} + 0.1$$

So $$S_{21} = 0.9486$$

$\therefore$ $$S_{21} = S_{12} = 0.9486 \qquad ...(5)$$

We have,

$$\left.\begin{aligned} S_{12}^2 + S_{23}^2 + S_{24}^2 &= 1 \\ S_{23}^2 + S_{34}^2 + 10^{-4} &= 1 \\ S_{24}^2 + S_{34}^2 + 10^{-1} &= 1 \end{aligned}\right\} \text{Using property (1) on rows I, II and III of } S\text{-matrix}$$

After solving,

$$\left.\begin{aligned} S_{34} = S_{43} &= 0.9486 \\ S_{24} = S_{42} &= 0.01 \\ S_{23} = S_{32} &= 0.3162 \end{aligned}\right\} \qquad ...(6)$$

Thus, the S-parameters are,

$$S = \begin{bmatrix} 0 & 0.9486 & 0.01 & 0.3162 \\ 0.9486 & 0 & 0.3162 & 0.01 \\ 0.01 & 0.3162 & 0 & 0.9486 \\ 0.3162 & 0.01 & 0.9486 & 0 \end{bmatrix}$$

Example 6.2. The impedance matrix of a certain lumped-elements network is given by

$$[Z_{ij}] = \begin{bmatrix} 4 & 2 \\ 2 & 4 \end{bmatrix}$$

Determine the scattering matrix by using S-parameter theory and indicate the values of the components:

$$[S_{ki}] = \begin{bmatrix} S_{11} & S_{12} \\ S_{21} & S_{22} \end{bmatrix}$$

Solution.
The scattering matrix equation is,

$$[S_{ki}] = [Z_{kj} - \delta_{kj}][Z_{ij} + \delta_{ij}]^{-1} = \frac{b_k}{a_i}$$

where

$$[Z_{ij}] = \begin{bmatrix} 4 & 2 \\ 2 & 4 \end{bmatrix}$$

$$b = [Z_{kj} - \delta_{kj}] = \begin{bmatrix} Z_{11} - 1 & Z_{12} \\ Z_{21} & Z_{22} - 1 \end{bmatrix} = \begin{bmatrix} 3 & 2 \\ 2 & 3 \end{bmatrix}$$

where

$$\delta_{kj} = 1 \text{ if } k = j$$

$$\delta_{kj} = 0 \text{ if } k \neq j$$

Or

$$a = [Z_{ij} + \delta_{ij}] = \begin{bmatrix} Z_{11} + 1 & Z_{12} \\ Z_{21} & Z_{22} + 1 \end{bmatrix} = \begin{bmatrix} 5 & 2 \\ 2 & 5 \end{bmatrix}$$

We know

$$[S_{ki}] = b \times a^{-1} \text{ and } a^{-1} = \begin{bmatrix} \frac{A_{11}}{A} & \frac{A_{21}}{A} \\ \frac{A_{12}}{A} & \frac{A_{22}}{A} \end{bmatrix}$$

where

$$A = \begin{vmatrix} Z_{11} + 1 & Z_{12} \\ Z_{21} & Z_{22} + 1 \end{vmatrix} = \begin{vmatrix} 5 & 2 \\ 2 & 5 \end{vmatrix} = 25 - 4 = 21$$

A_{ij} is the cofactor a_{ij} of determinant A

$$A_{ij} = \begin{bmatrix} 5 & -2 \\ -2 & 5 \end{bmatrix}$$

$$\therefore \quad a^{-1} = \begin{bmatrix} \frac{5}{21} & \frac{-2}{21} \\ \frac{-2}{21} & \frac{5}{21} \end{bmatrix}$$

Therefore

$$[S_{ki}] = b \times a^{-1} = \begin{bmatrix} 3 & 2 \\ 2 & 3 \end{bmatrix} \times \begin{bmatrix} \frac{5}{21} & \frac{-2}{21} \\ \frac{-2}{21} & \frac{5}{21} \end{bmatrix}$$

$$[S_{ki}] = \begin{bmatrix} \frac{11}{21} & \frac{4}{21} \\ \frac{4}{21} & \frac{11}{21} \end{bmatrix}$$

Example 6.3. A signal of power 64 mW is fed into one of the collinear ports of a lossless *H*-plane tee. Determine the powers in the remaining ports when other ports are terminated by means of matched loads.

Solution. Let the collinear ports be port (1) to which signal of 64 mW is fed. Then the other ports are port (2) (other collinear ports) and port (3) (*H*-armport) that are terminated in matched loads.

$$\therefore \quad a_2 = a_3 = 0 \text{ and } a_1 = 64 \text{ mW}$$

The *S*-matrix for *H*-plane tee.

$$[S] = \begin{bmatrix} \frac{1}{2} & -\frac{1}{2} & \frac{1}{\sqrt{2}} \\ -\frac{1}{2} & \frac{1}{2} & \frac{1}{\sqrt{2}} \\ \frac{1}{\sqrt{2}} & \frac{1}{\sqrt{2}} & 0 \end{bmatrix}$$

i.e.

$$\begin{bmatrix} b_1 \\ b_2 \\ b_3 \end{bmatrix} = \begin{bmatrix} \frac{1}{2} & -\frac{1}{2} & \frac{1}{\sqrt{2}} \\ -\frac{1}{2} & \frac{1}{2} & \frac{1}{\sqrt{2}} \\ \frac{1}{\sqrt{2}} & \frac{1}{\sqrt{2}} & 0 \end{bmatrix} \begin{bmatrix} 64 \\ 0 \\ 0 \end{bmatrix}$$

$$b_1 = \text{Power at port (1)} = \frac{1}{2}a_1 - \frac{1}{2}a_2 + \frac{1}{\sqrt{2}}a_3$$

$$= \left(\frac{1}{2}\right)^2 |a_1| = 16 \text{ mW}$$

Similarly,

$$b_2 = 16 \text{ mW}$$

and $b_3 = 32$ mW

$$b_1 = 16 \text{ mW}, \; b_2 = 16 \text{ mW}, \; b_3 = 32 \text{ mW}$$

it is seen that $b_3 = b_1 + b_2$.

Example 6.4. A helical slow-wave structure has a pitch P of 2 mm and a diameter of 4 cm. Calculate the wave velocity in the axial direction of the helix.

Solution. Given

Helix pitch $P = 2 \text{ mm} = 2\times10^{-3}$ m

Helix diameter $d = 4$ cm

Air filled helix

So wave velocity is

$$v_p = \frac{CP}{\sqrt{P^2 + (\pi d)^2}}$$

$$= \frac{3\times10^8 \times 2\times10^{-3}}{\sqrt{(2\times10^{-3})^2 + (\pi\times4\times10^{-2})^2}}$$

$$v_p = 4.78\times10^6 \text{ m/sec.}$$

Example 6.5. In a *H*-plane tee junction, 20 mW power is applied to port (3) that is perfectly matched to the junction. Calculate the power delivered to the load 60Ω and 75Ω connected to port (1) and port (2).

Solution. The [*S*] for *H*-plane tee with port (3) matched to the junction is

$$[S] = \begin{bmatrix} \frac{1}{2} & -\frac{1}{2} & \frac{1}{\sqrt{2}} \\ -\frac{1}{2} & \frac{1}{2} & \frac{1}{\sqrt{2}} \\ \frac{1}{\sqrt{2}} & \frac{1}{\sqrt{2}} & 0 \end{bmatrix}$$

Power at port (3) gets equally divided into port (1) and (2) *i.e.*, 20 mW power at port (1) delivers 10 mW at port (1) and (2).

Power reflected from port (1) and port (2) are

$$\frac{1}{2}[\rho_1 b_1]^2 \text{ and } \frac{1}{2}[\rho_2 b_2]^2.$$

∴ Power delivered to load Z_1 at port (1) = 60 Ω is

$$\rho_1 = \frac{1}{2}[b_1]^2 - \frac{1}{2}[\rho_1 b_1]^2 = \frac{1}{2}[b_1]^2\left[1-|\rho_1|^2\right]$$

Similarly, power delivered to load Z_2 at port (2) = 75 Ω

$$\rho_2 = \frac{1}{2}[b_2]^2 - \frac{1}{2}[\rho_2 b_2]^2 = \frac{1}{2}[b_2]^2\left[1-|\rho_2|^2\right]$$

Now taking the characteristics impedance of the line as 50 Ω

$$|\rho_1| = \frac{|60-50|}{|60+50|} = \frac{1}{11} \text{ and } |\rho_1|^2 = \frac{1}{121}$$

$$\rho_2 = \left|\frac{75-50}{75+50}\right| = \frac{1}{5}, |\rho_2|^2 = \frac{1}{25}$$

∴ $$P_1 = 10\times 10^3\left[1-\frac{1}{121}\right] = 9.92 \text{ mW}$$

$$P_2 = 10\times 10^3\left[1-\frac{1}{25}\right] = 9.6 \text{ mW}$$

REVIEW QUESTIONS AND PROBLEMS

1. What are scattering coefficients of a microwave device? Discuss the scattering properties of multiport junction and hence obtain the scattering matrix of a magic T. What are its uses? Explain the use of magic T (*i*) as duplexer; (*ii*) as a mixer for a superheterodyne receiver.
2. Explain the construction and working of two choke type movable short circuit used in microwave measurements.
3. Explain the use of magic tee for a four-port circulator.
4. Explain the operation of a circulator.
5. What is meant by Faraday rotation? Explain with the help of appropriate diagram the working of an isolator.
6. What are Ferrites? What properties do they have? Describe in detail a circulator and give its applications.
7. Explain the operation of Faraday Rotation Isolator and give its applications.
8. What are Ferrites? Derive the permeability matrix for ferrites. Using the matrix explain the phenomenon of Faraday rotation. Explain the working of gyrator.
9. Discuss propagation of microwave energy in ferrites. Obtain an expression for Faraday rotation. What is a reciprocal effect?

10. What do you understand by Faraday rotation? Obtain an expression for it. Is it a reciprocal effect? Show that a lossless, matched, nonreciprocal, three-port microwave junction is a three-port circulator. Hence describe the essential features of a practical three-port circulator.

11. What is Faraday rotation? Show that it is a non-reciprocal effect. Hence explain the construction and working of a rotation isolator. State few desired properties of ferrites.

12. What are Ferrites? Prove that for a ferrite, the permeability matrix is given by,

$$[\mu] = \begin{bmatrix} \mu & -jk & 0 \\ jk & \mu & 0 \\ 0 & 0 & \mu_0 \end{bmatrix}$$

where symbols have their usual meanings. Using this matrix, explain the rotation of plane of polarization of a plane electromagnetic wave propagation through the ferrite.

13. Show that scattering matrix for entirely symmetrical (120°) series. *T*-junction is given by

$$[S] = \frac{1}{3}\begin{bmatrix} 1 & 2 & 2 \\ 2 & 1 & -2 \\ 2 & -2 & 1 \end{bmatrix}$$

14. Show that scattering matrix for series *T*-junction matched at arm 3 is given by

$$[S] = \frac{1}{2}\begin{bmatrix} 1 & 1 & \sqrt{2} \\ 2 & 1 & -\sqrt{2} \\ \sqrt{2} & -\sqrt{2} & 0 \end{bmatrix}$$

15. Show that scattering matrix for shunt *T*-junction matched at arm 3 is given by

$$[S] = \frac{1}{2}\begin{bmatrix} -1 & 1 & \sqrt{2} \\ 1 & -1 & \sqrt{2} \\ \sqrt{2} & \sqrt{2} & 0 \end{bmatrix}$$

16. Show that scattering matrix for hybrid ring junction (rate-race, refer to Fig. 6.11) is given by

$$[S] = -\frac{1}{\sqrt{2}}\begin{bmatrix} 0 & 0 & 1 & 1 \\ 0 & 0 & 1 & -1 \\ 1 & 1 & 0 & 0 \\ 1 & -1 & 0 & 0 \end{bmatrix}$$

17. Show that the scattering matrix for an ideal unidirectional attenuator (isolator) without loss in the direction from arm 1 into arm 2 is given by

$$[S] = \begin{bmatrix} 0 & 0 \\ 1 & 0 \end{bmatrix}$$

18. Show that the scattering matrix for an ideal Y circulator is given by

$$[S]=\begin{bmatrix}0 & 0 & 1\\ 1 & 0 & 0\\ 0 & 1 & 0\end{bmatrix}$$

19. Show that the scattering matrix for a section of uniform lossless guide with an electric length $\theta = 2\pi L/\lambda_g$ or matched reciprocal phase shifter providing a θ phase shift is given by

$$[S]=\exp^{-j\theta}\begin{bmatrix}0 & 1\\ 1 & 0\end{bmatrix}$$

20. Prove that the input impedance of a waveguide section terminated in a matched load has a real normalized value of unity.
21. (*a*) What is magic tee? Describe the properties of magic tee, giving its *S*-matrix.
    (*b*) Show a waveguide with cylindrical post and describe its behaviour. How can it be used, when it is inserted half way into the waveguide?
22. (*a*) Enumerate the properties of *S*-parameters.
    (*b*) Formulate the *S*-parameter matrix of 4-port circular.
23. (*a*) Why 'Ferrites' are used in microwave passive devices? Explain.
    (*b*) Scattering matrix is a unitary matrix. Prove this statement.
24. Explain the construction, operation and applications of the following microwave components
    (*a*) Circular (*b*) Gyrator
25. (*a*) What are microwave ferrites? Explain the working of ferrite isolator.
    (*b*) Derive the *S*-matrix of a directional coupler in standard form.
26. (*a*) What are ferrite devices? Explain how Faraday rotation is utilized in the construction of a 4-port circulator.
    (*b*) What are the advantages of scattering matrix representation over impedance and admittance matrix representations?
27. (*a*) Derive the *S*-matrix of a magic tee.
    (*b*) Discuss the properties and applications of a Gyrator.

CHAPTER 7

Microwave Tubes—Klystrons and Reflex Klystrons

7.1 INTRODUCTION TO MICROWAVE TUBES

The microwave vacuum tubes that have enjoyed the greatest technological success in microwave generation and amplification are those that employ usefully the transit-time effects for their operation. These are (*i*) the klystron (*ii*) the magnetron (*iii*) the various travelling wave tubes (TWTs). The schemetic structure of these tubes are shown in Fig. 7.1, two cavity klystron, Fig. 7.2 magnetron and Fig. 7.3 travelling wave tube.

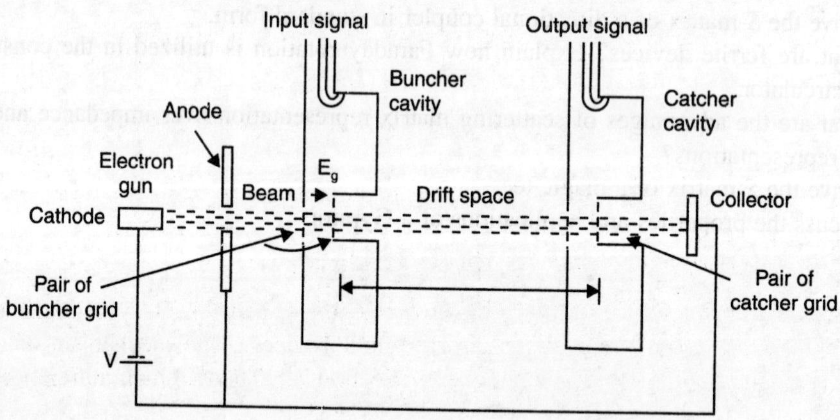

Fig. 7.1 Two-cavity klystron

The **mechanism of operation** of these tubes may differ in detail but involves transfer of power from a source of dc voltage to a source of alternating voltage by means of a density modulated stream of electrons. This is accomplished by accelerating electrons in a direct electric field and retarding them in alternating electric field thus delivering the energy to ac field. The involved **processes** are (*i*) beam production (*ii*) velocity modulation (*iii*) conversion to density modulation (*iv*) transfer of energy and (*v*) collection of electrons. In any electronic device these processes must take place, either separately (as in case of two-cavity klystron) or simultaneously (as in case of travelling wave tube).

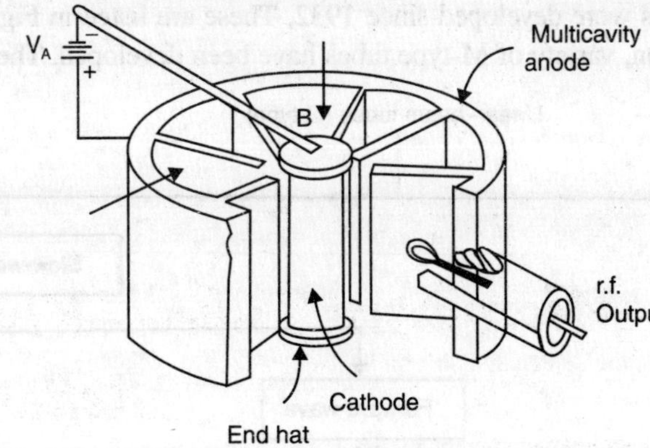

Fig. 7.2 Magnetron

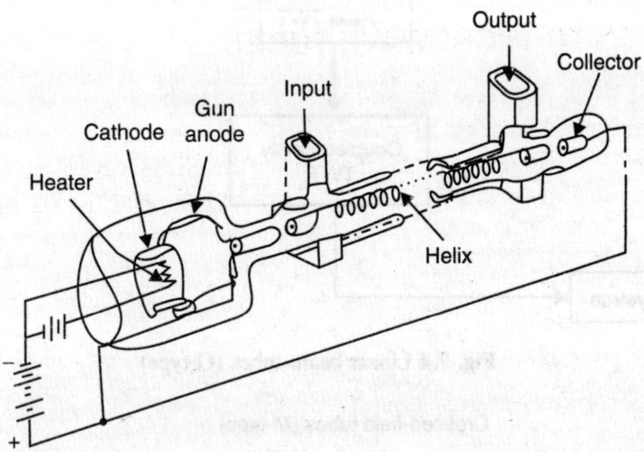

Fig. 7.3 Travelling wave tube (TWT)

7.2 CLASSIFICATION OF MICROWAVE TUBES

Microwave tubes can be broadly classified into two categories, namely, O-Type linear tubes (TWT, Klystrons) and M-Tubes tubes (magnetrons, crossed-field devices). The mechanism of operation in both the cases is essentially the same as mentioned in Section 7.1. The two main differences in the two types of tubes are listed in Table 7.1.

TABLE 7.1

S. No.		O-Type tubes	M-Type tubes
1.	Energy conversion	Kinetic energy to r.f. energy	Potential energy to r.f. energy
2.	Role of Magnetic field	For focussing of the electron beam only	Magnetic and electric fields are perpendicular to each other. Magnetic field plays direct role in r-f interaction process.

Variety of O-type tubes were developed since 1932. These are listed in Fig. 7.4. Since 1921, when Hull invented the magnetron, variety of M-type tubes have been developed. These are listed in Fig. 7.5.

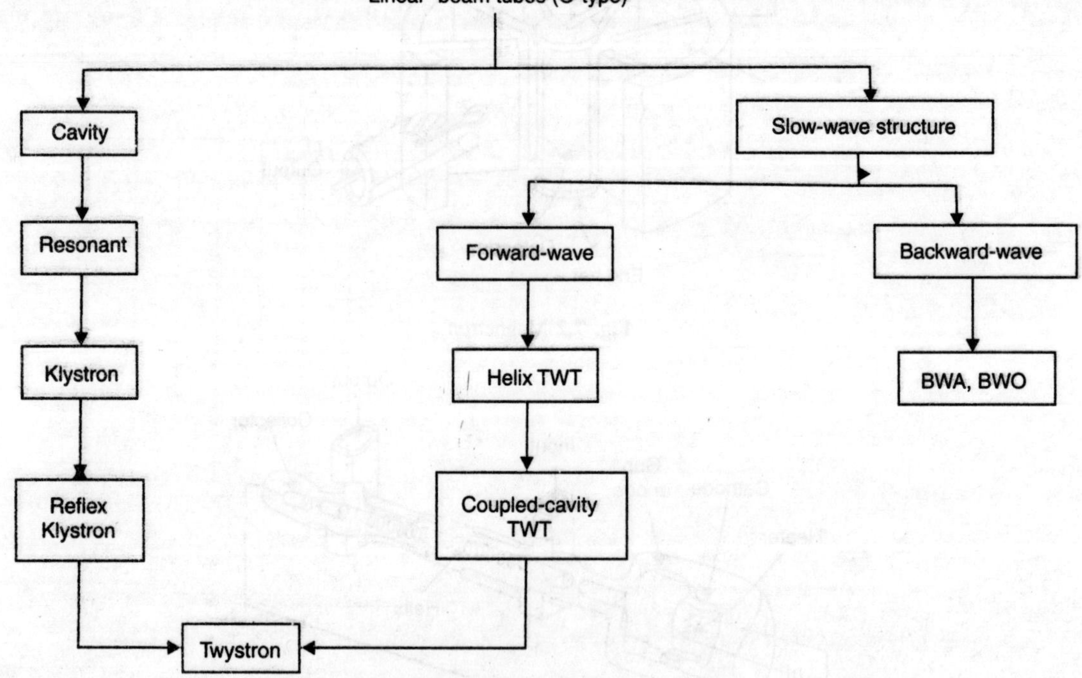

Fig. 7.4 Linear beam tubes (O-type)

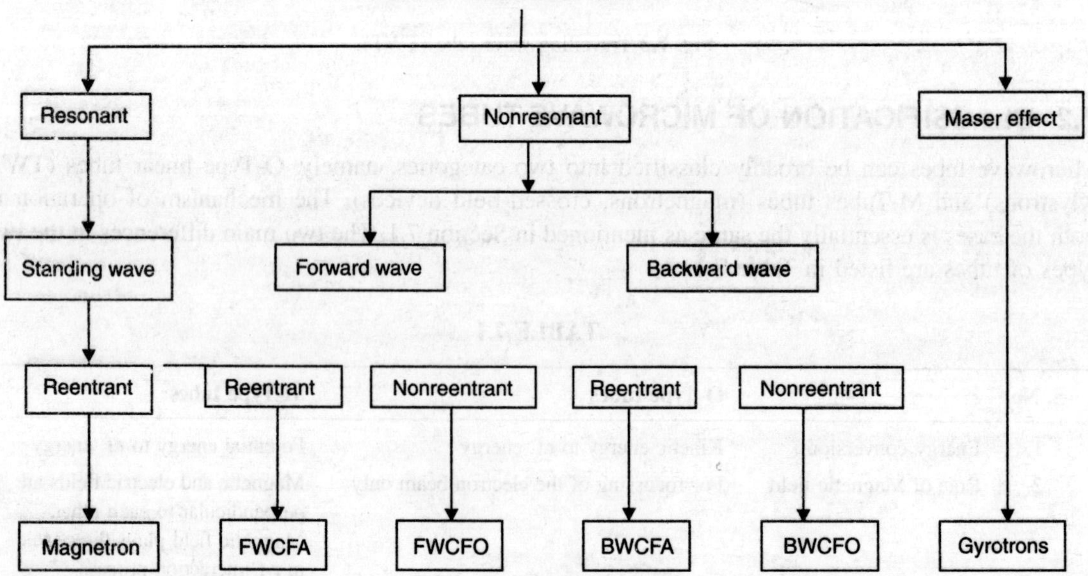

Fig. 7.5 M-type Tubes

7.3 PERFORMANCE OF TUBES AT GLANCE

Figures 7.6 and 7.7 summarize the frequency versus power performance of magnetrons, reflex klystrons, klystrons and helix travelling wave tubes. From the Fig. 7.6 we find that magnetrons are capable of very high CW power outputs in the range of several kilowatts. On the other hand reflex klystrons are capable of very low power outputs in the range of few miliwatts. From the graph of Fig. 7.7, we see that klystrons are capable of peak powers in the megawatt range and helix TWTs are capable of power in the range of several hundred watts.

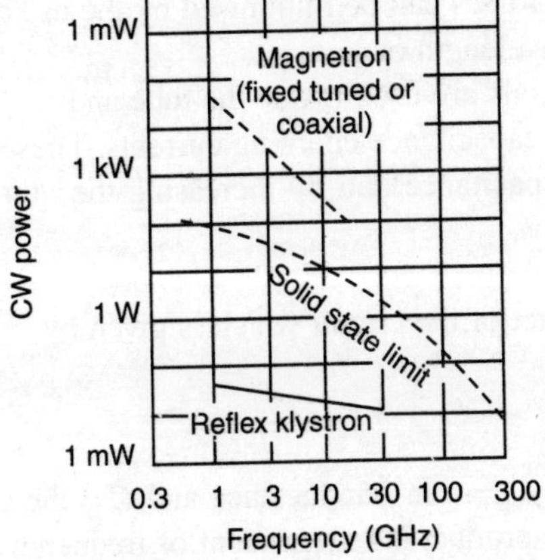

Fig. 7.6 Frequency versus power performance of reflex klystrons and magnetrons (oscillators)

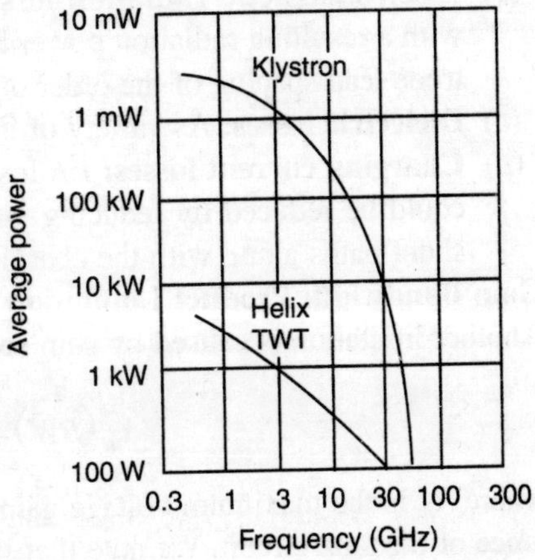

Fig. 7.7 Frequency versus powers performance of klystrons and helix TWTs (amplifiers)

Travelling wave tubes are characterised by very wide band width ranging from 30 to 120% and efficiency in the range of 20 to 40%. Two cavity klystrons are narrow bandwidth amplifiers, can produce gain about 20 dB and power efficiency in range of 30–50%. Klystrons have very low AM and FM noise. Magnetrons are very noisy, however they have AM efficiency of 80% or more.

7.4 NOISE IN MICROWAVE TUBES

Noise consists of unwanted signals having a random or incoherent nature, which limits the usefulness of microwave tubes and their associated circuits.

The major source of noise in a microwave tube is the cathode where random emission current (short noise) and random emission velocities (velocity fluctuation) results in initial noise current and noise velocity on the electron beams which are amplified and appear as output noise. Thermal noise and partition noise are other sources of noise in microwave tubes. There are number of ways to reduce the noise in microwave tubes.

O-Type devices are superior to M-Type devices with regard to noise behaviour.

7.5 LIMITATIONS AND LOSSES OF CONVENTIONAL TUBES AT MICROWAVE FREQUENCIES

With increasing frequency (>1GHz) the conventional tubes (triodes, tetrodes and pentodes) are less useful in amplification and generation because of two important limitations (1) Parasitic circuit elements such as interelectrode-capacitances and lead inductances and (2) Effects due to transit time of electrons between electrodes. These two effects have been analyzed in sections 7.5.1 and 7.5.2 respectively.

7.5.1 Other Limitations

In addition to these two effects there are **RF Losses**

(1) **Increase of Power Losses** resulting from

(*a*) **Skin effect losses:** As the frequency is increased, the current flows more and more on the surface of the conductor because of the skin-effect. This results in ohmic losses (I^2R) in lead wires as the effective resistance R increases. This can be minimised by the use of conductors having low resistivity and large surface areas: planer electrodes.

(*b*) **Electromagnetic radiation losses from the circuit:** Lead wires become small antennas with a resulting radiation power loss. Radiation losses can be minimised by the use of electrode lead spacing of the order of 1/100 of a wavelength or less.

(*c*) **Dielectric losses:** As variety of insulating materials are used inside the tube and

(*d*) **Charging current losses:** I^2R losses caused by capacitance charging currents. These losses could be reduced by reducing interelectrode capacitance and by increasing the number of shunt paths along with the charging currents flow.

(2) **Gain Bandwidth Product Limitation**

Another limitation is caused by gain bandwidth product of the circuit which is given by

$$A_m(BW) = \frac{g_m}{C}$$

where A_m is the maximum voltage gain at resonance, g_m = transconductance and C = the capacitance of the tank circuit. We note that gain bandwidth product is independent of frequency. For a given tube a higher gain can be achieved only at the expense of bandwidth or higher bandwidth can be achieved only at the expense of gain. This restriction is applicable to a resonant circuit only.

7.5.2 Effect of Parasitic Circuit Elements (Circuit-reactance limitation effect)

With increasing frequency of operation of an amplifier or an oscillator that uses conventional vacuum tubes it is found that the lead inductance* and interelectrode capacitances** of the tube become comparable to the external inductance and capacitance of resonator circuit and represent impedance that load the tube so as to affect the operation adversely. These effects are known as **circuit-reactance limitation effects**. To gain better understanding of these effects, we analyse the triode circuit shown in Fig. 7.8*a*. Its equivalent circuit is shown in Fig. 7.8(*b*) under the assumption that the interelectrode capacitances (C_{gk} and C_{gp}) and cathode lead inductance L_k are the only parasitic elements.

Input Impedance

Since $$C_{gp} << C_{gk}$$

and $$\omega L_k << \frac{1}{\omega C_{gk}},$$

where C_{gk} is the interelectrode capacitance between cathode and grid and C_{gp} is the interelectrode capacitance between grid and plate.

Input voltage V_{in} can be written as

$$V_{in} = V_g + V_k \qquad ...(7.1)$$

* A lead of 2.5 inc. in length and 100 mills in diameter is equivalent to an inductance = 0.5 μ H.

** Interelectrode capacitance C_{gk} is of the order of 0.5μμ*F*.

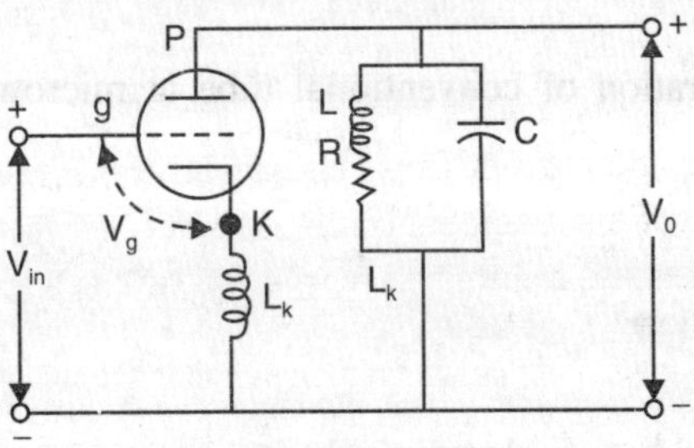

Fig. 7.8 (*a*) Triode circuit

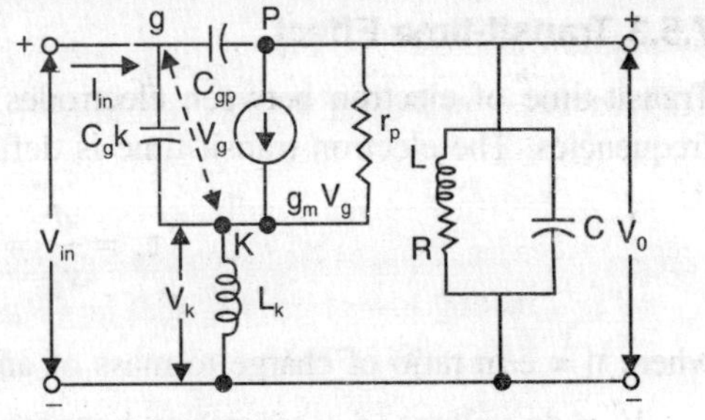

Fig. 7.8 (*b*) Equivalent circuit

where V_k in the cathode voltage given by

$$V_k = j\omega L_k g_m V_g \quad \text{...(7.2)}$$

The input current is given by

$$I_{in} = j\omega C_{gk} V_g \quad \text{...(7.3)}$$

From Eqns. (7.1) and (7.2),

$$V_{in} = I_{in} \frac{(1 + j\omega L_k g_m)}{j\omega C_{gk}} \quad \text{...(7.4)}$$

∴ The input admittance is given by

$$Y_{in} = \frac{I_{in}}{V_{in}} = \frac{j\omega C_{gk}}{(1 + j\omega L_k g_m)} \quad \text{...(7.5)}$$

In practical circuit $\omega L_k g_m \ll 1$ hence

$$Y_{in} \approx j\omega C_{gk}(1 - j\omega L_k g_m)$$

$$\approx j\omega C_{gk} + \omega^2 C_{gk} L_k g_m \quad \text{...(7.6)}$$

Equation 7.6 shows that the input conductance $\omega^2 L_k C_{gk} g_m$ increases with the square of the frequency. It also means that real part of the impedance is inversely proportional to the square of the frequency. As the frequency increases it loads the tube and when the frequencies are above 1 GHz the real part of the impedance becomes small enough to nearly short the signal. Consequently the output power decreases rapidly.

Load Impedance

Considering the resonant circuit of Fig. 7.8(*a*) in output circuit, it can be shown that real part of the load impedance is equal to $\frac{1}{R\omega_0^2 C^2}$ *i.e.*, the real part of the load impedance is inversely proportional to the square of the frequency. Consequently, the output power is decreased rapidly.

Thus it is clear that parasitic elements limit the operation of conventional tubes at high frequencies. These effects could be minimized and high frequency limit of operation can be extended by incorporating the grid-cathode and grid-plate capacitances into resonant and or by planer construction of electrodes and reduced lead and tube dimensions.

7.5.3 Transit-time Effect

Transit-time of electron between electrodes limits the operation of conventional tube at microwave frequencies. The electron transit-time is defined as

$$\tau_g = \frac{d}{v_0} = \frac{d}{\sqrt{2\eta V_0}}$$

where η = e/m ratio of charge to mass of an electron

V_0 = dc voltage, d = separation between cathode and grid, v_0 = electron velocity.

At lower frequencies transit-time (or angle) is a small fraction of the period of oscillations and is therefore insignificant. At microwave frequencies the transit-time τ_g is larger compared to the time period of the microwave signal. At lower frequencies an electron leaves the cathode and travels to the plate during the positive part of the grid voltage *i.e.,* voltage and current are in phase. At higher frequencies, the electron would leave the cathode when the grid is positive but before the electron passes the grid, the grid voltage might be negative or might even go through several cycles. Thus electron stream will be velocity modulated. This reduces the magnitude of mutual conductance and causes to have an imaginary component. The output voltage and current are no longer in phase because of a phase lag in the current produced by the finite transit time. This situation indicates that the output power is decreased.

From above discussion it is clear that the transit-time effect can be minimized by (*i*) decreasing the interelectrode spacing and (*ii*) by increasing the operating voltage. The latter results into increase in power dissipation at the anode while the decrease of interelectrode spacing increases the interelectrode capacitance.

7.5.4 Lighthouse and Disk-seal Tubes

The high frequency limitations due to parasitic circuit elements and transit-time were minimized by radical changes in spatial configuration, as in the light house tube in which the tube elements were made parts of the external circuit *i.e.,* incorporating interelectrode capacitances into resonant cavities and reducing interelectrode spacing. Figure 7.9 shows the lighthouse tube.

The problem of lead inductance and interelectrode capacitance has been solved in disk-seal tubes (Fig. 7.10) by making the electrodes in the form of metal disks which are sealed to the glass envelope. Several devices were constructed that could overcome the frequency limitations to some extent.

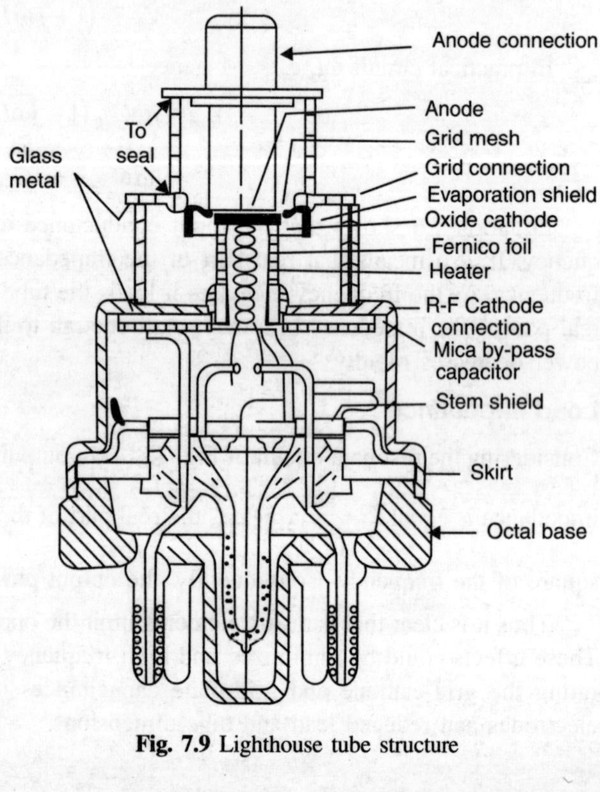

Fig. 7.9 Lighthouse tube structure

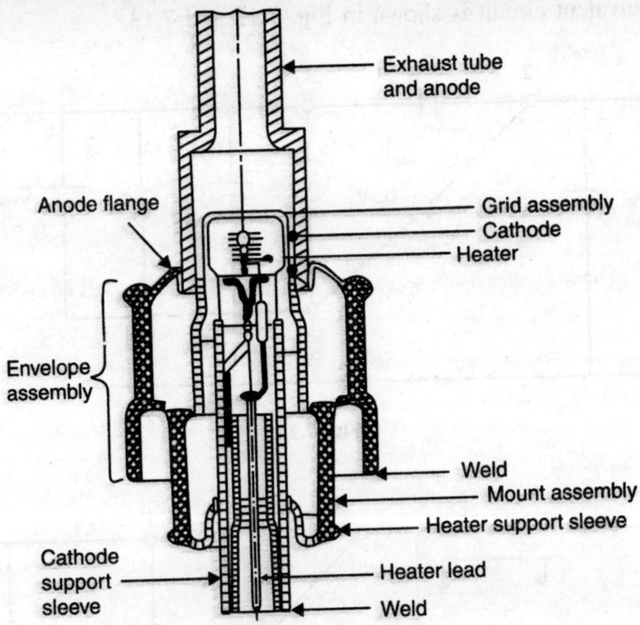

Fig. 7.10 Disk-seal power triode

Example 7.1. A vacuum pentode tube has five grids: a cathode, a control grid, a screen grid, a suppressor grid, and an anode plate as shown in Fig. 7.11.

(*a*) Draw the equivalent circuit.

(*b*) Drive an expression for the input impedance Z_{in} in term of the angular frequency ω and the circuit parameters.

(*c*) Determine the transit-angle effect.

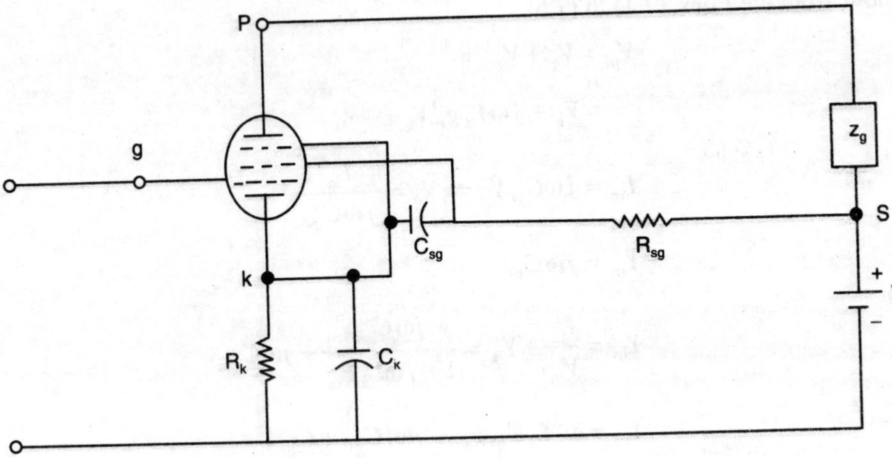

Fig. 7.11

Solution. (*a*) The equivalent circuit is shown in Fig. 7.12 and 7.13.

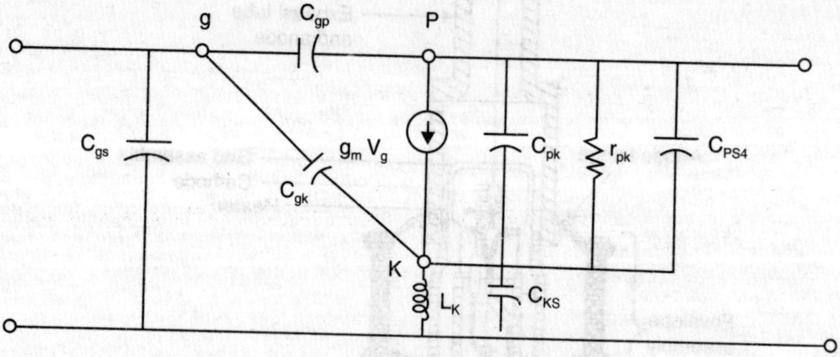

Fig. 7.12

or

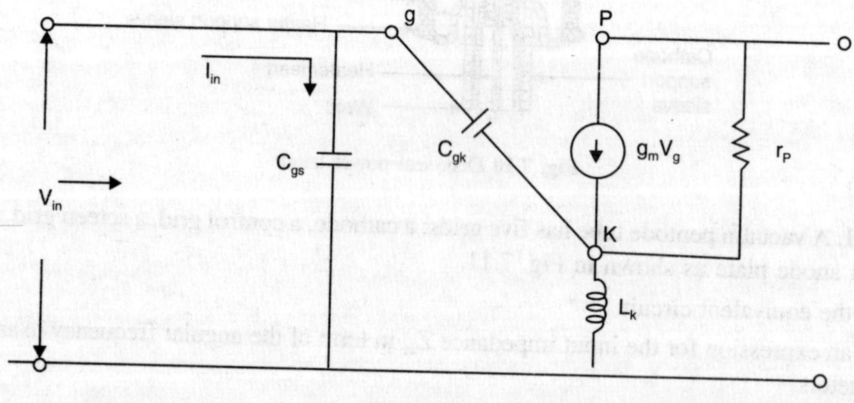

Fig. 7.13

(*b*) We know, from the Eqns. (7.1) to (7.6)

$$V_{in} = V_g + V_k$$

$$= V_g + j\omega L_k g_m V_g$$

or $$I_{in} = j\omega C_{gk} V_g \Rightarrow V_g = \frac{I_{in}}{j\omega C_{gk}}$$

and $$Y'_{in} = j\omega C_{gs}$$

or $$Y_{in} = \frac{I_{in}}{V_{in}} + Y'_{in} = \frac{j\omega C_{gk}}{1 + j\omega L_k g_m} + j\omega C_{gs}$$

$$Y_{in} = \omega^2 L_k C_{gk} g_m + j\omega\left(C_{gk} + C_{gs}\right)$$

$\therefore$ $$Z_{in} = \frac{1}{\omega^2 L_k C_{gk} g_m} - j\frac{\left(C_{gk} + C_{gs}\right)}{\omega^3 L_k^2 C_{gk}^2 g_m^2}$$

(*c*) From the sec. (7.5.3),
the electron transit-time,

$$\tau_g = \frac{d}{v_0}$$

So the transit-angle effect is,

$$\theta_g = \omega\tau_g = \frac{\omega d}{v_0}$$

Where d-Separation between cathode and grid

$v_0 = 0.593 \times 10^6 \sqrt{V_0}$ is the velocity of electron

V_0 = d.c. bias voltage.

7.6 TWO CAVITY KLYSTRON

Perhaps the simplest tube from a pedagogical point of view and which has also enjoyed the greatest technological success and widely used is the two-cavity klystron. In this type the various electronic processes as mentioned in section 7.1 take place in separate portions of the structure. Therefore the two cavity klystron will be described and analyzed first (In other words we will develop the principles underlying the operation of the two-cavity klystron first). Later on we shall describe in brief multicavity klystrons and reflex klystrons in detail.

7.6.1 Structure/Building Blocks

A schematic diagram of a two-cavity klystron amplifier is shown in Fig. 7.1. It consists of (*i*) cathode-anode configuration (*i.e.*, electron gun) to produce an accelerated electron beam; (*ii*) buncher (input) cavity, (*iii*) catcher (output) cavity, (*iv*) field-free drift space between two input and output cavities (These are reentrant cavities and discussed in Chapter 4), (*v*) pair of buncher grid—the first gap,

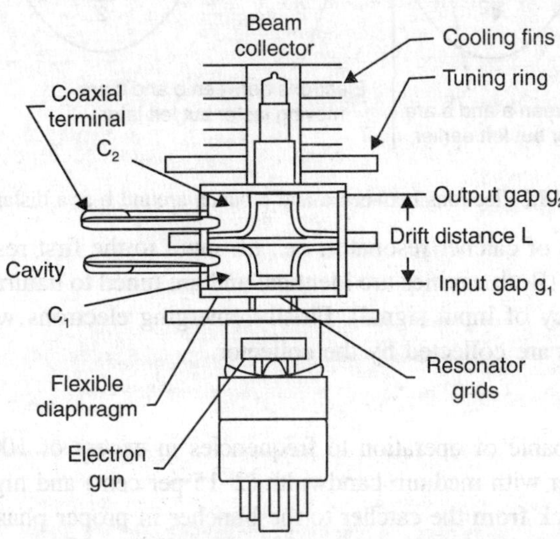

Fig. 7.14 Constructional details of two-cavity klystron

(*vi*) pair of catcher grid-the second gap, and (*vii*) collector. There is input coupling loop in buncher (input) cavity and output coupling loop in catcher (output) cavity. Whole assembly is in vacuum envelope with vacuum sealed coaxial fittings for external connections. In addition, there may be an applied dc magnectic field that is parallel to the beam to reduce spreading of the beam as it propagates from the source to the collector. The constructional details of two-cavity klystron are shown in Fig. 7.14.

7.6.2 Mechanism of Operation

The accelerated electron beam passes through a gap in the anode into a region between a pair of closely spaced buncher grids (first gap) across which an r.f. voltage exists, then into a drift space, and then through another pair of catcher grids (second gap) to the collector. The buncher (input) cavity is kept generally at the same d.c. potential as the anode. The electrons emerging from buncher grids (gap one) are velocity modulated due to presence of an r.f. voltage and conversion of velocity modulated electrons into density modulation (bunching) while drifting in drift region is effected and is called **drift-space bunching** (Fig. 7.15). It is clear from the Fig. 7.15 that those electrons that leave the first gap with higher velocities catch up with those left earlier with slower velocities, thus forming bunch. This phenomenon, commonly known as bunching, is the result of the velocity modulation produced at the first gap (buncher grids). When these bunches enter the second gap (catcher grids) during the retarding phase of r.f. field of the catcher resonator, then most of the electrons lose their kinetic energy in crossing

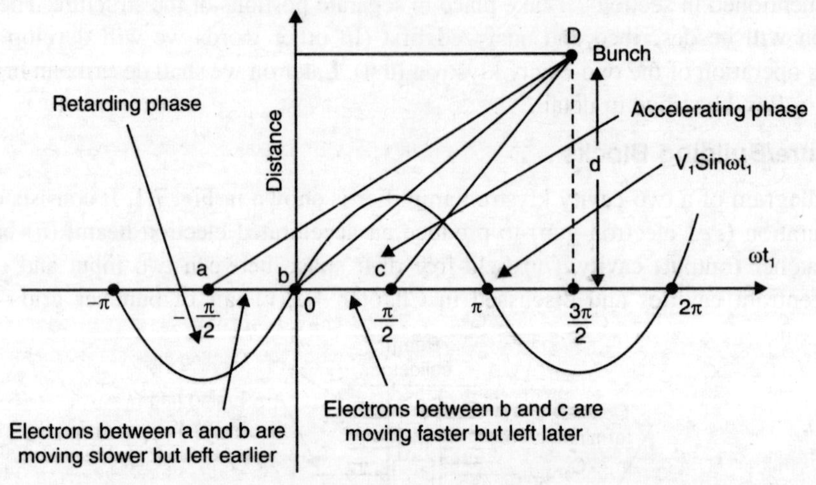

Fig. 7.15 Electrons between a and c bunch around b at a distance d

the second gap to r.f. field of catcher resonator, *i.e.*, r.f. input to the first resonator (buncher cavity) is amplified at catcher cavity (Both cavities are identical and are tuned to natural frequency of the cavities which is also the frequency of input signal). Finally emerging electrons with reduced velocity from second gap (catcher grids) are collected by the collector.

Performance

Two-cavity klystron is capable of operation to frequencies in excess of 100 GHz. As an amplifier it offers pulse and CW power with medium bandwidth (2–15 per cent) and high gain (15–70 dB). When sufficient energy is fed back from the catcher to the buncher in proper phase and if the loop gain has a magnitude of unity with a phase shift of multiple of 2π, it is double resonator klystron oscillator.

7.6.3 Small Signal Theory

7.6.3 (a) Velocity Modulation Process and Applegate Diagram

Electrons in the electron beam produced in a cathode anode region and accelerated by a d.c. anode voltage V_0, passing through a pair of buncher grids (gap one), across which an r.f. voltage $V_1 \sin \omega t$ exist (ω is the frequency of the input signal and it is also the resonant frequency of the first cavity) are either accelerated or decelerated depending on the part of the cycle during which they cross the gap (Fig. 7.15). The electrons, which are accelerated emerge with higher velocities than the entering velocity v_0 and those, which are decelerated emerge with velocities lower than the entering velocity v_0. Some electrons pass through at zero field and hence suffer no change in velocity (Fig. 7.15). This variation of velocities of the electrons in the electron beam is known as **velocity modulation.**

The analysis is carried out with the following assumptions:

1. Electrons leave the cathode with zero velocity and the electron beam have uniform density in the cross-section of the beam.
2. Space charge effects are ignored.
3. The magnitude (V_1) of input microwave signal is much smaller than the dc accelerating voltage (V_0), *i.e.*, $V_1 \ll V_0$.
4. r.f. field is perfectly longitudinal.

(i) Transit-time across the first gap is negligible

It is assumed that the gap spacing and frequency is such that the transit-time through gap (buncher grids) is a small fraction of the cycle and is negligible.

The velocity of the electrons before entering the buncher grid is given by

$$v_0 = \sqrt{2\eta V_0} = 5.93 \times 10^5 \sqrt{V_0} \text{ m/s} \qquad ...(7.7)$$

where $\eta = e/m$ is the magnitude of the electron charge to mass ratio 1.76×10^{11} C/kg and V_0 is the anode-cathode potential difference in volts.

(*i*) Let	$V_0 = 1600 \text{ V}$
	$v_0 = 5.93 \times 10^5 \times 40 = 2.37 \times 10^7 \text{ m/s.}$
(*ii*) If	$V_0 = 900 \text{ V}$
	$v_0 = 5.93 \times 10^5 \times 30$
	$= 1.779 \times 10^7 \text{ m/s}$

The energy of an electron emerging from the gap at time t_1 is

$$eV = e(V_0 + V_1 \sin \omega t_1)$$

$$= eV_0 \left(1 + \frac{V_1}{V_0} \sin \omega t_1\right) \qquad ...(7.8)$$

where $V = V_0 + V_1 \sin \omega t_1$ is the instantaneous voltage at time t_1 (buncher cavity is at V_0 dc voltage).

This energy is equal to kinetic energy $\frac{1}{2}mv^2$, where v is the velocity of electron at time t_1.

$$\therefore \quad \frac{1}{2}mv^2 = eV_0\left(1+\frac{V_1}{V_0}\sin\omega t_1\right) \qquad v = \sqrt{2\eta V_0\left(1+\frac{V_1}{V_0}\sin\omega t_1\right)}$$

$$v = v_0\sqrt{\left(1+\frac{V_1}{V_0}\sin\omega t_1\right)} = v_0\sqrt{1+\alpha\sin\omega t_1} \qquad \text{...(7.9)}$$

where $v_0 = \sqrt{2\eta V_0}$ is the injected velocity and $\alpha = \dfrac{V_1}{V_0}$ and is known as **depth of modulation.**

As $V_1 \ll V_0$ *i.e.,* $\alpha \ll 1$, then Eqn. 7.9 may be written as

$$v = v_0\left(1+\frac{1}{2}\alpha\sin\omega t_1\right) \qquad \text{...(7.10}a\text{)}$$

It is noted from Eqn. 7.10 that successive electrons are not travelling with the same velocity when they leave the first gap. The electrons, which leave the gap at a time corresponding to

$$\omega t_1 = 2\pi n, \qquad n = 0, 1, 2, 3$$

will suffer no change in velocity from the injected velocity v_0.

The electrons, which leave at a time in corresponding to

$$\omega t_1 = +\pi/2$$

the maximum accelerating phase of buncher voltage, will be moving with the maximum velocity

$$\mathbf{v}_{\max} = \mathbf{v}_0\left(\mathbf{1}+\alpha/\mathbf{2}\right)$$

The electrons which leave at a time corresponding to

$$\omega t_1 = -\pi/2$$

the minimum accelerating phase of buncher voltage, will be moving with the minimum velocity

$$\mathbf{v}_{\min} = \mathbf{v}_0\left(\mathbf{1}-\alpha/\mathbf{2}\right)$$

Thus the electrons which leave during the time

$$\pi/2\omega < t_1 < 0$$

will slow down (fall back) and those which leave during

$$0 < t_1 < \pi/2\omega$$

will move fast (catch up Fig. 7.15).

The Eqn. 7.10 is the **equation of velocity modulation** at cavity frequency ω.

(ii) Finite transit-time across the first gap

When the transit-time or transit angle across the gap is NOT zero, the velocity modulation Eqn. 7.10(a) is modified to

$$v = v_0\left[1+\frac{1}{2}\alpha\beta\sin\omega t_1\right] \qquad \text{...(7.10}b\text{)}$$

In above equation

$$\beta = \frac{\sin \omega\tau_g/2}{\omega\tau_g/2} = \frac{\sin \theta_g/2}{\theta_g/2} \qquad ...(7.11)$$

is called the **Beam Coupling Coefficient**; where $\tau_g = d/v_0$ is the electron transit time and $\theta_g = \omega\tau_g$ is the transit angle through the gap d. The Eqn. 7.11 is obtained by calculating the work done by an electron while travelling through the gap

$$d = 1 \text{ mm}, \; v_0 = 1.88 \times 10^7 \text{ m/s}, \; f = 3 \text{ GHz}$$

$$\tau_g = \frac{1\times10^{-3}}{1.88\times10^7} = \frac{1}{1.88} \times 10^{-10} \text{ s(transit-time)}$$

$$\omega\tau_g = \theta_g = \text{transit angle} = \frac{2\pi(3\times10^9)\times10^{-3}}{1.88\times10^7} = 1 \text{ rad.}$$

$$\beta = \frac{\sin\theta_g/2}{\theta_g/2} = \frac{\sin\frac{1}{2}}{\frac{1}{2}} = 0.952 \approx 1$$

We see that when magnitude of transit-time and transit angle is negligibly small, β is approaching 1, it implies a perfect beam coupling.

It is clear that finite transit-time or transit angle is equivalent to a reduction in the magnitude of depth of modulation. In other terms it is reduction in the modulating voltage V_1.

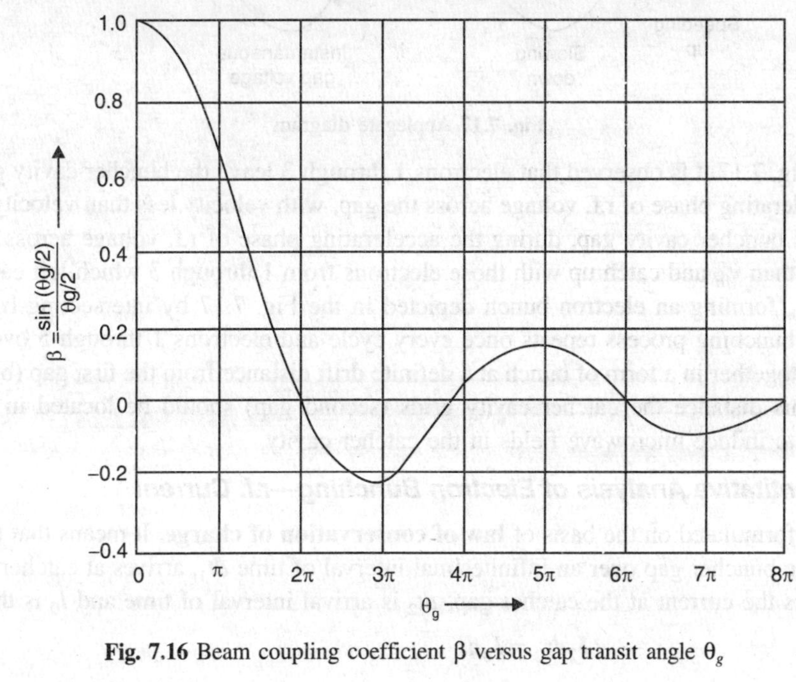

Fig. 7.16 Beam coupling coefficient β versus gap transit angle θ_g

The value of the beam coupling coefficient β depends on gap transit angle. Fig. 7.16 shows the plot of the beam-coupling coefficient β as a function of θ_g, the gap transit angle. It is observed that increasing the gap transit angle θ_g, decreases the coupling between the electron beam and the bunching cavity *i.e.*, the velocity modulation of the beam for a given microwave signal is decreased [in the region of practical interest where $\theta_g \to 0$].

The velocity modulated electron beam while drifting in field free drift space between two cavities is converted into density modulation. This is discussed in the following section:

7.6.3 (b) Bunching Process

The electrons passing through the first gap at different instants t_1 travel at different velocities (Eqn. 7.12) in the drift region. Those electrons that leave the first gap with higher velocities ($v > v_0$) catch up with those that left earlier with slower velocities ($v < v_0$). This is depicted in Fig. 7.17 in which the location of electron is plotted as function of time. This distance-time diagram is called as **Applegate diagram.**

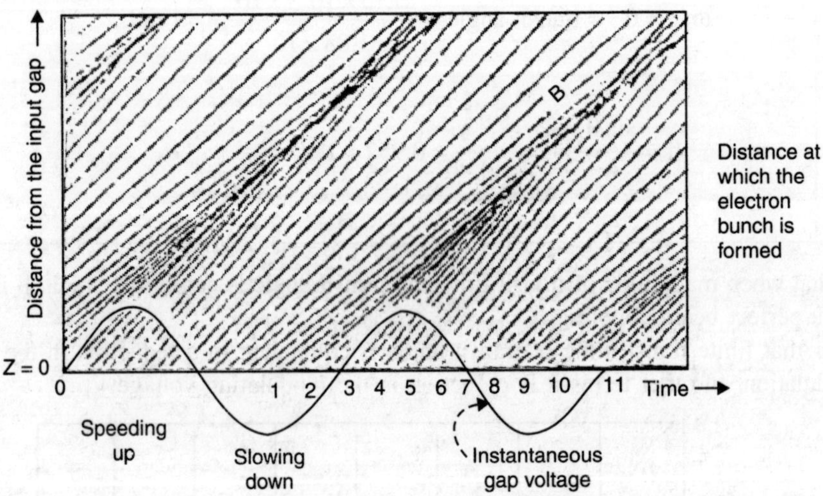

Fig. 7.17 Applegate diagram

From the Fig. 7.17 it is observed that electrons 1 through 3 leave the buncher cavity gap (first gap) during the decelerating phase of r.f. voltage across the gap, with velocity less than velocity v_0, electrons 3 to 5 leave the buncher cavity gap, during the accelerating phase of r.f. voltage across the gap, with higher velocity than v_0, and catch up with those electrons from 1 through 3 which left earlier with less velocity than v_0, forming an electron bunch depicted in the Fig. 7.17 by intersecting bundle of lines marked *B*. The bunching process repeats once every cycle and electrons 1 through 5 over one-half of the cycle come together in a form of bunch at a definite drift distance from the first gap (buncher grids). Obviously at this distance the catcher cavity grids (second gap) should be located in other for the electron bunch to induce microwave fields in the catcher cavity.

7.6.3 (c) Quantitative Analysis of Electron Bunching—r.f. Current

The analysis is formulated on the basis of **law of conservation of charge.** It means that the net charge $I_0 dt_1$, leaving the buncher gap over an infinitesimal interval of time dt_1, arrives at catcher cavity gap as $I_2 dt_2$, where I_2 is the current at the catcher gap, dt_2 is arrival interval of time and I_0 is the dc current.

$$\therefore \qquad I_2 dt_2 = I_0 dt_1$$

or
$$I_2 = I_0 \frac{dt_1}{dt_2} \qquad ...(7.12)$$

It is clear that t_2 is the arrival time for an electron for which departure time is t_1.

$$t_2 = t_1 + \frac{L}{v} \qquad ...(7.13)$$

where L is the length of the drift region. Substituting the value of v from Eqn. 7.10(*b*) in 7.31, we get

$$t_2 = t_1 + \frac{L}{v_0\left[1+\frac{\alpha\beta}{2}\sin\omega t_1\right]}$$

$$\approx t_1 + \frac{L}{v_0}\left[1-\frac{\alpha\beta}{2}\sin\omega t_1\right]$$

$$\approx t_1 + T_0\left[1-\frac{\alpha\beta}{2}\sin\omega t_1\right] \qquad ...(7.14)$$

as $\alpha \ll 1$, *i.e.*, $V_1 \ll V_0$ and $T_0 = \frac{L}{v_0}$ is the dc transit-time.

From Eqn. 7.14, we may write

$$dt_2 = dt_1 - \frac{\alpha\beta}{2}\omega T_0 \cos\omega t_1 dt_1$$

$$= dt_1\left[1-\frac{\alpha\beta}{2}\theta_0\cos\omega t_1\right]$$

or
$$\frac{dt_1}{dt_2} = \frac{1}{\left[1-\frac{\alpha\theta_0\beta}{2}\cos\omega t_1\right]} \qquad ...(7.15)$$

where $\theta_0 = \omega T_0$ the d.c. transit angle between the cavities.

For $L = 4$ cm, $v_0 = 1.88 \times 10^7$ m/s, $f = 3$ GHz
$\theta_0 = \omega T_0 = \frac{\omega L}{v_0} = \frac{2\pi(3\times10^9)\times4\times10^{-2}}{1.88\times10^7} = 40$ rad.

Substituting the value of $\frac{dt_1}{dt_2}$ from Eqn. 7.15 in Eqn. 7.12, we get

$$I_2 = \frac{I_0}{\left[1-\frac{\alpha}{2}\theta_0\beta\cos\omega t_1\right]}$$

$$= \frac{I_0}{[1 - X\cos\omega t_1]} \qquad ...(7.16)$$

$$\approx I_0[1 + X\cos\omega t_1]$$

where $X = \frac{\alpha\theta_0\beta}{2}$ is a dimensionless quantity called the **Bunching Parameter.** It is seen from Eqn. 7.16 that bunching parameter (X) depends on α, the drift angle $\theta_0 = \frac{\omega L}{v_0}$ and beam coupling coefficient β. The plot of output current (Eqn. 7.16) versus time for various values of X is shown in Fig. 7.13.

It gives one pulse per cycle centred about the electron that crosses the input gap at $\omega t_1 = 2n\pi$, n = 0, 1, 2, 3,.... It is seen from the Fig. 7.18 that for values $X > 1.0$ the catcher current is rich in the harmonics of the input frequency which is the resonance frequency of buncher cavity. This property is exploited in operating a klystron as an **harmonic generator.**

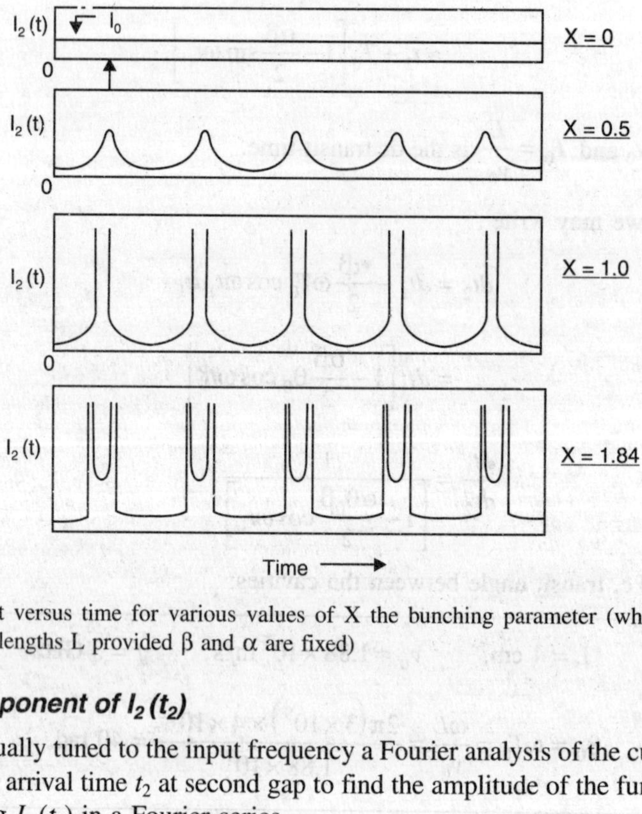

Fig. 7.18 Output current versus time for various values of X the bunching parameter (which would correspond to different drift lengths L provided β and α are fixed)

Fundamental Component of $I_2(t_2)$

Since the output is usually tuned to the input frequency a Fourier analysis of the current I_2 at the catcher gap as function of the arrival time t_2 at second gap to find the amplitude of the fundamental component is in order. Expanding $I_2(t_2)$ in a Fourier series

we have
$$I_2 = \sum_{n=0}^{\infty}\left[a_n\cos(n - \omega t_2) + b_n\sin(n\omega t_2)\right] \qquad ...(7.17)$$

where the coefficients

$$a_0 = \frac{1}{2\pi}\int_{-\pi}^{\pi} I_2 d(\omega t_2) = \frac{1}{2\pi}\int_{-\pi}^{\pi} I_0 d(\omega t_0) = I_0$$

$$a_n = \frac{1}{\pi}\int_{-\pi}^{\pi} I_2 \cos n\omega t_2 d(\omega t_2)$$

$$= \frac{1}{\pi}\int_{-\pi}^{\pi} I_0 \cos n(\omega t_1 + \theta_0 - X\sin\omega t_1) d(\omega t_1)$$

$$= 2I_0 J_n(nX)\cos n\theta_0$$

and

$$b_n = \frac{1}{\pi}\int_{-\pi}^{\pi} I_2 \sin n\omega t_2 d(\omega t_2)$$

$$= \frac{1}{\pi}\int_{-\pi}^{\pi} I_0 \sin n(\omega t_1 + \theta_0 - X\sin\omega t_1) d\omega t_1$$

$$= 2I_0 J_n(nX)\sin n\theta_0$$

where $J_n(n\text{X})$ is the nth order Bessel function of the first kind (Fig. 7.19).

The value of a_0, a_n and b_n were obtained by first converting these integrals into functions to t_1 instead of t_2 by using the Eqns. 7.12 and 7.14 and then evaluating the integrals using integral tables* in terms of the Bessel functions $J_n(nX)$.

The output current I_2 can therefore be written as

$$I_2 = I_0 + \sum_{n=1}^{\infty} 2I_0 J_n(nX)\cos n(\omega t_2 - \theta_0) \quad \text{...(7.18)}$$

From Eqn. 7.14 $\omega t_2 = \omega t_1 + \omega T_0\left[1 - \frac{\alpha\beta}{2}\sin\omega t_1\right]$

$$= \omega t_1 + \theta_0 - \frac{\alpha\beta\theta_0}{2}\sin\omega t_1$$

$$\therefore \quad \omega t_1 \approx \omega t_2 - \theta_0, \quad \left(\text{as } \frac{\alpha\beta\theta_0}{2}\sin\omega t_1 \text{ is negligible}\right)$$

The fundamental component of the beam current at the frequency of the input gap (n = 1) is

$$I_f = 2I_0 J_1(X)\cos(\omega t_2 - \theta_0) \quad \text{...(7.19)}$$

$$= 2I_0 J_1(X)\sin(\omega t_2 + \pi/2 - \theta_0)$$

* Dwight, H.B., **Tables of Integrals and Other Mathematical Data,** MacMillan Company, New York, 1947.

The magnitude of $|I_f|$ is $2I_0J_1(X)$. This fundamental component I_f has its maximum amplitude at $X = 1.841$ (Fig. 7.19).

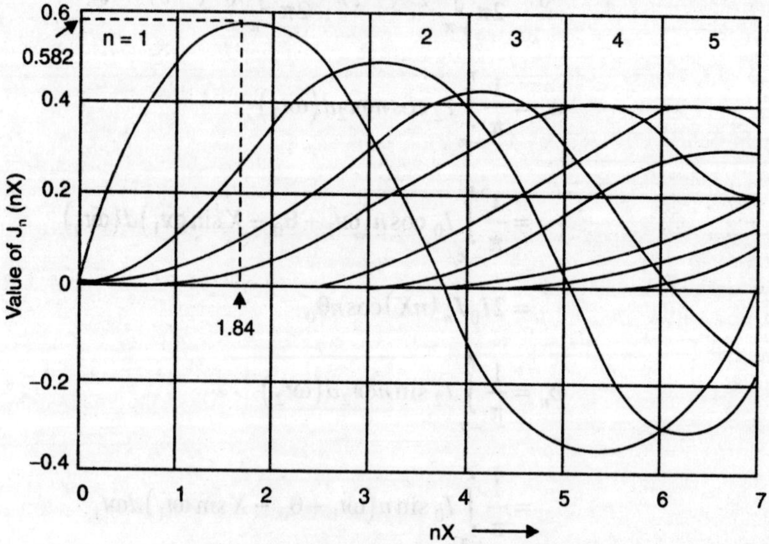

Fig. 7.19 Bessel function $J_n(nX)$ versus nX

7.6.3 (d) Optimal Length of Drift Region—Maximum Fundamental Component of Current

The optimum distance L at which the maximum fundamental component occurs is computed.

$$X = 1.841 = \frac{\theta_0 \alpha \beta}{2}$$

$$= \frac{\omega L_{opt}}{v_0} \frac{V_1 \beta}{2V_0} \qquad \text{...(7.20)}$$

$$\therefore \qquad L_{opt} = 3.682 \times \frac{v_0 V_0}{\omega \beta V_1} \qquad \text{...(7.20}a\text{)}$$

$$L_{opt} \propto \frac{1}{V_1}$$

The value of L_{opt} thus specifies the point at which the output cavity (catcher cavity) must be located. It is noted from Eqn. 7.20 that (*i*) since $L_{opt} \propto \frac{1}{V_1}$ any variation in the input signal level would require a change in the drift length to maintain maximum gain; (*ii*) it would also appear that a very long drift space enables enormous gain to be obtained. However, debunching force combined with finite mass of electrons results in the amount of debunching increasing with the time the electron spent in the drift region.

$L_{opt} = 4$ cm, $\quad V_0 = 1000$ V, $\quad \therefore \; v_0 = 1.88 \times 10^7$ m/s,

$\beta = 0.952, \quad f = 3$ GHz

From Eqn. 7.20,

$$\therefore \quad V_{1\max} = \frac{3.682 \times v_0 V_0}{\omega\beta L_{opt}} = \frac{3.682 \times 1.88 \times 10^7 \times 10^3}{2\pi(3\times10^9)\times 0.952 \times 4 \times 10^{-2}} \approx 0.952 \text{ V}$$

It is input gap voltage to give maximum voltage V_2.

7.6.3 (e) Transfer of Energy—Expressions for Output Power, Efficiency

When the bunched electron beam passes through the retarding phase of catcher gap voltage, its kinetic energy is transfered to the field of the catcher cavity. These electrons emerging with reduced velocity are finally collected by the collector. For a two-cavity klystron amplifier both cavities (buncher and catcher) are tuned to the signal frequency and the beam current at the catcher cavity is given by $2I_0 J_1(X) \cos(\omega t_2 - \theta_0)$. Due to finite transit-time of the beam through the catcher cavity gap, the effective beam current is reduced by the beam coupling coefficient β (assumed identical buncher and catcher cavities *i.e.*, β is same for both cavities) to $2\beta I_0 J_1(X) \cos(\omega t_2 - \theta_0)$. The magnitude of the fundamental component of the beam current is (it is induced current in the catcher cavity)

$$|I_f| = \beta 2 I_0 J_1(X) \qquad \text{...(7.21)}$$

and V_2 is fundamental component of catcher gap voltage. The output power delivered to the catcher and the load is given by

$$P_{out} = \beta 2 I_0 \frac{J_1(X)}{\sqrt{2}} \times \frac{V_2}{\sqrt{2}}$$

$$= V_2 \beta I_0 J_1(X) \qquad \text{...(7.22)}$$

$$\therefore \text{ Efficiency of klystron} \quad \eta = \frac{\text{Output power}}{\text{Input power}} = \frac{P_{out}}{P_{in}}$$

$$= \frac{V_2 \beta I_0 J_1(X)}{2 I_0 V_0} \qquad \text{...(7.23)}$$

$$= \frac{V_2}{V_0} \beta J_1(X) \qquad \text{...(7.24)}$$

If the coupling is perfect, $\beta = 1$, then

$$\eta = \frac{V_2}{V_0} J_1(X) \qquad \text{...(7.25)}$$

We know from Fig. 7.19 that when $X = 1.841$, $J_1(X) = 0.582$, the maximum value.

$$\eta = \frac{V_2}{V_0} 0.582 \qquad \text{...(7.26)}$$

when $V_2 \to V_0$, *i.e.*, $\frac{V_2}{V_0} \to 1$ then the maximum electronic efficiency is about 58%. Figure 7.20 shows the variation of electronic efficiency of a klystron as function of catcher cavity transit angle θ_g since the efficiency is a function of β (β depends upon catcher gap transit angle).

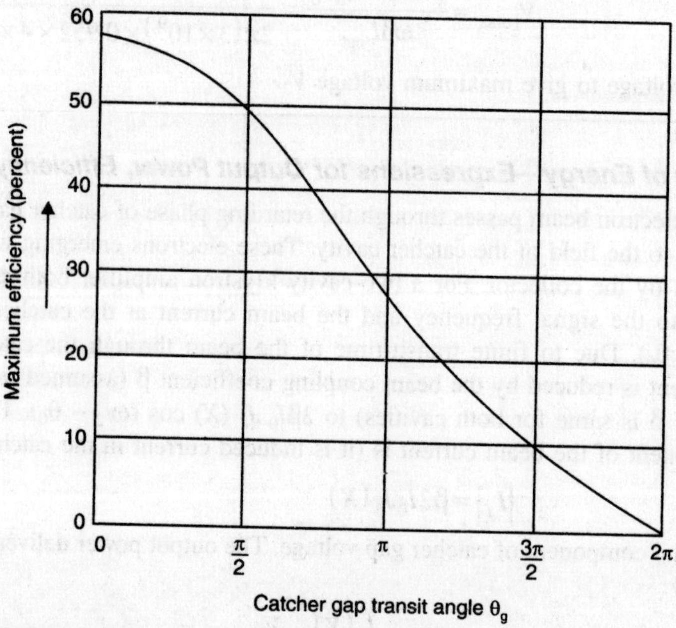

Fig. 7.20 Maximum efficiency of klystron versus catcher gap transit angle θ_g

In most of the klystron, the **electronic efficiency ranges between 15% to 30%.**

7.6.3 (f) Mutual Conductance of a Klystron Amplifier

Mutual conductance (g_m) is defined as a ratio of induced (output) current to input voltage (r.f.)

$$|g_m| = \frac{|I_f|}{V_1} = \frac{2\beta I_0 J_1(X)}{V_1} \quad ...(7.27)$$

V_1 in terms of bunching parameter X is

$$V_1 = \frac{2V_0}{\beta\theta_0}X, \quad \left[X = \frac{\alpha\theta_0\beta}{2} = \frac{V_1\theta_0\beta}{2V_0}\right]$$

Substituting the value of V_1 in Eqn. 7.27, we get

$$|g_m| = \frac{2\beta I_0 J_1(X)}{2V_0 X}\beta\theta_0$$

$$= \beta^2\theta_0 \frac{I_0}{V_0}\frac{J_1(X)}{X}$$

The normalized mutual conductance

$$\frac{|g_m|}{g_0} = \beta^2\theta_0 \frac{J_1(X)}{X} \qquad ...(7.28)$$

where $g_0 = \dfrac{I_0}{V_0}$ is the d.c. beam conductance.

For small signal parameter

$$J_1(X) \approx X/2$$

$$\therefore \qquad g_m = \beta^2\theta_0 \frac{I_0}{2V_0}$$

or normalized mutual conductance

$$\frac{|g_m|}{g_0} = \frac{\beta^2\theta_0}{2}$$

7.6.3 (g) Voltage Gain

The voltage gain of a klystron amplifier is defined as

$$A_v = \left|\frac{V_2}{V_1}\right| \qquad ...(7.29)$$

A cavity is equivalent to LC circuit with losses because of (*i*) the resistance in the walls in the cavity; and (*ii*) the useful load coupled to cavity. Figure 7.21 shows an output equivalent circuit.

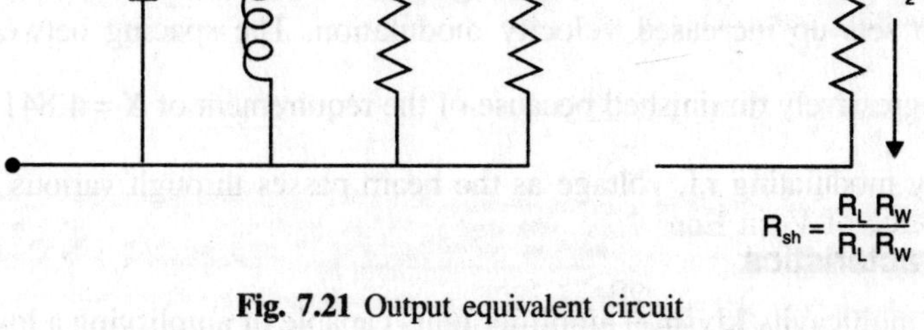

Fig. 7.21 Output equivalent circuit

The fundamental component of current induced in the catcher cavity has a magnitude = $\beta 2 I_0 J_1(x)$

$$\therefore \qquad |V_2| = \beta 2 I_0 J_1(X) R_{sh}$$

$$\therefore \qquad A_v = \left|\frac{V_2}{V_1}\right| = \frac{\beta^2\theta_0 2 I_0 J_1(X)}{2V_0 X} R_{sh}, \quad \left(\text{since } |V_1| = \frac{2V_0}{\beta\theta_0} X\right)$$

$$= \beta^2 \theta_0 \frac{J_1(X)}{R_0 X} R_{sh} \qquad ...(7.30)$$

where $R_0 = V_0/I_0$ is the d.c. beam resistance and R_{sh} is the effective shunt resistance (Fig. 7.21).

$$A_v = g_m R_{sh}, \quad \left(\text{Since } g_m = \beta^2 \theta_0 \frac{I_0}{V_0} \frac{J_1(X)}{X} \right) \qquad ...(7.31)$$

When $\beta = 1$, $\theta_0 = 40$ rad, $R_{sh} = 30\ \text{k}\Omega$, $\frac{J_1(X)}{(X)} = \frac{0.582}{1.841}$

and $R_0 = 40\ \text{k}\Omega$

calculate voltage gain. From Eqn. 7.30, we get

$$A_v = \beta^2 \theta_0 \frac{J_1(X)}{R_0 X} \cdot R_{sh}; \text{ Substituting the values}$$

$$= \frac{40 \times 0.582 \times 30 \times 10^3}{40 \times 10^3 \times 1.841} \approx 9.5.$$

The preceding analysis merely uses the principle of conservation of charge and energy and equates the energy lost by the electrons to the energy gained by the electromagnetic field and gives surprisingly good quantitative accuracy. However, it lacks in describing the detailed mechanism of the electron-field interaction within the cavity actually taking place.

7.7 MULTICAVITY KLYSTRON AMPLIFIER

For achieving high overall gain, several two-cavity tubes are connected in cascade, feeding the output of each of the tubes to input of the following one. Instead of two cavities, one can use chain of cavities in one-vacuum envelope tube as shown in Fig. 7.22. The first cavity is again used to provide the initial velocity modulation of the beam. The last cavity in a chain is used as the output cavity. Each of the intermediate cavities, placed at a distance of the bunching parameter X of 1.841 away from the previous cavity, acts as buncher with electron beam inducing a more enchanced r.f. voltage than the previous cavity, which in turn sets up increased velocity modulation. The spacing between the consecutive cavities would be progressively diminished because of the requirement of $X = 1.841$ $\left(L_{\text{opt}} \propto \frac{1}{V_{\text{r.f.}}} \right)$ and an increasing velocity modulating r.f. voltage as the beam passes through various cavities.

Performance Characteristics

There are varieties of multicavity klystron amplifier units capable of amplifying a low level signal to the desired values. Power gain of 50 dB to 60 dB can be achieved with a chain of cavities in cascade. Multicavity klystrons are capable of **greater bandwidth with less gain,** when operated with the cavities stagger tuned, as compared to two-cavity klystron's bandwidth, which is limited on account of high Q's of the resonators. There are number of commercially available multicavity klystron amplifier units capable of delivering **tens of milliwatts up to several hundred kilowatts of amplified output power.**

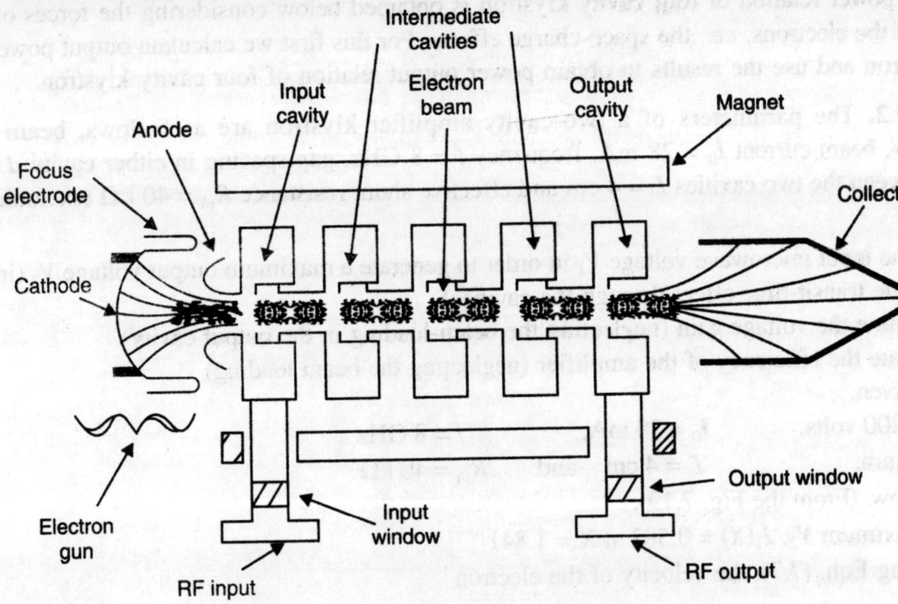

Fig. 7.22 Schematic diagram of a multicavity klystron

External cavity tube

The r.f. circuit (cavities) of most klystrons is entirely within the vacuum envelope and are called internal cavity tubes. There are tubes in which the r.f. cavities are distinct from the vacuum envelope and are known as external cavity tubes. Figure 7.23 shows a typical external cavity.

In multicavity klystron mostly magnetic beam focussing (permanent magnet or solenoid) is used to keep the electron beam uniformly small.

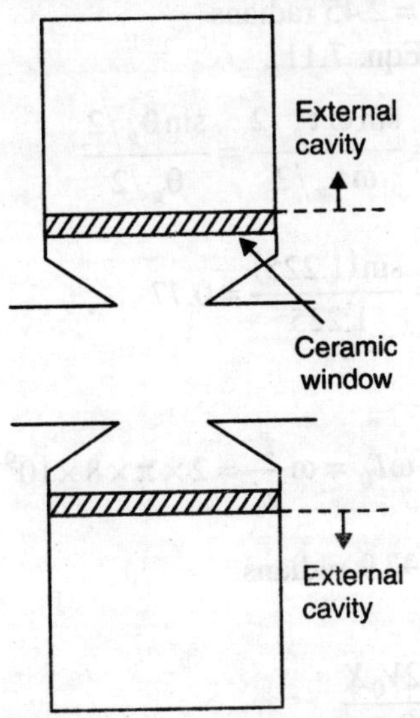

Fig. 7.23 Typical external cavity

Output power relation of four cavity klystron is obtained below considering the forces of mutual repulsion of the electrons, *i.e.*, the space-charge effects. For this first we calculate output power of two cavity klystron and use the results to obtain power output relation of four cavity klystron.

Example 7.2. The parameters of a two-cavity amplifier klystron are as follows, beam voltage $V_0 = 1200$ V, beam current $I_0 = 28$ mA, frequency $f = 8$ GHz, gap spacing in either cavity $d = 1$mm, spacing between the two cavities $L = 4$ cm and effective shunt resistance $R_{sh} = 40$ kΩ (excluding beam loading) so,

(*a*) Find the input microwave voltage V_1 in order to generate a maximum output voltage V_2 (including the finite transit-time effect through the cavities)
(*b*) Determine the voltage gain (neglecting the beam-loading in the output cavity)
(*c*) Calculate the efficiency of the amplifier (neglecting the beam loading) [IES]

Solution. Given,

$V_0 = 1200$ volts, $I_0 = 28$ mA, $f = 8$ GHz
$d = 1$mm, $L = 4$ cm and $R_{sh} = 40$ kΩ

(*a*) We know, (From the Fig. 7.19),
For maximum V_2, $J_1(X) = 0.582$ at $X = 1.841$
By using Eqn. (7.7), the velocity of the electron

$$v_0 = 0.593 \times 10^6 \sqrt{v_0} \text{ m/s}$$

So

$$v_0 = 0.593 \times 10^6 \sqrt{1200}$$

$$= 20.54 \times 10^6 \text{ m/s.}$$

We know the gap transit angle is,

$$\theta_g = \frac{\omega d}{v_0} = \frac{2\pi \times 8 \times 10^9 \times 10^{-3}}{20.54 \times 10^6}$$

$$= 2.45 \text{ radians}$$

the beam coupling coefficient (Eqn. 7.11),

$$\beta = \frac{\sin \omega\tau_g/2}{\omega\tau_g/2} = \frac{\sin\theta_g/2}{\theta_g/2}$$

$$\beta = \frac{\sin(1.225)}{1.225} = 0.77$$

and the dc transit angle is

$$\theta_0 = \omega T_0 = \omega\frac{L}{v_0} = 2 \times \pi \times 8 \times 10^9 \times \frac{4 \times 10^{-2}}{20.54 \times 10^6}$$

$$= 97.9 \text{ radians.}$$

By using Eqn. (7.20),

$$V_{1\max} = \frac{2V_0 X}{\beta_i \theta_0}$$

$$= \frac{2 \times 12 \times 10^2 \times 1.841}{0.77 \times 97.9}$$

$$V_{1\max} = 58.62 \text{ volts.}$$

(*b*) From the Eqn. (7.30), we get

$$A_v = \frac{\beta_0^2 \theta_0 J_1(X)}{R_0 X} R_{sh},$$

$$R_0 = \frac{V_0}{I_0} = \frac{1200}{28 \text{ mA}} = 42857$$

$$\therefore \qquad A_v = \frac{(0.77)^2 (97.9)(0.582) 40 \times 10^3}{42857 \times 1.841}$$

$$= 17.1$$

Gain in dB = 24.66 dB.

(*c*) The efficiency of the amplifier from the Eqn. (7.23),

$$\eta = \frac{V_2 \beta I_0 J_1(X)}{2 I_0 v_0}$$

or $$I_2 = 2I_0 J_1(X) = 2 \times 28 \times 10^{-3} \times 0.582 = 32.6 \times 10^{-3} \text{ A}$$

and $$V_2 = \beta_0 I_2 R_{sh} = 0.77 \times 32.6 \times 10^{-3} \times 40 \times 10^3 = 1003.83 \text{ V}$$

so $$\eta = \frac{\beta_0 I_2 V_2}{2 I_0 V_0} = \frac{(0.77)(32.6 \times 10^{-3})(1003.83)}{2 \times 28 \times 10^{-3} \times 1200}$$

$$= 37.5\%$$

Example 7.3. A two cavity amplifier klystron has the following parameters, voltage gain A_v = 15 dB, input power P_{in} = 5 mW, total shunt impedance of input cavity R_{sh} = 30 kΩ, load impedance at output cavity R_e = 40 kΩ. Determine:

(*a*) The input voltage (rms)

(*b*) The output voltage (rms)

(*c*) The power delivered to the load in watts.

Solution. Given A_v = 15 dB = 5.623, $R_{sh\text{ in}}$ = 30 kΩ,

P_{in} = 5 mW, $R_{sh\text{ out}}$ = 40 kΩ and

R_e = 40 kΩ

(*a*) We know

$$P_{in} = \frac{V_1^2}{R_{sh}}$$

$$\therefore \qquad V_1 = \sqrt{P_{\text{in}} R_{sh}}$$

$$= \sqrt{5 \times 10^{-3} \times 30 \times 10^3}$$

$$= 12.25 \text{ V(rms)}$$

(*b*) Using equation (7.29), we get

$$V_2 = V_1 A_v$$

$$= (12.25)\,(5.623)$$

$$V_2 = 68.87 \text{ V(rms)}$$

(*c*) We know,

$$P_{\text{out}} = \frac{V_2^2}{R_e} = \frac{(68.87)^2}{40 \times 10^3}$$

$$= 118.57 \text{ mW}$$

Example 7.4. A two-cavity klystron operates at 5 GHz with dc beam voltage 10 kV, cavity gap 2 mm. For a given input RF voltage, the magnitude of the gap voltage is 100 volts. Calculate the transit-time at the cavity gap, the transit angle, and the velocity of the electrons leaving the gap.

Solution.

DC beam velocity $v_0 = 0.593 \times 10^6 \sqrt{V_0}$

$$= 0.593 \times 10^6 \sqrt{(10 \times 10^3)} \text{ m/s}$$

$$= 0.593 \times 10^8 \text{ m/s}$$

Gap transit-time $\tau_g = \dfrac{d}{v_0} = \dfrac{2 \times 10^{-3}}{0.593 \times 10^8} = 33.7 \times 10^{-12}$ sec.

The gap transit angle $\theta_g = \omega t_g = 2\pi \times 5 \times 10^9 \times 33.7 \times 10^{-12}$ rad

$$= 1.059 \text{ rad} = 60.7 \text{ deg.}$$

The beam coupling coefficient

$$\beta_1 = \frac{\sin \theta_g / 2}{\theta_g / 2}$$

$$= 0.505/0.5295 = 0.95337$$

The velocity of electron leaving the input cavity gap

$$v(t) = v_0 \left[1 + \frac{\beta_1 V_1}{2V_0} \sin(\omega t_1) \right]$$

$$= 0.593 \times 10^8 \left[1 + (0.954 \times 100) / (2 \times 10 \times 10^3) \sin(\omega t_1) \right]$$

$$= 0.593 \times 10^8 \, [1 + 0.00477 \sin(\omega t_1)]$$

is varying sinusoidally at the input cycle.

The minimum velocity

$$v(t)_{max} = v_0 \, (1 + \alpha/2) = 0.593 \times 10^{10} \, (1 + 0.00477)$$
$$= 0.5958 \times 10^8 \text{ m/s}$$

The minimum velocity

$$v(t)_{min} = v_0 \, (1 - \alpha/2) = 0.593 \times 10^8 \, (1 - 0.00477)$$
$$= 0.5902 \times 10^8 \text{ m/s}$$

Example 7.5. A two-cavity klystron operates at 10 GHz with $I_0 = 3.6$ mA, $V_0 = 10$ kV. The drift space length is 2 cms and the output cavity total shunt conductance is G_{sh} = 20 μmho and beam coupling coefficient β = 0.92. Find the maximum voltage and power gain.

Solution. Maximum voltage gain

$$A = \frac{\beta^2 \theta_0 I_0 J_1(X)_{max}}{X V_0 G_{sh}}$$

dc beam velocity $v_0 = 0.593 \times 10^6 \sqrt{V_0}$

$$= 0.593 \times 10^6 \sqrt{10 \times 10^3}$$
$$= 0.593 \times 10^8 \text{ m/s}$$

Transit angle in drift space

$$\theta_0 = \frac{\omega L}{v_0}$$

$$= \frac{2\pi \times 10 \times 10^9 \times 2 \times 10^{-2}}{0.593 \times 10^8}$$

$$= 21.19 \text{ rad.}$$

$$A_{max} = \frac{0.92 \times 0.92 \times 21.19 \times 3.6 \times 0.582 \times 10^{-3}}{1.841 \times 10 \times 10 \times 20 \times 10^{-6}} = 102.1$$

Example 7.6. An identical two-cavity klystron amplifier operates at 4 GHz with $V_0 = 1$ kV, $I_0 = 22$ mA, cavity gap 1 mm, drift space 3 cms. If dc beam conductance and catcher cavity total equivalent conductance are 0.25 × 10^{-3} mhos and 0.3 × 10^{-4} mhos, respectively, calculate

(*a*) the beam coupling coefficient, dc transit angle in the drift space and the input cavity voltage magnitude for maximum output voltage.

(*b*) voltage gain and efficiency, neglecting the beam loading.

Solution.

(*a*) dc beam velocity $v_0 = 0.593 \times 10^6 \sqrt{V_0} = 0.59 \times 10^6 \sqrt{1 \times 10^3}$

$$= 1.88 \times 10^7 \text{ m/s}$$

Gap transit angle $\theta_g = \frac{\omega d}{v_0} = \frac{2\pi \times 4 \times 10^9 \times 1 \times 10^{-3}}{1.88 \times 10^7}$

$= 1.337$ rad $= 76.6$ deg.

The beam coupling coefficient

$$\beta_1 = \beta_2 = \frac{\sin\frac{\theta_g}{2}}{\frac{\theta_g}{2}} = \sin 39.3°/0.6685 = 0.927$$

dc transit angle in the drift space

$$\theta_0 = \frac{\omega L}{v_0} = \frac{2\pi \times 4 \times 10^9 \times 3 \times 10^{-2}}{1.88 \times 10^7} \text{ rad} = 40.11 \text{ rad}$$

For maximum output voltage $X = 1.84$, $J_1(X) = 0.582$, so that the input cavity gap voltage magnitude

$$V_1 = \frac{2V_0X}{\beta_1\theta_0} = \frac{2 \times 10^3 \times 1.841}{0.927 \times 40.11} \text{ volts} = 99 \text{ V}$$

(*b*) The voltage gain

$$A_v = \frac{\beta^2\theta_0 J_1(X) I_0}{XV_0G_{sh}}$$

$$= \frac{0.927 \times 0.927 \times 40.11 \times 0.582 \times 22 \times 10^{-3}}{1.841 \times 0.55 \times 10^{-4} \times 10^3}$$

$$= 4.36$$

$$= 12.8 \text{ dB}$$

Catcher voltage $V_2 = A_v \times V_1 = 4.36 \times 99 = 431.64$ V

Power efficiency

$$\eta = \frac{P_{rf}}{P_{dc}} = \frac{\beta_2 I_0 J_1(X) V_2}{I_0 V_0} = \frac{\beta_2 J_1(X) V_2}{2V_0}$$

$$= \frac{0.927 \times 0.582 \times 431.64}{10^3}$$

$$= 23.28\%$$

7.8 REFLEX KLYSTRON

Two-cavity klystron can be made to oscillate by a positive feedback of power from output cavity to the input cavity as shown in Fig. 7.24. Two cavity klystron oscillators are usually not constructed due to operational difficulties. As the frequency of oscillations is varied, the resonant frequency of each cavity and the feedback path phase shift has to be readjusted for a positive feedback which is cumbersome and is seldom used. These disadvantages were overcome in a single cavity low power klystron oscillator (reflex klystron) in which beam feedback is used to produce output energy. This type is widely used as a local oscillator in superhetero-dyne operations, low-power FM transmitter and test signal source in the laboratory for microwave measurements. Reflex klystron are used primarily from 4 to 40 GHz. Power outputs of 1 W or less are common. The efficiency is about 20 to 30 percent.

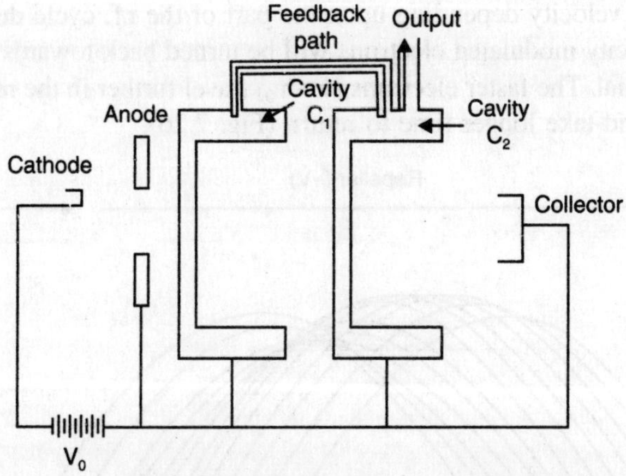

Fig. 7.24 Two-cavity klystron oscillator

7.8.1 Structure/Building Blocks

A sectional view of a typical reflex klystron is shown in Fig. 7.25. It consists of: (*i*) Cathode-anode configuration (electron gun) to produce an accelerated electron beam; (*ii*) resonator (cavity); (*iii*) flexible diaphragm; (*iv*) control grid; (*v*) pair of cavity grids (gap); (*vi*) repeller electrode, which is at a negative potential with respect to the cathode; and (*vii*) output coupling loop (iris). In addition to these there is a tuning lever and tension spring. The whole assembly is sealed in vacuum envelope and vacuum fittings for external connections.

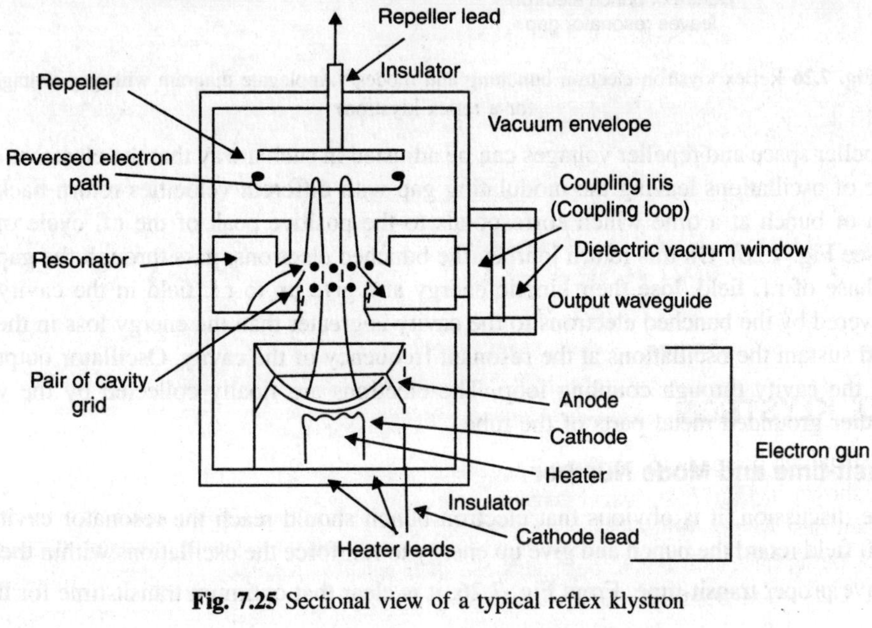

Fig. 7.25 Sectional view of a typical reflex klystron

7.8.2 Mechanism of Operation

The accelerated electron beam from electron gun while passing through cavity gap is velocity modulated due to presence of microwave noise field (r.f. field) at cavity resonant frequency, generated due to dc voltage in the cavity circuit. Electrons emerging from the gap are either accelerated or decelerated or there is no change in velocity depending upon the part of the r.f. cycle during which it crosses the gap. However these velocity modulated electrons will be turned back towards the cavity by the repeller due to its negative potential. The faster electrons ($v > v_0$) travel further in the repeller space as compared to slower one ($v < v_0$) and take longer time to return (Fig. 7.26).

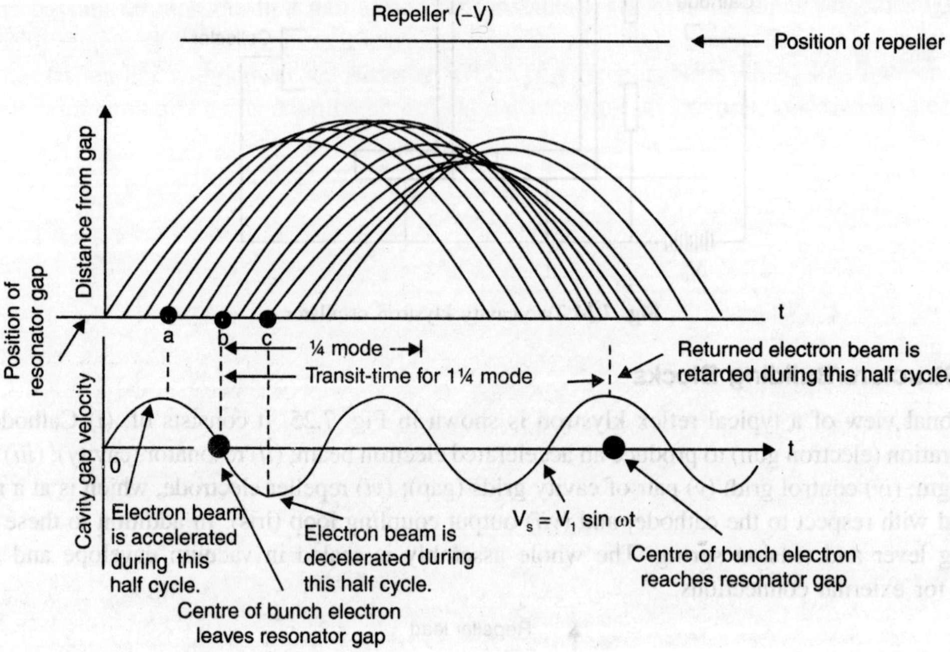

Fig. 7.26 Reflex klystron electron bunching and modes. (Applegate diagram with gap voltage for a reflex klystron)

The repeller space and repeller voltages can be adjusted in such a way that the electrons spread over a half cycle of oscillations leaving the modulating gap with different velocities return back to the gap in the form of bunch at a time which corresponds to the positive peak of the r.f. cycle of the cavity resonator (see Fig. 7.26). On this return journey the bunched electrons, pass through the gap during the retarding phase of r.f. field, lose their kinetic energy and give up to r.f. field in the cavity. When the energy delivered by the bunched electrons to the cavity is greater than the energy loss in the cavity, the cavity could sustain the oscillations at the resonant frequency of the cavity. Oscillator output energy is taken from the cavity through coupling loop. The electrons are finally collected by the walls of the cavity or other grounded metal parts of the tube.

7.8.3 Transit-time and Mode Number

From above discussion, it is obvious that electron bunch should reach the resonator cavity at a time when the r.f. field retard the bunch and give up energy to reinforce the oscillations within the cavity, *i.e.*, it should have proper transit-time. From Fig. 7.26, it is clear that optimum transit-time for the bunch to arrive at the cavity is $\left(n+\frac{3}{4}\right)$ cycles after the beam initially left the cavity.

∴ Transit-time is related to the frequency by the relation;

$$T_0' = \frac{n+3/4}{f} = \left(n+\frac{3}{4}\right)T \qquad \text{...(7.32)}$$

where $T = \frac{1}{f}$ time period of r.f. voltage across the gap. In terms of transit angle,

$$\theta_0' = \omega T_0' = 2\pi f\left(n+\frac{3}{4}\right)T = \left(2\pi n+\frac{3}{2}\pi\right) \text{ where } n = 0, 1, 2, 3,... \qquad \text{...(7.33)}$$

When n = 1, 2, 3,........(as in some books)
Eqn. 7.33 for transit angle would be

$$\theta_0' = \omega T_0' = 2\pi f\left(n-\frac{1}{4}\right)T = \left(2\pi n-\frac{\pi}{2}\right)$$

From Eqns. 7.32 and 7.33, it is clear that reflex klystron can be operated at different drift times (round trip transit-time), which correspond to different values of n. These different transit-times are referred to as modes and are labelled to values of n, *i.e.*,

$$\text{mode number } N_n = \left(n+\frac{3}{4}\right) \qquad \text{...(7.34)}$$

where n = 0 is known as 3/4 mode, n = 1 is known as 1¾ mode and so on.

For fixed value of anode voltage V_0 and resonant frequency the relationship between repeller voltage and mode number is (see Eqn. 7.45*a*)

$$\frac{\left(|V_{R_1}|+V_0\right)}{\left(|V_{R_2}|+V_0\right)} = \frac{N_2}{N_1} \qquad \text{...(7.35)}$$

where $|V_{R_1}|$ and $|V_{R_2}|$ are the magnitudes of repeller voltages required to operate the klystron in mode number N_1 and N_2 respectively. Substitution of mode number equation into repeller voltage relationship (Eqn. 7.35) yields

$$\frac{\left(|V_{R_1}|+V_0\right)}{\left(|V_{R_2}|+V_0\right)} = \frac{n+1+\frac{3}{4}}{n+\frac{3}{4}} = \frac{n+1.75}{n+0.75}. \qquad \text{...(7.36)}$$

for two adjacent modes of operation. Figure 7.27 shows the variation of power output and frequency versus repeller voltage ($-V_R$). These are known as mode curves. For n = 0 $\left(i.e., T_0' = \frac{3}{4}T\right)$, the gain mechanism is usually not strong enough to overcome system losses. **Higher order modes ($n > 1$) are present at lower voltages (Fig. 7.27).**

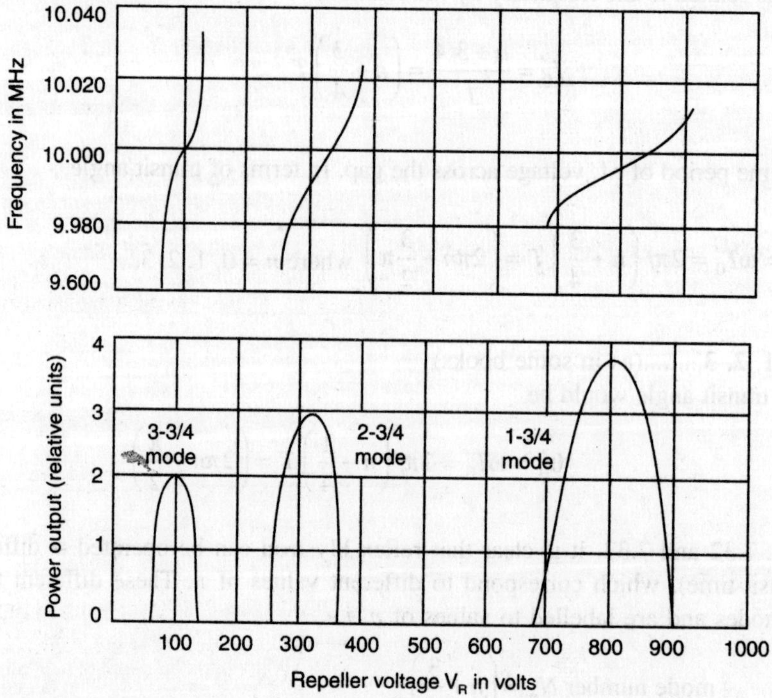

Fig. 7.27 Variation of power output and frequency versus repeller voltage (mode curves)

7.8.4 Tuning*

Reflex klystron oscillator can be electronically tuned (1 percent, see Fig. 7.27) as well as mechanically tuned (up to 30 percent) by suitable tuning mechanism (changing the space between the grids of the resonator).

7.8.5 Quantitative Analysis

7.8.5 (a) Velocity Modulation

The analysis of a reflex klystron is similar to that of a two-cavity klystron. The accelerated electron beam with velocity,

$$v_0 = 0.593 \times 10^6 \sqrt{V_0} \qquad \text{...(7.37)}$$

where V_0 is the accelerating voltage, enters the gap and leaves with a velocity v at instant t_1 (neglecting the transit-time across the gap)

$$v = v_0\left(1 + \frac{V_1}{V_0}\sin\omega t_1\right)$$

---

* Sisodia, M.L., and G.S. Raghuvanshi, **Basic Microwave Techniques and Laboratory Manual,** New Age International (P) Ltd., (Chapter 4).

$$= v_0\left(1+\frac{1}{2}\alpha\sin\omega t_1\right) \qquad ...(7.38)$$

where, $\alpha = \dfrac{V_1}{V_0}$ is the depth of modulation and V_1 is the r.f. voltage.

With the finite transit-time across the gap, electrons entering with velocity v_0 at instant t_1 leave the gap with velocity

$$v = v_0\left(1+\frac{\beta\alpha}{2}\sin\omega t_1\right) \qquad ...(7.39)$$

where $\beta = \dfrac{\sin\omega\tau_g/2}{\omega\tau_g/2}$ is the beam coupling coefficient, τ_g is the electron transit-time across the gap, $\omega\tau_g = \theta_g$ is the electron transit angle across the gap.

The electron leaving gap at a different instant t_1 travels with different velocities in repeller space.

The electron, which leaves the gap at a time t_1 corresponding to $\omega t_1 = 2\pi n$, $n = 0, 1, 2, 3,...$ will suffer no change in velocity.

7.8.5 (b) Bunching, Arrival Time and Modes

Suppose the voltage difference between the resonator and repeller is $(V_R - V_0)$ and the distance between the two is L. It is measured from the exit plane of the gap, *i.e.*, $x = 0$ (Fig. 7.28).

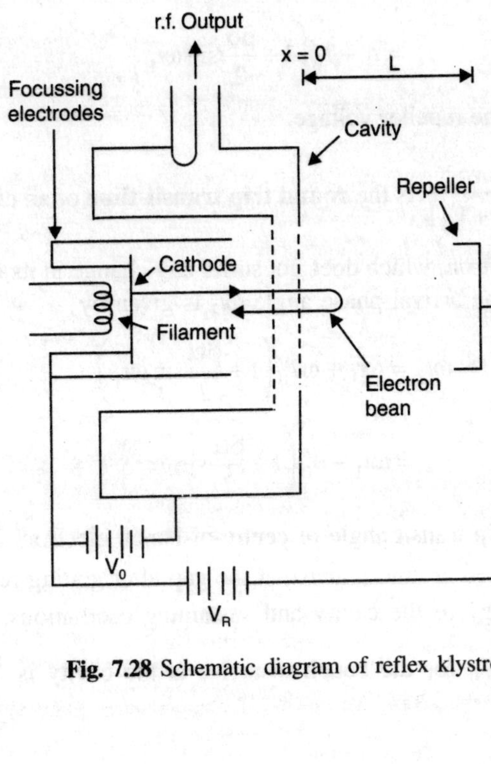

Fig. 7.28 Schematic diagram of reflex klystron

The force on the electron in the repeller space is given by $m\dfrac{d^2x}{dt^2} = \dfrac{e(V_R - V_0)}{L}$

$$\therefore \quad \frac{d^2x}{dt^2} = \frac{\eta(V_R - V_0)}{L} \qquad \text{...(7.40)}$$

where $\eta = e/m$ is the magnitude of the electronic charge to mass ratio = 1.759×10^{11} C/kg

Integrating Eqn. (7.40) twice, we obtain

$$x = \frac{\eta}{2}\frac{(V_R - V_0)}{L}(t - t_1)^2 + v(t - t_1) \qquad \text{...(7.41)}$$

where $t = t_1$ is the instant when the electron leaves with a velocity v. The electron is retarded as it approaches the negative repeller electrode it than reverses its direction of travel (Fig. 7.21) and continues to travel towards the resonator gap and return back at the gap $x = 0$ at the instant $t = t_2$ *i.e.*, **electron leaving $x = 0$ at t_1 return back to $x = 0$ at t_2.**

Thus the total transit-time of the electron in the repeller space (round trip transit-time) is

$$t_2 - t_1 = \frac{-2Lv}{\eta(V_R - V_0)} \qquad \text{...(7.42)}$$

Substituting Eqn. 7.39 in Eqn. 7.42, the arrival time t_2 for the electron which left the gap at t_1 is

$$t_2 = t_1 - \frac{2Lv_0}{\eta(V_R - V_0)}\left(1 + \frac{\beta\alpha}{2}\sin\omega t_1\right)$$

$$= t_1 + T_0'\left(1 + \frac{\beta\alpha}{2}\sin\omega t_1\right) \qquad \text{...(7.43)}$$

where $|V_R|$ is the magnitude of the repeller voltage.

where $T_0' = -\dfrac{2Lv_0}{\eta(V_R - V_0)} = \dfrac{2Lv_0}{(|V_R| + V_0)}$, is the **round trip transit-time** of an electron that leaves the gap when r.f. voltage is zero *i.e.*, electron, which does not suffer any change in its dc velocity v_0. It is known as **centre-of-bunch electron.** The arrival phase angle ωt_2 is given by

$$\omega t_2 = \omega t_1 + \omega T_0'\left(1 + \frac{\beta\alpha}{2}\sin\omega t_1\right)$$

$$= \omega t_1 + \theta_0'\left(1 + \frac{\beta\alpha}{2}\sin\omega t_1\right) \qquad \text{...(7.44)}$$

where $\theta_0' = \omega T_0'$ is the round trip transit angle of centre-of-bunch electron. It is clear (Fig. 7.26) that electrons between *a* and *c* centered around *b* arrive at the gap after getting reflected from repeller at a retarding phase, delivering energy to the cavity and sustaining oscillations. As discussed in Section 7.8.3 the optimum transit-time T_0' for the bunch to arrive at the cavity is $\left(n + \dfrac{3}{4}\right)T$ after the beam initially left the cavity, *i.e.*,

$$T_0' = \left(n + \frac{3}{4}\right)T$$

The optimum transit angle $$\theta_0' = 2\pi\left(n + \frac{3}{4}\right) = \left(2\pi n + \frac{3\pi}{2}\right).$$

Thus there are several values of T_0' or θ_0' satisfying the oscillation condition. These values correspond to **various modes** of a reflex klystron (Fig. 7.27).

7.8.5c Relation between Accelerating Voltage (V_0), Repeller Voltage (V_R) and Repeller Space (L)

The optimum value of T_0' is

$$T_0' = \frac{2Lv_0}{\eta\left(|V_R| + V_0\right)} = \left(n + \frac{3}{4}\right)T \qquad \text{...(7.45)}$$

for oscillation condition.

Substituting the value of $v_0 = \sqrt{2\eta V_0}$ in Eqn. 7.45 and arranging

$$\frac{V_0}{\left(|V_R| + V_0\right)^2} = \frac{\eta}{8L^2}\left(n + \frac{3}{4}\right)^2 \cdot T^2, \quad \left(T = \frac{1}{f}\right)$$

$$\frac{V_0}{\left(|V_R| + V_0\right)^2} = \frac{\eta}{8L^2 f^2}\left(n + \frac{3}{4}\right)^2 \qquad \text{...(7.45a)}$$

$$|V_R| = \frac{6.744 \times 10^{-6}\, fL\sqrt{V_0}}{\left(n + \frac{3}{4}\right)} - V_0 \qquad \text{...(7.45b)}$$

Smaller value of n corresponds to higher value of V_R. Thus a particular mode of operation can be selected by the choice of repeller voltage (Eqn. 7.45*b*).

For a selected gap-repeller spacing L, the various values of the repeller voltage V_R for oscillations can be obtained using Eqn. (7.45*b*), for a given value of V_0. Alternatively if the repeller voltage is fixed, the gap repeller spacing L can be determined.

7.8.5 (d) Power Output and Efficiency

The power output and efficiency of a reflex klystron can be determined in the same manner as that of a two-cavity klystron in Section 7.6.3. From the charge conservation law the net charge $I_0\, dt_1$ leaving the resonator cavity gap over an interval of time dt_1 arrives back to cavity gap over an interval of time dt_2 inducing current I_2. Thus

$$I_2 dt_2 = I_0 dt_1 \quad \text{(Charge conservation)}$$

$$I_2 = I_0 \frac{dt_1}{dt_2} \qquad \text{...(7.46)}$$

From Eqn. 7.43, we get

$$I_2 = \frac{I_0}{\left[1 + \frac{\beta\alpha\theta_0'}{2}\cos\omega t_1\right]}$$

$$\approx I_0\left[1 - \frac{\beta\alpha\theta_0'}{2}\cos\omega t_1\right] \quad ...(7.47)$$

$$\approx I_0\left[1 + X'\cos\omega t_1\right] \quad ...(7.48)$$

where $X' = -\frac{\theta_0'\alpha\beta}{2} = -\theta_0'\frac{\beta V_1}{2V_0}$ is **bunching parameter** and β is **beam coupling** coefficient and for small cavity gap $\beta \approx 1$, *i.e.*, transit-time across the gap is negligible.

X' the bunching parameter has a negative sign with respect to the bunching parameter X of a two-cavity klystron amplifier. The beam current injected into the cavity gap from the repeller space flows in the negative direction. The fundamental component of the current induced in the cavity by the density modulated return electron beam is given in similar way as in case of two-cavity klystron.

$$I_f = 2I_0J_1(X')$$

where $J_1(X')$ is the Bessel function of the first kind (see Fig. 7.29).

The dc power supplied by the beam voltage V_0 is

$$P_{dc} = V_0I_0$$

and the r.f. power delivered to the load is given by

$$P_{\text{r.f.}} = \frac{V_1I_f}{2} = V_1I_0J_1(X') \quad ...(7.49)$$

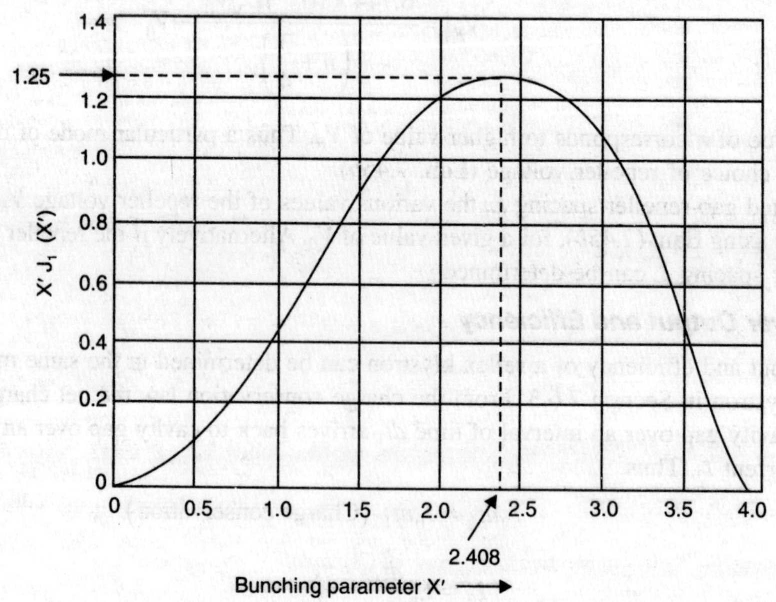

Fig. 7.29 X' $J_1(X')$ versus X'

The ratio of V_1 over V_0 is given by

$$\frac{V_1}{V_0} = \frac{2(X')}{(2\pi n + 3\pi/2)}$$

$$\therefore \quad V_1 = \frac{2(X')V_0}{(2\pi n + 3\pi/2)} \qquad ...(7.50)$$

Substituting the value of V_1 from Eqn. 7.50 in Eqn. 7.49, we get

$$P_{\text{r.f.}} = \frac{2I_0V_0(X')J_1(X')}{2\pi(n+3/4)} \qquad ...(7.51)$$

$$P_{\text{r.f.}} = \frac{I_0V_0(X')J_1(X')}{\pi(n+3/4)} \qquad ...(7.52)$$

From Fig. 7.29, it is clear that the factor $(X')J_1(X')$ reaches the maximum value of 1.25 at $X' = 2.408$ and $J_1(X') = 0.52$. Thus optimum maximum r.f. power would be

$$P_{\text{r.f.(max)}} = \frac{0.3986\ I_0V_0}{(n+3/4)}$$

$$\approx \frac{0.3986 I_0V_0}{N} \qquad ...(7.53)$$

where $N = (n + 3/4)$, $n = 0, 1, 2, 3,...$

Electronic efficiency from Eqn. 7.53 is given below:

$$\text{Efficiency } = \eta = \frac{P_{\text{r.f.}}}{P_{\text{d.c.}}} = \frac{(X')J_1(X')}{\pi(n+3/4)} \qquad ...(7.54)$$

Maximum or optimum efficiency

$$\eta_{max} = \frac{0.3986}{(n+3/4)} \qquad ...(7.55)$$

The mode $n = 1$ *i.e.*, $1\frac{3}{4}$ mode leads to maximum r.f. power output (Eqn. 7.53) and efficiency (Eqn. 7.55).

$$P_{\text{r.f.max}} = 0.227\ V_0I_0$$

$$\eta_{max} = 22.7\%$$

Example 7.7. A reflex klystron is to be operated at frequency of 10 GHz, with dc beam voltage 400 V, repeller space 0.1 cm for $1\frac{3}{4}$ mode. Determine optimum (maximum) value of power and corresponding repeller voltage for beam current of 30 mA.

Solution.

From Eqn. 7.53

Given:

$$I_0 = 30\times10^{-3}\text{ A}$$
$$V_0 = 400\text{ V}$$
$$n+\frac{3}{4} = 1\frac{3}{4} = 1.75$$
$$f = 10\times10^{9}\text{ Hz}$$
$$L_{opt} = 0.1\times10^{-2}$$
$$= 10^{-3}\text{ m}$$

$$P_{\text{r.f.(max)}} = \frac{0.3986\times I_0V_0}{(n+3/4)}$$

$$= \frac{0.398\times400\times30\times10^{-3}}{1.75}$$

= 2.73 watts

From Eqn. 7.45*b*, we can write

$$|V_R| = \frac{6.744\times10^{-6} fL\sqrt{V_0}}{(n+3/4)} - V_0$$

$$= \frac{6.744\times10^{-6}\times10\times10^{9}\times10^{-3}\times\sqrt{400}}{1.75} - 400$$

$$= \frac{6.744\times2\times10\times10}{1.75} - 400$$

$$V_R \approx -370.44\text{ V}.$$

Example 7.8. Find the optimum length for $1\frac{3}{4}$ mode of a repeller space of reflex klystron which operates with centre frequency 9 GHz with dc beam voltage 600 V and repeller voltage $V_R = -250\ V$ and $\eta = 1.76\times10^{11}$ C/kg

Solution. From Eqn. (7.45*b*)

Given:

$$V_0 = 600\text{ V}$$
$$V_R = -250\text{ V}$$
$$f_c = 9\times10^{9}\text{ Hz}$$
$$N = 1\frac{3}{4} = 1.75$$

$$\frac{V_0}{(V_R - V_0)^2} = \frac{\eta}{8L^2}\frac{\left(n+\frac{3}{4}\right)^2}{f^2}$$

$$L^2 = \frac{\eta}{8}\frac{\left(n+\frac{3}{4}\right)^2}{f^2}\cdot\frac{(V_R - V_0)^2}{V_0}$$

or

$$L = \sqrt{\frac{\eta}{8}}\frac{\left(n+\frac{3}{4}\right)}{f}\cdot\frac{(V_R - V_0)^2}{V_0}$$

or

$$L = \frac{\sqrt{1.76\times10^{11}}}{\sqrt{8}}\times\frac{1.75}{9\times10^{9}}\times\frac{850}{\sqrt{600}}$$

$$= 10^{-3}\text{ m} = 0.1\text{ cm}$$

7.8.5 (e) Electronic Admittance

Figure 7.30 shows the equivalent circuit for the reflex klystron, where *C*, *L*, *R* represent equivalent lumped circuit components of the resonator cavity. $Y_e = G_e + jB_e$ is electronic admittance presented by bunches as the bunched electrons return to the cavity gap a little before the optimum transit-time $T_0' = \left(n + \frac{3}{4}\right)T$, the r.f. beam current lags behind the field and an inductive reactance is presented to the external circuit. If the bunched electrons return slightly after T_0', a capacitive reactance appears in the circuit. The condition for oscillations is satisfied when G_e becomes negative.

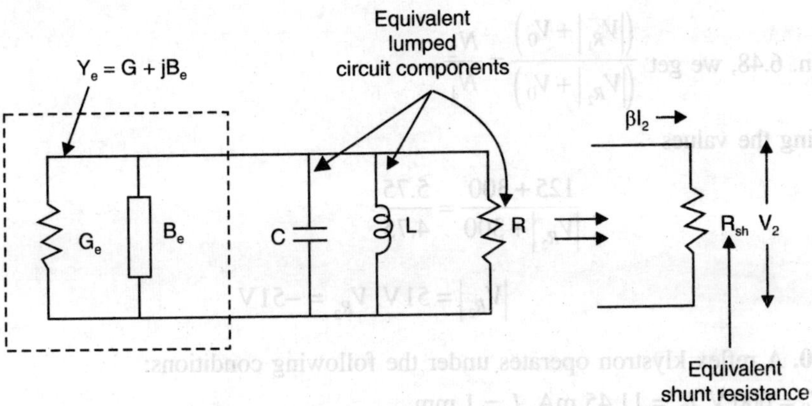

Fig. 7.30 Equivalent circuit of a reflex klystron

In Fig. 7.31, electronic admittance $Y_e = G_e + jB_e$ is plotted in the complex admittance plane. It shows the variation of G_e and B_e qualitatively. The origin in this plane corresponds to infinite repeller voltage.

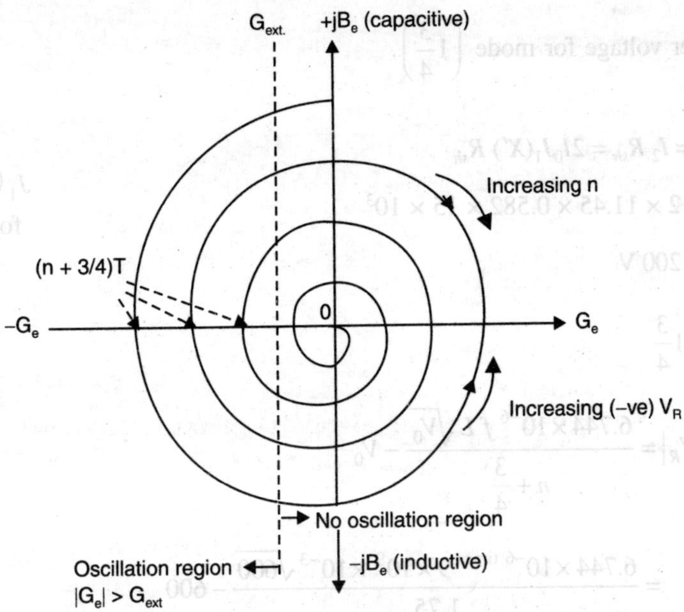

Fig. 7.31 Electronic admittance of a reflex klystron

A decrease in repeller voltage moves outwards along the spiral. Oscillations can take place whenever $|G|$ is greater than G_{ext}, the conductance of the external circuit. When jB_e is negative, oscillations take place at a frequency higher than the resonance frequency of the cavity, whereas positive values of jB_e correspond to a lower frequency than the resonator resonance frequency. This leads to the frequency variation with the repeller voltage as shown in Fig. 7.27.

Example 7.9. For 4.75 mode the repeller voltage is 125 V find the repeller voltage for 5.75 mode, the anode voltage $V_0 = 300$ V.

Solution.

From Eqn. 6.48, we get $\dfrac{\left(|V_{R_1}| + V_0\right)}{\left(|V_{R_2}| + V_0\right)} = \dfrac{N_2}{N_1}$

Substituting the values

$$\therefore \quad \frac{125 + 300}{|V_{R_2}| + 300} = \frac{5.75}{4.75}$$

$$|V_{R_2}| = 51\,\text{V},\ V_{R_2} = -51\,\text{V}$$

Example 7.10. A reflex klystron operates under the following conditions:

$V_0 = 600$ V, $I_0 = 11.45$ mA, $L = 1$ mm.

$R_{sh} = 15$ kΩ, $f = 9$ GHz

The tube is oscillating at f at the peak of $n = 1\frac{3}{4}$ mode. Assume $\beta = 1$. Find (*a*) the microwave gap voltage and repeller voltage for mode $\left(1\frac{3}{4}\right)$.

Solution.

(*a*) $V_2 = I_2 R_{sh} = 2I_0 J_1(X') R_{sh}$

$= 2 \times 11.45 \times 0.582 \times 15 \times 10^3$

≈ 200 V

$I_0 = 11.45$ mA

$J_1(X') = 0.582$ for $X' = 2.408$

$R_{sh} = 15 \times 10^3\ \Omega$

$L = 10^{-3}$ m

(*b*) V_R for mode $1\frac{3}{4}$

$$|V_R| = \frac{6.744 \times 10^{-6} f L \sqrt{V_0}}{n + \frac{3}{4}} - V_0$$

$$= \frac{6.744 \times 10^{-6} \text{ times } 9 \times 10^9 \times 10^{-3} \sqrt{600}}{1.75} - 600$$

$$= -250\,\text{V}.$$

The reflex klystron is a low-power generator and operating frequency extending up to millimetre region.

Example 7.11. A reflex klystron operates at the peak mode of $n = 2$ with beam voltage $V_0 = 300$ V, beam current $I_0 = 20$ mA and signal voltage $V_1 = 40$ V. Determine the input power, output power in watts and efficiency.

Solution. Given $V_0 = 300$ V, $I_0 = 20$ mA, $V_1 = 40$ V and $n = 2$

The input power is,

$$P_{in} = P_{dc} = V_0 I_0 = 300 \times 20 \times 10^{-3} = 6 \text{ watts}$$

The output power is (using Eqn. 7.49),

$$P_{ac} = V_1 I_0 \beta_1 J_1(X') \text{ let } \beta_i = 1$$

where

$$X' = \frac{\beta_i V_1}{2V_0}\theta_0' = \frac{V_1}{2V_0}\omega T_0'$$

$$= \frac{V_1}{2V_0}(2n\pi - \pi/2)$$

$$= \frac{40}{2 \times 300}(2 \times 2 \times 3.14 - 3.14/2)$$

$$X' = 0.734$$

By using Fig. (7.29), we get

$$J_1(0.734) = 0.345$$

$$\therefore \quad P_{ac} = 40 \times 20 \times 10^{-3} \times 1 \times 3.45$$

$$= 0.276 \text{ W}$$

The efficiency is

$$\eta = \frac{P_{ac}}{P_{dc}} = \frac{0.276}{6} = 4.6\%.$$

Example 7.12. A reflex klystron operates under the following conditions; $V_0 = 500$ V, $R_{sh} = 20$ kΩ, $f_r = 8$ GHz and $L = 1$ mm is the spacing between repeller and cavity. The tube is oscillating at f_r at the peak of the $n = 2$ mode or 1¾ mode. Assume that the transit-time through the gap and the beam loading effect can be neglected.

(*a*) Find the value of repeller voltage V_R.

(*b*) Find the direct current necessary to give microwave gap voltage of 200 V.

(*c*) Find the electronic efficiency.

Solution. Given $V_0 = 500$ V, $R_{sh} = 20$ kΩ, $n = 2$

$f = 8$ GHz, $L = 1$ mm.

(*a*) Using the Eqn. 7.45(*a*), the repeller voltage,

$$\frac{V_0}{\left(|V_R| + V_0\right)^2} = \frac{\eta}{8L^2}\frac{(n+3/4)^2}{f^2} = \frac{(2\pi n - \pi/2)^2}{8\omega^2 L^2}\frac{e}{m}$$

$$= \frac{(4\pi - \pi/2)^2}{8\times(2\pi\times8\times10^9)^2(10^{-3})^2}(1.759\times10^{11})$$

$$\frac{V_0}{(|V_R|+V_0)^2} = 1.053\times10^{-3}$$

$\therefore$
$$|V_R| = \sqrt{\frac{500}{1.05\times10^{-3}}} - 500$$

$$|V_R| = 189 \text{ volts.}$$

(*b*) Let $\beta_1 = \beta_0 = 1$,

We know
$$X' = \frac{\beta_i V_1}{2V_0}(2n\pi - \pi/2)$$

$$X' = 2.2.$$

or using Fig. (7.29), we get

$$J_1\,(2.2) = 0.556.$$

Then the direct current is

$$I_0 = \frac{V_1}{2J_1(X')R_{sh}} = \frac{200}{2(0.556)20\times10^3}$$

$$= 8.99 \text{ mA}$$

(*c*) The electronic efficiency is (using Eqn. 7.54).

$$\eta = \frac{X'J_1(X')}{\pi\left(n+\frac{3}{4}\right)} = \frac{2.2\times0.556}{3.14\left(2+\frac{3}{4}\right)}$$

$$\eta = 22.23\%.$$

Example 7.13. A reflex klystron operates at the peak of the $n = 2$ mode. The dc power input is 40 mW and $v_1/v_0 = 0.278$. If 20% of the power delivered by the beam is dissipated in the cavity walls, find the power delivered to the load.

Solution.

Given $\eta = 2$, $\frac{V_1}{V_0} = 0.278$

$P_{dc} = 40$ mW $B_i = 1$ is assumed

By using the Eqn. (7.32) we get

$$T_0' = \left(n+\frac{3}{4}\right)T,$$

So $\omega T_0 = 2\pi n - \pi/2 = 11.00$

and $X' = \dfrac{\beta_i V_1}{2V_0}\omega T_0' = \dfrac{1}{2}(.278)11$

$= 1.53$

So by using Fig. (7.29),

$$J(1.53) = 0.562$$

So the efficiency is using Eqn. (7.54),

$$\eta = \frac{2X'J_1(X')}{2\pi n - \dfrac{\pi}{2}} = \frac{2 \times 1.53 \times 0.562}{4\pi - \pi/2}$$

$$= 16\%$$

And the total power output is,

$$P_{out} = \eta P_{dc}$$

$$= 6.25 \text{ mW.}$$

and the power at the load is,

$$P_{load} = 6.25 \times 0.8 = 5.00 \text{ mW.}$$

Example 7.14. A reflex klystron is operating at 4 GHz with the following parameters V_R = 2000 V, V_0 = 500 V, L = 2 cm, n = 0 *i.e.*, $\dfrac{3}{4}$ mode. Calculate the change in frequency for a 4% change in repeller voltage ($\eta = e/m = 1.76 \times 10^{11}$ C/kg).

Solution. From Eqn. 7.45*a*

$$\frac{V_0}{(V_R - V_0)^2} = \frac{\eta\left(n + \dfrac{3}{4}\right)^2}{8L^2 f^2} \quad ...(7.56)$$

$$(V_R - V_0)^2 = \frac{8L^2 f^2 V_0}{\eta\left(n + \dfrac{3}{4}\right)^2} \quad ...(7.57)$$

Differentiating with respect to f

$$2(V_R - V_0)\frac{dV_R}{df} = \frac{16L^2 f V_0}{\eta\left(n + \dfrac{3}{4}\right)^2} \quad ...(7.58)$$

$$\frac{dV_R}{df} = \frac{8L^2 f}{\eta\left(n + \dfrac{3}{4}\right)^2} \times \frac{V_0}{(V_R - V_0)} \quad ...(7.59)$$

Substituting the value from Eqn. 7.56

$$\frac{dV_R}{df} = \frac{8L^2 f V_0}{\eta\left(n+\frac{3}{4}\right)^2} \cdot \frac{\sqrt{\eta\left(n+\frac{3}{4}\right)^2}}{8L^2 f^2 V_0} \qquad \text{...(7.60)}$$

$$\frac{dV_R}{df} = \frac{L}{\left(n+\frac{3}{4}\right)}\sqrt{\frac{8V_0}{\eta}} \qquad \text{...(7.61)}$$

Substituting the values in 7.61

$$\frac{dV_R}{df} = \frac{2\times 10^{-2}}{3/4}\sqrt{\frac{8\times 500}{1.76\times 10^{11}}} = 4\times 10^{-6} \qquad df = \frac{dV_R}{4\times 10^{-6}} = \frac{0.04\times 2000}{4\times 10^{-6}} = 20 \text{ MHz.}$$

Example 7.15. In a certain reflex klystron for a given value of gap-repeller electrode spacing and accelerating potential, 5% change in repeller voltage produces 15% change in operating frequency when the ratio of repeller (V_R) to operating frequency is 1 volt/MHz. If the ratio *VR/f* is changed to 1.5 V/MHz. What would be percentage change in (f) for 5% change in (V_R)?

Solution.

$$\frac{dV_R}{df} = \frac{L}{\left(n+\frac{3}{4}\right)}\sqrt{\frac{8V_0}{\eta}}$$

Which gives

$$\frac{df}{f}(dV_R/V_R) = \left(\frac{V_R}{f}\right)\frac{\left(n+\frac{3}{4}\right)}{L}\sqrt{\frac{\eta}{8V_0}} = k\left(\frac{V_R}{f}\right)$$

Where $k = \text{constant} = \left(\frac{n+3/4}{L}\right)\sqrt{\frac{\eta}{8V_0}}$ $\quad k\left(\frac{V_R}{f}\right) = \frac{1}{5}$ $\quad \therefore\ k = \frac{1}{5}\left(\frac{f}{V_R}\right) = \frac{1}{5}$

New value of $\frac{V_R}{f} = 1.5$ volts/MHz

$$\therefore \text{ Therefore, } \frac{\text{Percentage change in } (f)}{\text{Percentage change in } (V_R)} = 1.5k = 1.5\times\frac{1}{5} = 0.3 = \frac{3}{10}$$

or percentage changes $(f) = \frac{3}{10}\times$ percentage change in $V_R = \frac{3}{10}\times 5 = 1.5\%$.

REVIEW QUESTIONS AND PROBLEMS

1. What are limitations of conventional tubes at microwave frequencies? Describe construction and working of a two cavity klystron amplifier. Explain in detail bunching process and obtain expression for bunching parameter. Compare two cavity klystrons with TWT as microwave amplifier.
2. Describe the mechanism of velocity modulation in a two cavity klystron and hence obtain an expression for the bunched beam current. Also find out condition for maximum power output.
3. What is velocity modulation? How is it achieved in a two cavity klystron? Describe the construction and working of a reflex klystron.
4. Describe the phenomenon of velocity modulation in two-cavity klystron and explain how microwave power is amplified by the interaction of velocity modulated beam with electromagnetic field in the two-cavity klystron.
5. Describe the mechanism of velocity modulation in a two cavity klystron and hence obtain an expression for the bunched beam current. Also find out condition for maximum power output.
6. A two-cavity klystron amplifier has the following parameters:
   $V_0 = 1000$ V, $f = 5$ GHz, Cavity gap = 1 mm.
   Find: (*i*) the gap transit angle; and (*ii*) the optimum length of the drift region.
7. A two-cavity klystron amplifier has the folllowing parameters:
   $V_0 = 1200$ V, $I_0 = 25$ mA, $R_0 = 30$ kΩ, $f = 10$ GHz, $d = 1$ mm, $L = 4$ cm, $R_{sh} = 30$ kΩ Calculate: (*i*) the input voltage for maximum output voltage; (*ii*) the voltage gain in decibles; (*iii*) efficiency.
8. A klystron amplifier is operated with a beam voltage of 3 kV. If the coupling coefficient is 0.9 and the magnitude of the signal voltage at the input cavity gap is 100 V, find the velocities of the electrons leaving the input gap.
9. A two-cavity klystron amplifier has beam current 400 mA. An output cavity coupling coefficient 0.95 and the magnitude of the voltage across the output cavity gap is 75 V. Calculate the magnitude of the fundamental induced current and the maximum power output.
10. A two-cavity klystron is operated at 10 GHz with $V_0 = 1200$ V. $I_0 = 30$ mA, $d = 1$ mm. $L = 1$ cm and $R_{sh} = 40$ k-ohm. Neglecting beam loading calculate (*a*) input RF voltage V_1 for a maximum output voltage, (*b*) voltage gain, (*c*) efficiency, (*d*) beam loading conductance, (*e*) the electron velocity, and (*f*) the dc electron transit-time.
11. A klystron amplifier having two similar cavities with a spacing $L = 40$ mm, with cavity gap distance of 1 mm, operates under d.c. accelerating voltage $V_0 = 1$ kV. The d.c. beam current $I_0 = 25 \times 10^{-3}$ A. The effective shunt resistance R_{sh} of the cavity is $R_{sh} = 30 \times 10^3 \Omega$. The d.c. beam resistance R_{dc} (beam) $= 2 \times 10^4$ Ω. Calculate the voltage gain and efficiency of the amplifier.
12. A double-resonator klystron is tuned to a frequency of 3500 MHz. The gap between the buncher grids and catcher grids is 0.12 cm and the length of the drift space is 2.1 cm. The beam current is 24 mA, and the ratio V_2/V_0 is 0.35. If the two resonators are assumed to oscillate in time phase, calculate the dc accelerating voltage, buncher voltage, and catcher voltage required for maximum power output.
13. Explain with the help of a big clear labelled diagram the construction and principle of working of two cavity klystron amplifier. Explain bunching process using Applegate diagram. State two of its main uses.
14. What is velocity modulation? How is it achieved in a two cavity klystron? Describe the construction and working of a reflex klystron.

15. Explain the phenomenon of "Bunching" in a Reflex Klystron. Derive an expression for time taken by electron (to) return to resonator and show that it is inversely proportional to the reflector voltage. What is the ideal time to in terms of resonator parameters.

16. Describe principle, construction and opertion of Reflex Klystron. Mention reflection protection arrangement for this device. Write its important applications.

17. Using an equivalent circuit representation of electron Bunch and Resonantor show how and under what conditions a reflex klystron will oscillate? Plot its output power and frequency against reflector voltage.

18. Explain construction and principle of working of two cavity klystron amplifier. Explain bunching process using Applegate diagram.

19. In a certain reflex klystron, for given values of gap repeller electrode spacing and accelerating potential, 5% change in repeller voltage produces a 1% change in operating frequency when the ratio of repeller voltage (V_r) to operating frequency (f) is 1 volt/MHz. If this ratio is changed to 1.5 volts/MHz, what would be the percentage change in operating frequency (f) of 5% change in (V_r)?

20. Explain the construction and working principle of Reflex Klystron. Which tube parameters decide the oscillation frequency and amplitude?

21. Calculate the resonant frequency of the resonator of a reflex klystron which operation with beam voltage $V_0 = 300$ volts, beam coupling coefficient $\beta = 0.9$ and gap spacing $d = 0.286$ μm.

22. Describe the phenomenon of velocity modulation in two-cavity klystron and explain how microwave power is amplified by the interaction of velocity-modulated beam with electromagnetic field in the two-cavity klystron.

23. A reflex klystron operates at the peak of the $n = 1$ or ¾ mode. The dc power input is 40 mW and the ratio of V_1 over V_0 is 0.278.

(*i*) Determine the efficiency of the reflex klystron.

(*ii*) Find the total output power in milli-watts.

(*iii*) If 20% of the power delivered by the electron beam is dissipated in the cavity walls, find the power delivered to the load.

24. A reflex-klystron has the following parameters:

$V_0 = 2500$ V, $L = 6$ mm, $f = 3$ GHz

Calculate the repeller voltage for which tube can oscillate in 1¾ mode and 3¾ mode.

25. A reflex klystron having an accelerated field of 300 volts oscillates at a frequency of 10 GHz with a retarding field of 500 volt. If its cavity is retuned to 9 GHz, what must be the new value of retarding field for oscillations is the same mode to take place?

26. A reflex klystron having an accelerated field of 300 volts oscillates at a frequency of 10 GHz with a retarding field of 500 volt. If its cavity is retuned to 9 GHz, what must be the new value of retarding field for oscillations to take place in the same mode?

27. Name different methods of generating microwave power. Describe with necessary theory, the working of a reflex klystron. Explain how frequency stabilization is achieved in this tube.

28. In a reflex klystron, calculate the magnitude of the acceleration in the resonator repeller space using the following data—

Positive accelerating voltage $V_0 = 400$ volts, Negative repeller voltage $V_r = -800$ volts

Resonator repeller spacing $S = 4 \times 10^{-3}$ meters. $e/m = 1.76 \times 10^{-11}$ e.m.u.

29. Using a clear labelled diagram, describe the construction and principle of operation of a Reflex Klystron. Describe repeller protection arrangements.

CHAPTER 8

Microwave Tubes: Helix Travelling Wave Tube

8.1 HELIX TRAVELLING WAVE TUBE—O-TYPE

In the last chapter the behaviour of velocity modulated cavity type microwave linear-tubes (O-type) were described. These tubes are characterized by high gain at low signal levels and limited bandwidth due to use of high resonant cavities. On the other hand, another type of O-type devices (see Table 7.2) are broadband devices which use a periodic slow-wave structure for continuous electron interaction with slow electromagnetic waves on a structure. These O-type devices include the Helix travelling-wave tube (TWT), the coupled cavity travelling wave tube, backward wave oscillator (BWO) and backward wave amplifier (BWA).

8.1.1 Performance and Application

Travelling-wave tubes (TWTs) are widely used as broadband, low and medium power microwave active devices. A large variety of travelling-wave tubes are available, that differ from one another in the type of periodic slow-wave structure (circuit) used in supporting a slow electromagnetic wave. **Helix travelling-wave amplifiers** are almost exclusively used for broadband applications, such as wideband ground and satellite communication systems, and anti jamming and countermeasure radar systems. Octave bandwidths (2 : 1) are possible with travelling wave tubes with power rating of several hundred watts. The power rating can be increased to several kilowatts by using a set of coupled cavities as an interaction region but with reduced bandwith. Power versus frequency performance of helix and coupled cavity travelling tubes, is summarized in Fig. 8.1. Gridded controlled travelling wave tube is another type of high power tube.

8.2 HELIX TRAVELLING-WAVE TUBES

In this section the structure/building blocks, mechanism of operation, and small signal analysis of helix travelling-wave tube are described.

8.2.1 Structure/Building Blocks

A schematic diagram of a typical helix-type travelling-wave tube is shown in Fig. 8.2 (*Pierce depicted TWT as in Fig.* 7.3). It consists of (*i*) a cathode anode configuration (electron gun Fig. 8.2) to produce an electron beam with a velocity close to the velocity of propagation of microwaves on the slow-wave

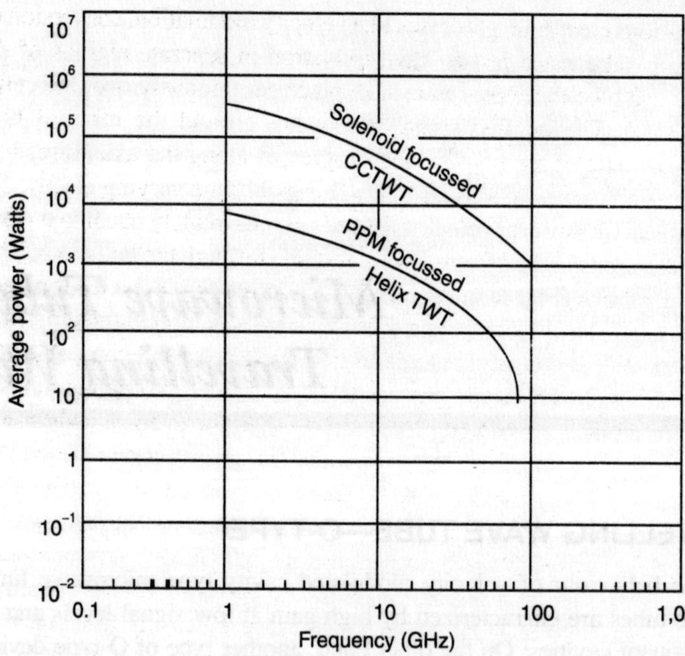

Fig. 8.1 Power capability versus frequency of helix and coupled cavity travelling wave tubes

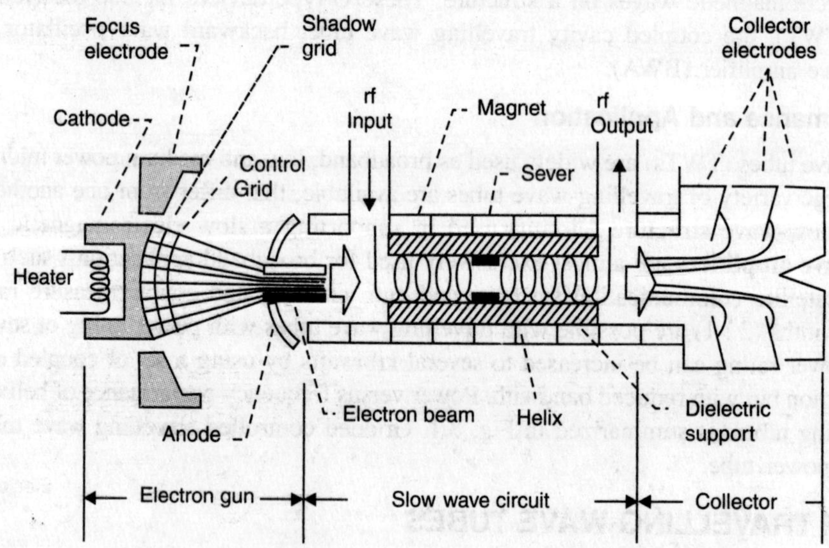

Fig. 8.2 Schematic diagram of TWT (see Fig. 7.3 also)

structure; (*ii*) a long helix slow-wave structure (nonresonant) capable of propagating microwaves and at the same time enclosing an electron beam; (*iii*) a collector which is generally kept at a dc potential lower than that of the helix; slow-wave structures are discussed in section 8.2.1(a) (*iv*) input and output couplers; and (*v*) an external cover to vacuum envelope magnetic focussing structure, usually a long solenoid/PPM.

8.2.2 Mechanism of Operation

We have seen that various electronic processes like velocity modulation, conversion to density modulation, transfer of energy etc. take place in two-cavity klystron in separate regions of the tube. On the other hand in travelling-wave tube these processes take place continuously over the entire length of the slow-wave structure (8.2.2*a*). The applied signal propagates around the turns of helix, and produces an electric field along the axis of the helix which progresses along the axial direction with phase velocity which is small compared to the velocity of light. This axial time varying electric field produces velocity modulation in an otherwise uniform velocity which is approximately equal to the phase velocity (actually slightly greater) of electrons in the electron beam moving through the helix axis. The velocity modulation tends to bunch the electrons (Fig. 8.3). As can be seen in the Fig. 8.3, electrons near the zero amplitude value are not accelerated at all; those in one-half cycle are accelerated (electric field directed against the electron flow accelerates); and those in the other half cycle are decelerated (electric field directed along the electron flow). The field obviously passess through zero at two points in each cycle. Consider the point A, electrons ahead of it are moving more slowly than the average and those behind it are moving faster than the average; then electrons tend to bunch themselves around point A. The zero-field point B (180° away) has a corresponding decrease in density associated with it. The action of the field is thus to produce bunching, similar to that of klystron, except that the electrons travel with the field, and as a consequence the field continually acts to increase the bunching *i.e.* velocity and density modulation occurs simultaneously in the same region. Since the **dc electron velocity is slightly greater than the wave velocity,** it results into a situation in which more electrons face the retarding field than the accelerating field; energy is thus lost by the electron beam and gained by the travelling-wave, resulting in an amplified amplitude signal. This cumulative interaction between the electrons and travelling waves along the axis can be discussed in terms of **space charge waves** that exist on the slow wave structure in the presence of the electronic beam. Signal Analysis is given in Annexure 8.1.

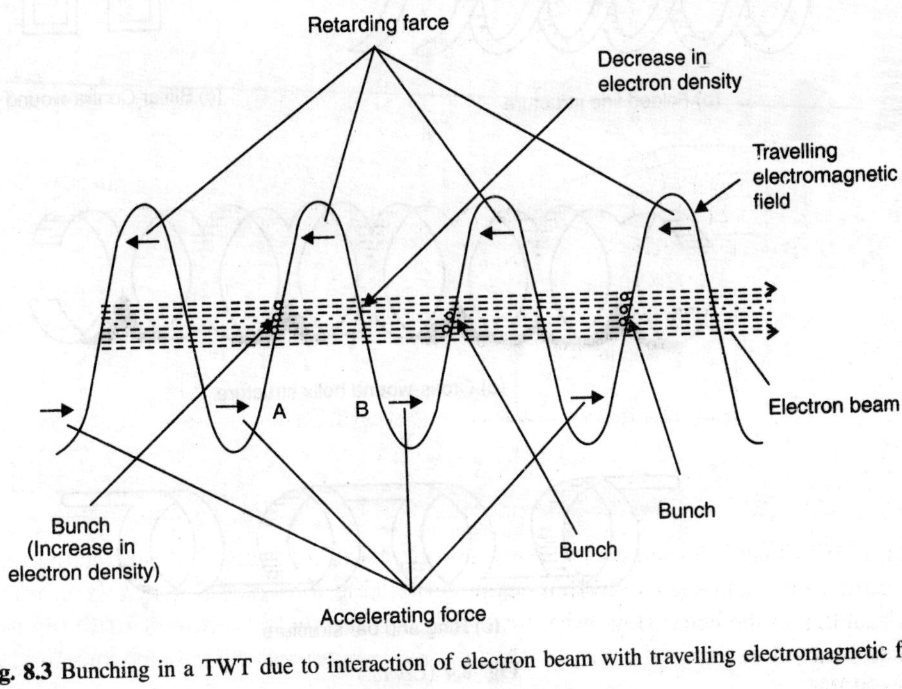

Fig. 8.3 Bunching in a TWT due to interaction of electron beam with travelling electromagnetic field

8.2.2a Types and Characteristics of Slow-wave Structure

Introduction

Slow-wave structures are integral part of many microwave tubes since their phase velocities can be made comparable to the velocity of an electron beam. This is a necessary condition for efficient interaction between the electron beam and the electromagnetic field existing in the structure. Several microwave devices like O-type, M-type, MASERs etc. make use of this interaction. Several types of slow-wave structures are shown in Fig. 8.4. In high power TWTs coupled cavity structures are used. Figure 8.5 depicts a typical coupled cavity structure.

The slow-wave structures are periodically loaded transmission line (waveguides) structures. In most cases the loading elements in transmission lines or waveguides are reactive elements. These are characterized by pass band and stop band similar to those of filters and support slow-wave propagation.

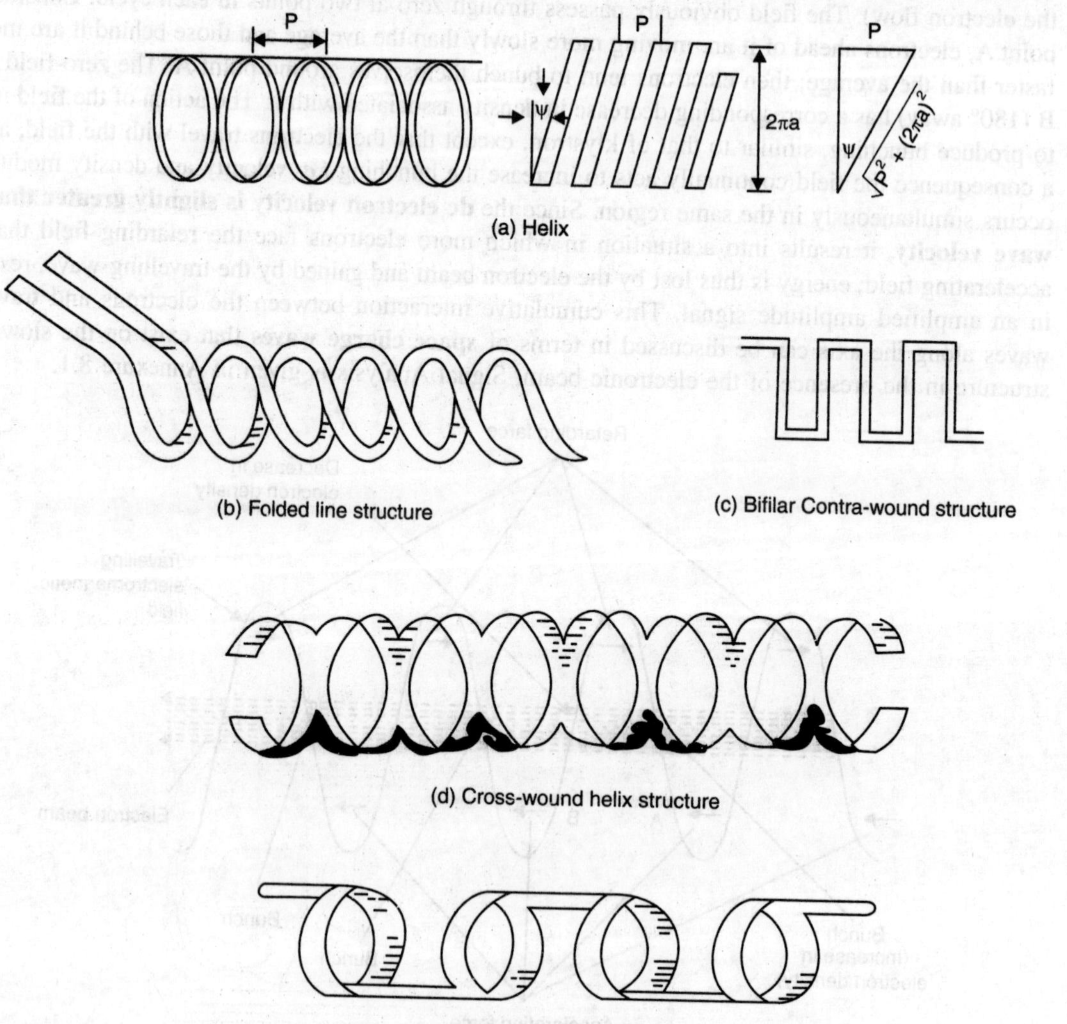

(a) Helix

(b) Folded line structure

(c) Bifilar Contra-wound structure

(d) Cross-wound helix structure

(e) Ring and bar structure

Fig. 8.4 *(Contd.)*

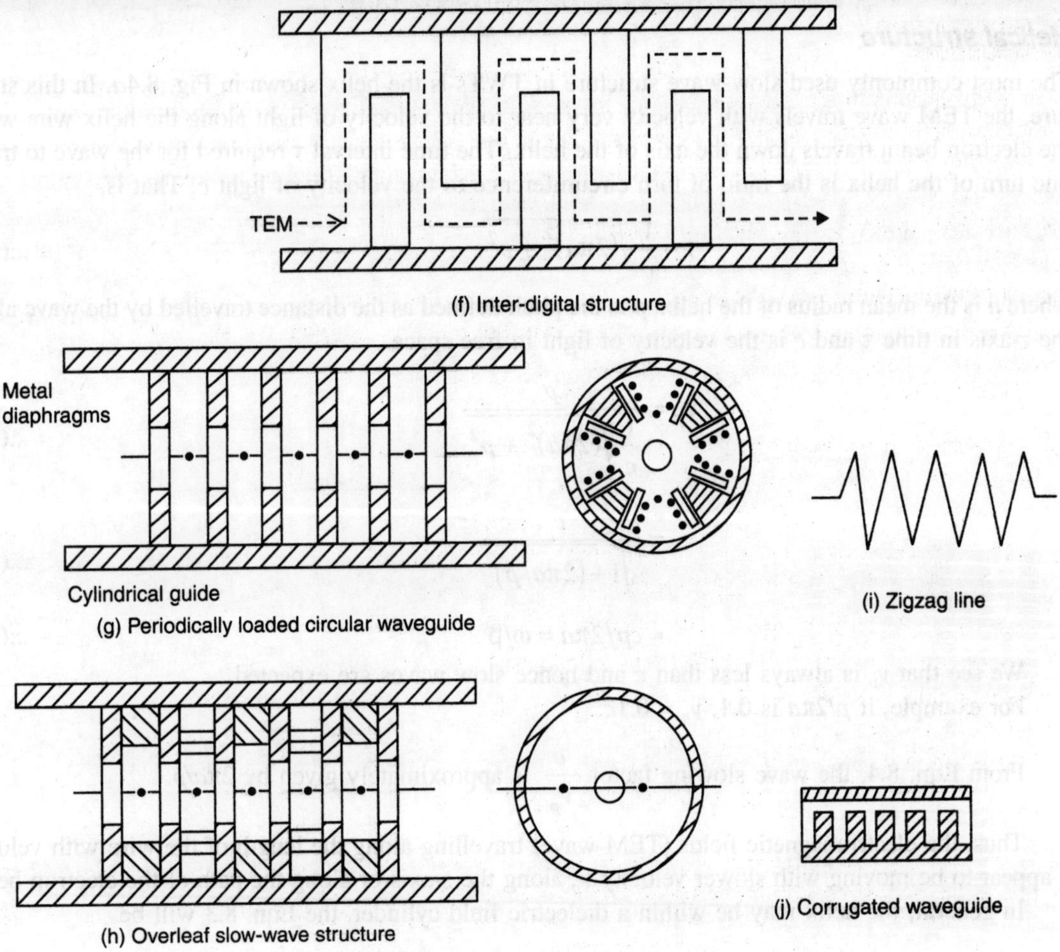

Fig. 8.4 Several types of slow-wave structures

These are usually analysed from both a circuit and field point of view. Below we discuss the general behaviour of periodically loaded transmission structures.

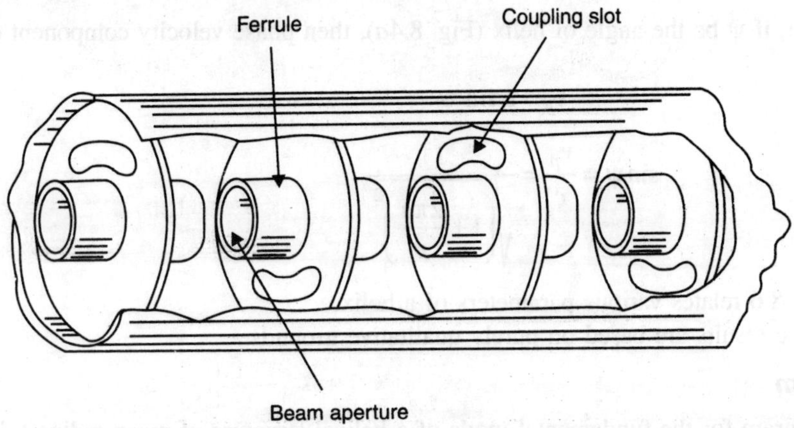

Fig. 8.5 A typical coupled cavity structure

Helical structure

The most commonly used slow-wave structure in TWTs is the helix shown in Fig. 8.4*a*. In this structure, the TEM wave travels with velocity very near to the velocity of light along the helix wire while the electron beam travels down the axis of the helix. The time interval τ required for the wave to travel one turn of the helix is the ratio of turn circumference to the velocity of light c. That is,

$$\tau = \frac{1}{c}\sqrt{(2\pi a)^2 + p^2} \qquad ...(8.1)$$

where a is the mean radius of the helix, p is the pitch defined as the distance travelled by the wave along the z-axis in time τ and c is the velocity of light in free space.

$$v_p = \frac{p}{\frac{1}{c}\sqrt{(2\pi a)^2 + p^2}} \qquad ...(8.2)$$

$$= \frac{c}{\sqrt{1 + (2\pi a/p)^2}} \qquad ...(8.3)$$

$$\approx cp/2\pi a = \omega/\beta \qquad ...(8.4)$$

We see that v_p is always less than c and hence slow waves are expected.

For example, if $p/2\pi a$ is 0.1, $v_p = 0.1c$.

From Eqn. 8.4, the wave slowing factor $\frac{c}{v_p}$ is approximately given by $2\pi a/p$.

Thus, the electromagnetic fields (TEM wave) travelling along the length of the wire with velocity c appear to be moving with slower velocity v_p along the z-axis or along the path of the electron beam.

In general, the helix may be within a dielectric field cylinder, the Eqn. 8.3 will be

$$v_p = \frac{1}{\sqrt{\mu\varepsilon\left(1 + \frac{2\pi a}{p}\right)^2}} \qquad ...(8.5)$$

Moreover, if ψ be the angle of helix (Fig. 8.4*a*), then phase velocity component of c along the z-axis will be

$$v_p = c \sin \psi$$

or

$$\sin \psi = \frac{v_p}{c} = \frac{1}{\sqrt{\left(1 + \frac{2\pi a}{p}\right)^2}} \qquad ...(8.6)$$

Equation 8.6 relates various parameters of a helix.

The above results are based on purely qualitative grounds.

ω–β diagram

The ω–β diagram for the fundamental mode of a helical structure of mean radius '*a*' and pitch '*p*' is shown in Fig. 8.6 which is obtained by using Eqn. 8.4. Helix is a periodic structure with period *p*, thus

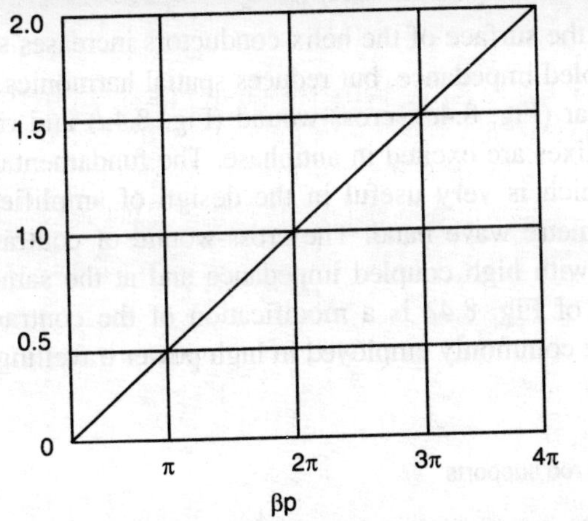

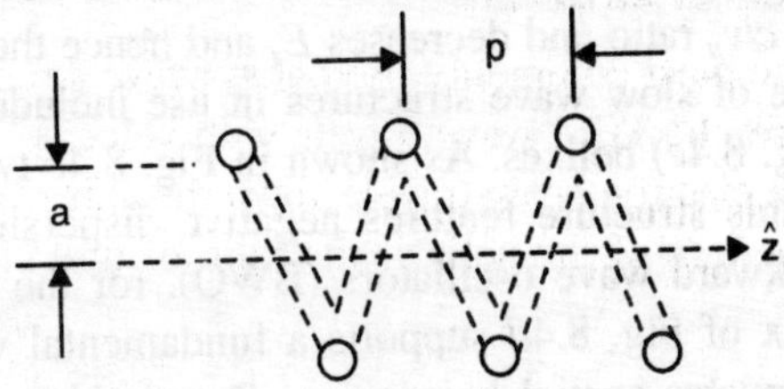

Fig. 8.6 ω–β diagram for the fundamental mode of a helical structure

its ω–β diagram repeats at $\beta p = \pm 2\pi, \pm 4\pi, \pm 6\pi + ...$intervals as any other periodic structure as shown in Fig. 8.6 and depicts the stop-band-passband characteristics. The dotted regions of the ω–β diagram (Fig. 8.7) of the helix are forbidden regions. In these regions $v_p > c$ and the helix radiates energy ($v_p > c$), (acts as a helical antenna). These regions are not useful for TWT purposes.

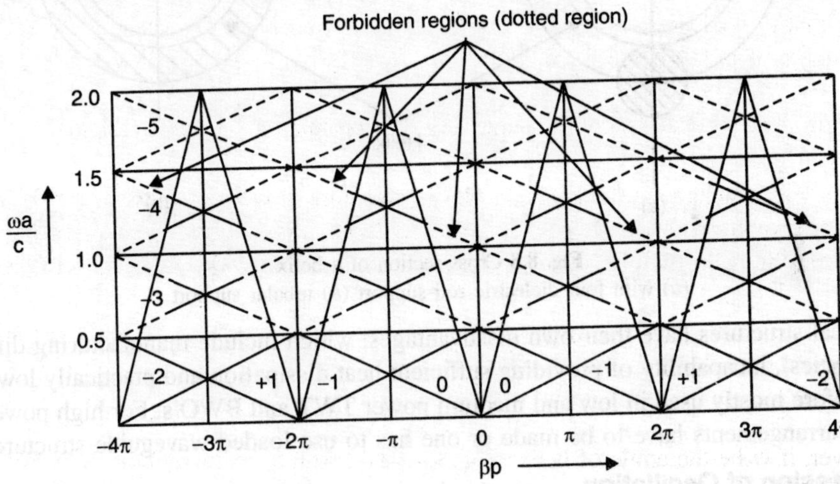

Fig. 8.7 ω–β diagram for a helix

Slow wave structures in use

The helix was first put to use in microwave travelling-wave tubes (TWT). Currently helical slow wave structures are commonly used in microwave TWT and other devices with non-resonant oscillatory systems. The main advantage of a helical slow wave system is its wideband characteristics. The phase velocity of a slow wave is almost precisely equal to the group velocity and remains practically constant within a comparatively wide range of frequencies—of the order of an octave and even wider. However, the actual designs of slow-wave helical structures differ from the model discussed in this section, because undesired spatial harmonics need to be reduced in amplitude. A typical structure is shown in Fig. 8.8. This mechanically rigid structure is obtained by mounting the helix on dielectric supports, rods or on a solid tube. The partial dielectric filling at the surface of the helix conductors increases slightly the c/v_p ratio and decreases E_z and hence the coupled impedance, but reduces spatial harmonics. Other type of slow wave structures in use include bifilar (Fig. 8.4*c*), cross-wound (Fig. 8.4*d*) and ring-bar (Fig. 8.4*e*) helixes. As shown in Fig. 8.4*c* two helixes are excited in antiphase. The fundamental wave in this structure features negative dispersion which is very useful in the design of amplifiers and backward wave oscillators (BWO), for the centimetre wave band. The cross-wound or contrawound helix of Fig. 8.4*d* supports a fundamental wave with high coupled impedance and at the same time attenuates spatial harmonics. The ring-bar helix of Fig. 8.4*e* is a modification of the contrawound structure. Ring-bar and contrawound structures are commonly employed in high power travelling-wave tubes (TWT).

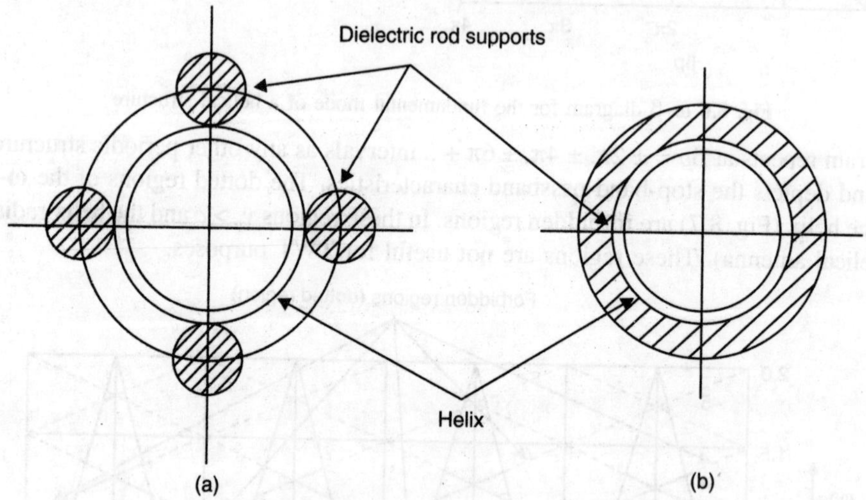

Fig. 8.8 Cross-section of a helix
(*a*) with four dielectric rod support (*b*) tubular support

Still, helical structures have their own disadvantages, which include manufacturing difficulties (at higher frequencies), incapability of providing sufficient heat dissipation and practically low c/v_p ratios. They are therefore mostly used in low and medium power TWT and BWO's. For high power operation either special arrangements have to be made or one has to use loaded waveguide structures.

8.2.3 Suppression of Oscillation

There is an **attenuator or sever** at the centre of the circuit which absorbs the reflecting wave from mismatch at the output end. Thus preventing reflections reaching the input causing oscillations. The attenuator or sever will attenuate both the forward (causing gain degradation) and reflected wave (preventing oscillations) on the helix without affecting the electron beam to start the microwave signal for further amplification in the next section. To minimize the gain degradation effect of the attenuator or sever, its length should be kept to a minimum and gain in the output section as high as possible.

8.3 THE NATURE OF FOUR PROPAGATION CONSTANTS AND GAIN CONSIDERATION

The small signal analysis given in Annexure 8.1 shows that four space charge waves are set upon the slow wave structure in the presence of the electron beam.

For the travelling wave tube, we are particularly interested in a wave that propagates in the direction of electron flow and is nearly of the electron speed, which will account for the observed gain. Making a reasonable assumption that the electron speed is equal to that of circuit wave in the absence of beam interaction, we have

$$-\gamma_0 = -j\beta_e$$

We can further assume that the propagation constant, in the presence of beam interaction γ differs from β_e by a small quantity ξ, we have

$$-\gamma = -j\beta_e + \xi \quad \text{...(8.7)}$$

With the assumption $\xi < \beta_e$, we obtain

$$\xi^3 = \frac{jZ_0 I_0 \beta_e^3}{4V_0} \quad \text{...(8.8)}$$

$$= -jC^3\beta_e^3 \quad \text{...(8.9)}$$

where $C = \left(\frac{I_0 Z_0}{4V_0}\right)^{1/3}$ is the travelling wave gain parameter first suggested by Pierce*

where

Z_0 = the helix to beam coupling impedance (characteristic impedance of the line)

I_0 = the beam current

V_0 = dc voltage equivalent to the beam velocity v_0

$\gamma_0 = j\sqrt{ZY}$ propagation constant of circuit in absence of beam

$\beta_e = \omega/v_0$, where v_0 is the dc beam velocity $\left(v_0 = \sqrt{2\eta V_0}\right)$

γ = the complex propagation constant in the presence of the beam.

$$\therefore \quad \xi = [-j]^{1/3}\beta_e C$$

$$= \delta_n \beta_e C$$

Where $\delta_n = (-j)^{1/3} = e^{-j[(\pi/2+2n\pi)/3]}$ (n = 0, 1 and 2) and its three roots are

$$\delta_1 = \frac{\sqrt{3}}{2} - j/2, \quad \text{(corresponding to } n = 0)$$

$$\delta_2 = -\frac{\sqrt{3}}{2} - j/2, \quad \text{(corresponding to } n = 1)$$

and

$$\delta_3 = j, \quad \text{(corresponding to } n = 2)$$

$$\therefore \quad -\gamma = -j\beta_e + \xi$$

$$= -j\beta_e + \delta_n \beta_e C$$

---

* Pierce, J.R., **Theory of the Beam-type Travelling Wave Tube**, Proc. I.R.E., 35, pp. 111–123, 1947.

or $$\gamma_n = j\beta_e - \delta_n \beta_e C$$

Thus the values of three propagation constants are

$$\gamma_1 = -\beta_e C\frac{\sqrt{3}}{2} + j\beta_e\left(1+\frac{C}{2}\right) \quad ...(8.10a)$$

$$\gamma_2 = \beta_e C\frac{\sqrt{3}}{2} + j\beta_e\left(1+\frac{C}{2}\right) \quad ...(8.10b)$$

$$\gamma_3 = j\beta_e(1-C) \quad ...(8.10c)$$

Thus, it is obvious that three waves are set up on the circuit in the presence of the electron beam. The waves corresponding to γ_1 and γ_2 propagate more slowly than the circuit wave when the beam is absent. From modified phase constant we can write—

$$\beta_e(1+C/2) = \frac{\omega}{v_0\Big/\left(1+\frac{C}{2}\right)} = \frac{\omega}{v_{p_1}} = \frac{\omega}{v_{p_2}},$$

i.e., $v_{p_1} < v_0$, Similarly $v_{p_2} < v_0$

Thus the first two waves have phase velocities slightly less than the beam velocity.

Amplitude of each of the wave varies as $e^{j\omega t - \gamma_n z}$. Substituting the values of γ_1 and γ_2 and arranging, we get the amplitudes of the waves with propagation constant γ_1 and γ_2 respectively varying as

$$e^{(\sqrt{3}/2)\beta_e C z} \cdot e^{j\left[\omega t - \beta_e\left(1+\frac{C}{2}\right)z\right]}, \quad \text{(Corresponding to } \gamma_1) \quad ...(8.11a)$$

and

$$e^{(-\sqrt{3}/2)\beta_e C z} \cdot e^{j\left[\omega t - \beta_e\left(1+\frac{C}{2}\right)z\right]}, \quad \text{(Corresponding to } \gamma_2) \quad ...(8.11b)$$

We note from Eqn. 8.11*a* that the wave corresponding to γ_1 is a growing wave with distance and energy flows from the electron beam to the wave whereas the wave corresponding to γ_2 (Eqn. 8.11*b*) is a decaying wave with distance and energy flows from the wave to electron beam.

Similarly, we find that the wave corresponding to γ_3 is unattenuated, but is faster than the first two waves. Its phase constant is

$$\beta_e(1-C) = \frac{\omega}{v_0/(1-C)}$$

$$= \frac{\omega}{v_{p_3}}, \text{ where } v_{p_3} = \frac{v_0}{(1-C)}, \; i.e., \; v_{p_3} > v_0$$

The amplitude of the wave with propogation constant γ_3 varies as $e^{j[\omega t - \beta_e(1-C)]}$, thus the wave is unattenuated and no energy exchange occurs between the wave and electron beam.

Fourth root wave

The propagation constant γ has four roots. The fourth root δ_4 corresponding to the backward travelling wave can be obtained by setting

$$\delta_4 = -jC^2/4$$

$$\gamma_4 = -j\beta_e - \beta_e C\delta_4$$

$$\therefore \quad \gamma_4 = -j\beta_e\left(1 - \frac{C^3}{4}\right) \qquad \text{...(8.12)}$$

This is a backward wave* set up at output and travels towards input end with a velocity slightly higher than the velocity of the electron beam. Usually the structure is perfectly matched so that there is no reflected backward wave. Attenuator is usually put at the centre of the tube which absorbs the reflected backward wave if any due to some mismatch.

Gain relation

From the above discussion it is clear that total circuit voltage is the sum of the three forward voltages corresponding to the three forward travelling waves. The input signal voltage at the input end ($z = 0$) naturally splits into three voltages that is

$$V(0) = V_1 + V_2 + V_3$$

With the proper boundary conditions** ($i_1 = 0$, $v_1 = 0$ and $V_1 = V(0)$ at $z = 0$) it can be shown that

$$V_1 = V_2 = V_3 = \frac{V(0)}{3} \qquad \text{...(8.13)}$$

and it looks logical, as at the input end there is no reason why at the start any wave may have more or less voltage than the other. Therefore, the amplitude of the growing wave

$$V(z) = \frac{V(0)}{3} e^{(\sqrt{3}/2)\beta_e Cz}$$

When $z = L$, the circuit length then

$$V(L) = \frac{V(0)}{3} e^{(\sqrt{3}/2)\beta_e CL}$$

$$= \frac{V(0)}{3} e^{\sqrt{3}\,\pi NC} \qquad \text{...(8.14)}$$

where $\beta_e L = 2\pi N$, N is the circuit length in terms of electronic wavelength, that is

$$N = \frac{L}{\lambda_e} \text{ and } \beta_e = \frac{2\pi}{\lambda_e}$$

The output power gain in decibel is

$$A_p = 10\log\left|\frac{V(L)}{V(0)}\right|^2 = 10\log\left[\frac{1}{3}e^{\sqrt{3}\pi NC}\right]^2$$

$$= -9.54 + 47.3\, NC \text{ dB.} \qquad \text{...(8.15)}$$

---

* This backward wave corresponding to Eqn. 8.12 is not the wave used in the backward wave devices.

** Pierce J.R., Travelling Wave Tubes, D. Van Nostrand Company, Inc., Princeton, N.J., 1950.

where NC is a numerical number. From Eqn. 8.15 it can be noted that power gain is proportional to the circuit length in electronic wavelength and the gain parameter C of the circuit. For larger C there is larger gain for the same physical length of the tube.

There is initial loss at the circuit input of 9.54 dB due to splitting of signal into three waves of equal magnitude and the growing wave voltage is only one-third of the total input voltage.

Performance

It is observed that with the increase of r.f. input power, r.f. output power attains a maximum value and finally starts decreasing (Fig. 8.9*a*).

The point at which r.f. output power attains maximum value is termed as **saturation point** and the gain at that point is known as saturation gain (Fig. 8.10*a*). For low r.f. inputs, the small signal gain is almost constant (Fig. 8.9*b*). **These tubes are known for their low noise, wide bandwidth and high power gain characteristics**. As such, as mentioned earlier, they have been extensively used in microwave communication systems, radar and electronic countermeasure (ECM) systems.

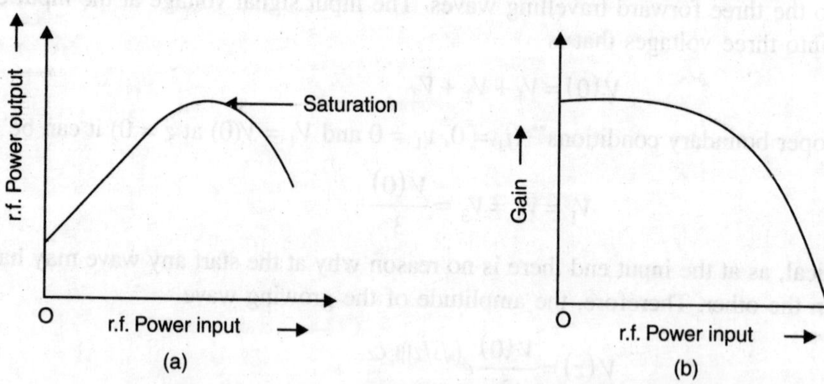

Fig. 8.9 Radiofrequency (r.f.) Power output and gain characteristics

Example 8.1. A helix travelling-wave tube operates at 4 GHz under a beam voltage $V_0 = 4$ kV, and a beam current $I_0 = 20$ mA. If the helix impedance (Z_0) is 100 ohm and circuit length $N = 30$, find the output power gain.

Solution. From Eqn. 7.34 the gain in decibel is given by

$$A_p = -9.54 + 47.3\ NC$$

and gain parameter
$$C = \left(\frac{I_0 Z_0}{4V_0}\right)^{1/3}$$

Substituting the value of I_0, Z_0, V_0, we find

$$C = \left[\frac{20 \times 10^{-3} \times 100}{4 \times 4 \times 10^3}\right]^{1/3}$$

$$= 0.5 \times 10^{-1}$$

$$\therefore \quad \text{Gain } A_p = -9.54 + 47.3 \times 0.5 \times 10^{-1} \times 30$$

$$= 61.41 \text{ dB.}$$

Example 8.2. A travelling-wave tube operates the following parameters, beam current I_0 = 50 mA, beam voltage V_0 = 2.5 kV, characteristic impedance of helix Z_0 = 7.75 Ω, circuit length N = 45, and frequency f = 8 GHz. Determine:

(*a*) The gain parameter *C*.
(*b*) The output power gain A_p in decibels.
(*c*) All four propagation constant.

Solution. Given $I_0 = 50$ mA, $V_0 = 2.5$ kV, $Z_0 = 7.75\ \Omega$, $N = 45$ and $f = 8$ GHz.

(*a*) Using the Eqn. (8.9), the gain parameter

$$C = \left(\frac{I_0 Z_0}{4V_0}\right)^{1/3}$$

$$= \left(\frac{50\times10^{-3}\times6.75}{4\times2.5\times10^{3}}\right)^{1/3}$$

$$C = 3.23 \times 10^{-2}$$

(*b*) From Eqn. (8.15) the gain in dB.

$$A_p = -9.54 + 47.3\ NC$$

$$= -9.54 + 47.3 \times 45 \times 3.23 \times 10^{-2}$$

$$A_p = 59.21 \text{ dB.}$$

(*c*) We know

$$\beta_e = \frac{\omega}{v_0} = \frac{2\pi\times8\times10^9}{0.593\times10^6\sqrt{V_0}} = \frac{2\pi\times8\times10^9}{0.593\times10^6\sqrt{2.5\times10^3}}$$

$$\beta_e = 1.69\times10^3 \text{ rads/m.}$$

Using equations (8.12) and (8.15), the propagation constant are,

$$\gamma_1 = -\beta_e C\frac{\sqrt{3}}{2} + j\beta_e\left(1+\frac{C}{2}\right)$$

$$= -1.69\times10^3\times3.23\times10^{-2}\times0.87 + j\,1.69\times10^3\times\left(1+1.61\times10^{-2}\right)$$

$$\gamma_1 = -47.49 + j\ 1717.2$$

$$\gamma_2 = \beta_e C\frac{\sqrt{3}}{2} + j\beta_e\left(1+\frac{C}{2}\right)$$

For high power operation coupled cavity travelling-wave tubes are available. In these, instead of helix, coupled cavity shown in Fig. 8.10 is used. It overcomes the thermal limitation of helix type TWT.

$$= 47.49 + j1717.2$$

$$\gamma_3 = j\beta_e(1 - C)$$

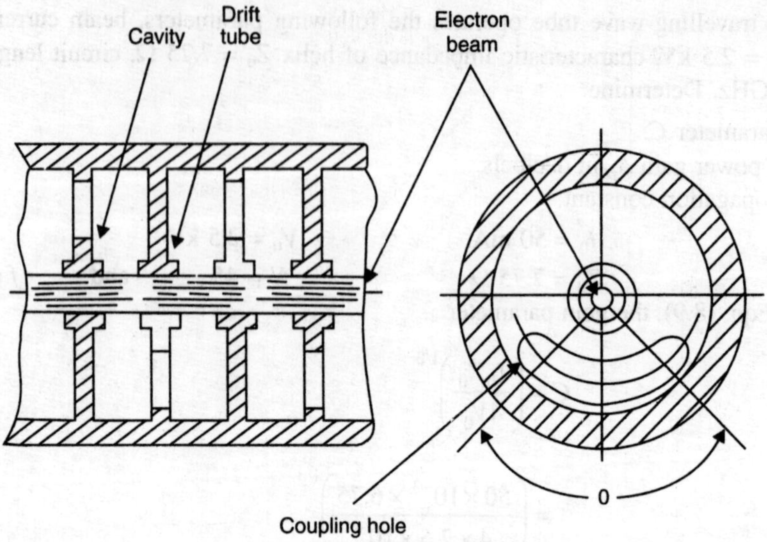

Fig. 8.10 Coupled cavity circuit

$$= j\,1.69 \times 10^3\,(1 - 3.23 \times 10^{-2})$$

$$= j\,1632.4$$

and

$$\gamma_4 = -j\beta_e\left(1 - \frac{C^3}{4}\right)$$

$$= -j\,1.69\times10^3\left(1 - \frac{(3.23\times10^{-2})^3}{4}\right)$$

$$= -j\,1690.$$

8.4 IMPROVED EFFICIENCY AND POWER

(*a*) **Derived helix:** In order to increase the efficiency, the helix pitch can be gradually decreased in output section. This slows the axial propagation of the r.f. signal in approximate synchronism with the decelerating electron beam. Such a circuit modification is known as a **velocity taper** and increases efficiency without degrading phase linearity. Helix derivative circuits are capable of significantly higher power and efficiency than the single helix circuits. Use of multistage depressed collector enhances the efficiency of the TWTs. Figure 8.11 shows comparative peak average power for helix and helix derived TWTs.

(*b*) **Depressed-Collector:** Use of single stage depressed collector and multistage depressed collector improves the overall efficiency of the TWTs. By depressing the voltage of the collector electrode to a value less than that of the circuit, the total power supplied to the TWT is reduced, resulting in an increase in overall efficiency, as defined by

$$\eta = \left[P_{\text{r.f.out}}\right] / \left[P_0 + P_h + P_{\text{sol}} - P_{\text{rec}}\right]$$

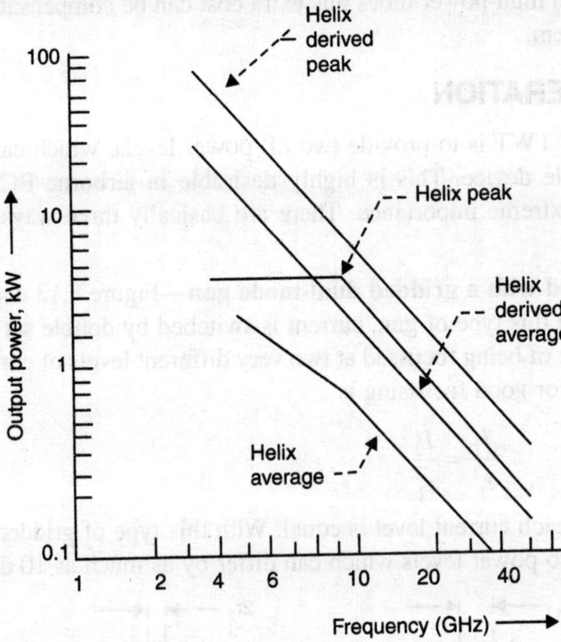

Fig. 8.11 Peak and average power for helix and helix derived TWTs

where

$P_{\text{r.f. out}}$ is the r.f. output power

P_0 is the input beam power (I_0V_0)

P_{sol} is the power supplied to the solenoid, if the TWT is solenoid focussed.

P_{rec} is the power recovered in the collector by decelerating electrons.

P_h is the heater power

Larger P_{rec} larger the efficiency

In a multistage depressed collector (Fig. 8.12), which collects spent beam electrons (which have spread of velocities) on several electrodes at different depressed voltages, large power is recovered. This results in increase in overall efficiency of a TWT to a much greater extent than in a single-stage depressed collector; one can see that a separate power supply for each collector section is required, this

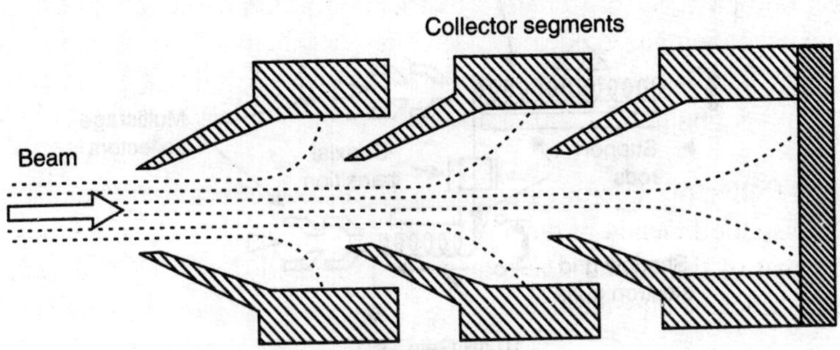

Fig. 8.12 Multistage depressed collector

will add to overall cost but in high-power tubes this extra cost can be compensated by the saving of the energy and the cooling system.

8.5 DUAL MODE OPERATION

The purpose of a dual-mode TWT is to provide two r.f. power levels, which can differ by as much as 10 dB; efficiently in a single device. This is highly desirable in airborne ECM systems, for where volume and weight are of extreme importance. There are basically three ways to achieve dual-mode operation.

1. **Beam current is varied with a gridded dual-mode gun**—Figure 8.13 depicts principle of dual-mode gun operation. In this type of gun, current is switched by double set of control grids which provide a beam capable of being focussed at two very different levels of current (for two r.f. power levels). The condition for good focussing is

$$\frac{I_1}{r_1^2} = \frac{I_2}{r_2^2}$$

i.e., current density at each current level is equal. With this type of gridded dual-mode gun TWT could be operated at two power levels which can differ by as much as 10 dB.

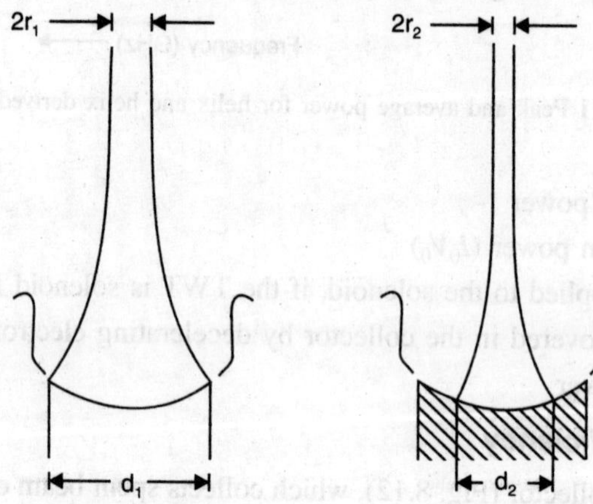

Fig. 8.13 Depicts principle of dual-mode gun operation

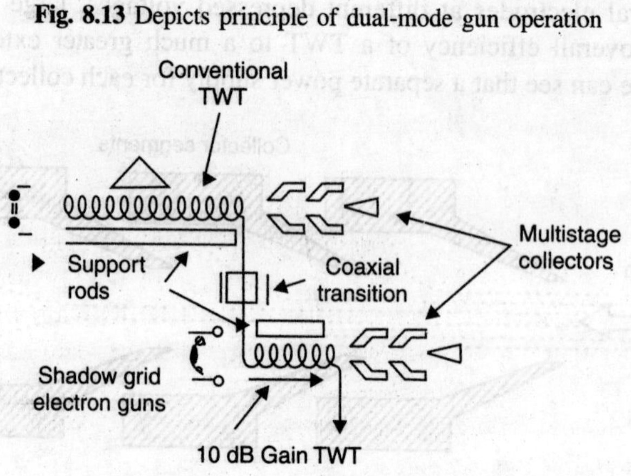

Fig. 8.14 Schematic drawing of a dual-beam, dual mode TWT

2. **Changing r.f. level**—In this method beam current and voltage are held constant and r.f. drive level is changed to produce two different r.f. power levels.
3. **Two tubes in one envelope**—In this method conventional TWT is in series with very short, low gain (10 dB), nearly lossless TWT, which is known as transparent tube (Fig. 8.14). In the low power mode, the beam in the transparent tube (10 dB device) is switched off and the signal from the conventional TWT passes through it with no amplification and little loss. In the high power mode, transparent tube is turned on to provide additional 10 dB gain. Although this device has the capability of providing a full 10 dB difference in two power levels, but it is more complex and heavier than the single device.

Higher overall efficiency is obtained by using a multistage depressed collector in dual-mode TWT.

REVIEW QUESTIONS AND PROBLEMS

1. Draw a labelled schematic diagram of Helix TWT and show that output power gain of TWT is G = –9.54 + 47.3 CN dB (Starting with the assumption that there are three travelling waves on TWT).
2. Describe with necessary theory the amplification of microwaves by O-type travelling-wave tube. In what respect is it superior to two cavity amplifier?
3. Discuss the enhancement of overall efficiencies of TWT.
4. Explain the bunching of electrons in TWT, with suitable sketch.
5. A TWT has the following parameters:
   $V_0 = 3$ kV, $I_0 = 4$ mA, $f = 10$ GHz, $Z_0 = 30$ ohm and $N = 50$ in wavelength. Calculate: (*i*) the gain parameter; and (*ii*) power gain in dB.
6. Calculate the interaction length for a TWT to give a 45 dB gain operating in the frequency band 5.9–7.4 GHz with beam voltage 7.5 kV, beam current 200 mA and interaction impedance $Z_0 = 16\ \Omega$.
7. Compute the gain parameter and output power gain of TWT which has the following parameters $V_0 = 15$ kV, $I_0 = 3$ A, $N = 10$ in wavelength, $Z_0 = 50\ \Omega$.
8. A travelling-wave tube (TWT) operates under the following parameters:
   Beam voltage $V_0 = 3$ kV, Beam current $I_0 = 30$ mA, characteristic impedance of helix $Z_0 = 10\ \Omega$, Circuit length = 50 in wavelength terms and Frequency 10 GHz. Determine the (*i*) gain parameter *C* and (*ii*) the power gain in dB.
9. Discuss construction, operation and practical consideration of helix type TWT. What are its different applications?
10. Describe the construction and working of a slow wave type helix travelling-wave tube. Mention different methods of focussing and state two of its important applications.
11. Explain various types of slow wave structures used for—Travelling-wave tube applications.
12. Describe the principle of operation of a travelling-wave tube. Explain why an attenuator is some time placed midway along the slow-wave structure. What is the main advantage of the TWT amplifier compared to the klystron amplifier?
13. A TWT operates under the following parameters:
    Beam voltage = 20 V
    Beam current = 25 A

Characteristic impedance of helix = 15 ohms.

Circuit length = 50, electronic wavelength, frequency = 9 GHz. Find: (*i*) gain parameter (*ii*) output power gain and (*iii*) value of β_e.

14. A helix travelling-wave tube operates at 4 GHz under a beam voltage of 10 kV and beam current 500 mA. If the helix impedance is 25 ohm and the interaction length is 20 cm. Find the output power gain in dB.

15. An O-type travelling-wave tube operates at 3 GHz. The slow-wave structure has a pitch angle of 6.2°. Determine the propagation constant of the travelling-wave in the tube. It is assumed that the tube is lossless.

16. Draw a labelled schematic diagram of Helix TWT and show that output power gain of TWT is

$$A_p = -9.54 + 47.3 \text{ NC dB}.$$

Annexure 8.1

The purpose of the following analysis is to determine the propagation constants of the space-charge waves that can exist on the slow-wave structure in the presence of an electron beam and to calculate the gain of the device from these constants.

(a) Electronic equation: r.f. current produced by the circuit field

R.F. current (convection current) induced by the circuit field in the electron beam at any point along the slow-wave structure (helix) is determined in terms of assumed electric field using the equation of continuity, the definition of current density and equation of motion. The equation obtained for convection current in terms of voltage is known as electronic equation.

The electron velocity $v(z)$, the volume charge density $\rho(z)$, the beam current $I(z)$ and the axial electric field $E(z)$ may be expressed as

$$v(z) = v_0 + v_1 e^{(j\omega t - \gamma z)} \quad \text{...(A8.1)}$$

$$\rho(z) = \rho_0 + \rho_1 e^{(j\omega t - \gamma z)} \quad \text{...(A8.2)}$$

$$I(z) = I_0 + i_1 e^{(j\omega t - \gamma z)} \quad \text{...(A8.3)}$$

and

$$E(z) = E_{z1} e^{(j\omega t - \gamma z)} \quad \text{...(A8.4)}$$

($E_0 = 0$, since dc potential V_0 is constant along z-direction $\therefore \dfrac{\partial V_0}{\partial z} = E_0 = 0$)

Here $\gamma = \alpha_e + i\beta_e$ is the propagation constant of the axial waves. v_1, ρ_1 and i_1 are r.f. velocity, charge density and beam current respectively and are much smaller than corresponding dc quantities v_0, ρ_0 and I_0.

The beam current is the product of volume charge density $\rho(z)$, the velocity $v(z)$ and beam area A. Thus $I(z)$ could be written as

$$I(z) = I_0 + i_1 e^{(j\omega t - \gamma z)}$$

$$= \rho(z)\, v(z) \times \text{beam area } A \quad \text{...(A8.5)}$$

Substituting for $\rho(z)$ and $v(z)$ from Eq. A8.2 and A8.1 respectively in Eq. A8.5 and arranging, we get

$$I(z) \approx \left[\rho_0 v_0 + (\rho_0 v_1 + \rho_1 v_0) e^{(j\omega t - \gamma z)} + \rho_1 v_1 e^{2(j\omega t - \gamma z)}\right] A \quad \text{...(A8.6)}$$

The last term in the Eqn. A8.6 is a second order term involving the product of two small quantities ρ_1 and v_1 and can be neglected. This yields ac (r.f.) component of the beam current (since $I_0 = \rho_0 v_0 A$).

$$i_1 = (\rho_0 v_1 + \rho_1 v_0) A \quad \text{...(A8.7)}$$

From the **current continuity equation**

$\nabla \cdot J = -\partial\rho/\partial t$, we obtain

$$\nabla \cdot \frac{i_1}{A} = -\frac{\gamma i_1}{A} = -j\omega\rho_1 \quad \text{...(A8.8)}$$

$$\therefore \quad \rho_1 = -\frac{j\gamma i_1}{\omega A} \quad \text{...(A8.9a)}$$

Since
$$i_1 = (\rho_0 v_1 + \rho_1 v_0) A$$
$$= \left(\rho_0 v_1 - \frac{j\gamma i_1}{\omega A} v_0\right) A, \text{ (using Eqn. A8.9}a)$$

$$\therefore \quad i_1\left[1 + \frac{j\gamma v_0}{\omega}\right] = \rho_0 v_1 A$$

The expression for i_1, therefore, is derived as

$$i_1 = \frac{\omega \rho_0 v_1 A}{(\omega + j\gamma v_0)} \qquad \text{...(A8.9}b)$$

$$= \frac{j\beta_e \rho_0 A}{(j\beta_e - \gamma)} v_1 \qquad \text{...(A8.10)}$$

where $\beta_e = \omega/v_0$ is defined as the phase constant of the velocity modulated electron beam and v_0 is the dc velocity of the electron beam.

From the equation of motion for the electron

$$\frac{dv(z)}{dt} = \eta E(z) = \eta \frac{\partial V}{\partial z} = \eta \frac{\partial V_1}{\partial z} \qquad \text{...(A8.11)}$$

since the dc potential V_0 is constant along the flow of the electron beam, so that $\frac{\partial V_0}{\partial z} \equiv 0.$

The left hand side of Eqn. A8.11 can be written as:

$$\frac{dv(z)}{dt} = \frac{dv_1}{dt} = \frac{\partial v_1}{\partial t} + \frac{\partial v_1}{\partial z}\frac{\partial z}{\partial t}$$
$$= j\omega v_1 - \gamma v_1 (v_0 + v_1)$$
$$\approx j\omega v_1 - \gamma v_1 v_0$$
$$\approx (j\omega - \gamma v_0) v_1$$

$\therefore$ The Eqn. A8.11 can be written as

$$(j\omega - \gamma v_0) v_1 = -\eta \gamma V_1$$

$$\therefore \quad v_1 = \frac{-\gamma \eta V_1}{v_0 (j\beta_e - \gamma)} \qquad \text{...(A8.12)}$$

Using Eqn. A8.12 in Eqn. A8.10 we have

$$i_1 = \frac{j\beta_e \rho_0 A}{(j\beta_e - \gamma)} \cdot \frac{-\gamma \eta V_1}{v_0 (j\beta_e - \gamma)}$$

$$i_1 = \frac{j I_0 \beta_e \gamma V_1}{2V_0 (j\beta_e - \gamma)^2} \qquad \text{...(A8.13)}$$

where $I_0 = -\rho_0 v_0 A$ and $v_0^2 = 2\eta V_0$. Equation A8.13 is called the **electronic equation.** It relates the r.f. beam current (convection current induce by the axial electric field) to the r.f. voltage on the slow-wave structure.

(b) Circuit equation : Electric field produced in the circuit by the r.f. current in the beam

The analysis is based on transmission line theory approach. Figure A8.1 depicts the equivalent circuit of the helix (helix is represented by lossless transmission line with series impedance Z per unit length and shunt admittance Y per unit length) in the vicinity of an electron beam which induces a current in the transmission line.

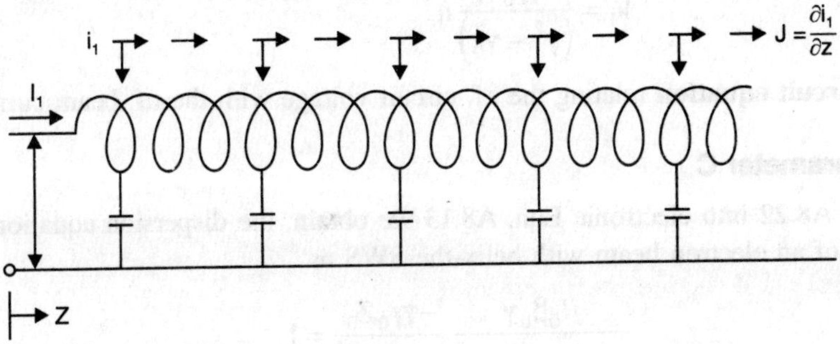

Fig. A8.1 Equivalent circuit of helix in the vicinity of an electron beam

From the transmission line relation,

$$\gamma_0 = j\sqrt{ZY} \qquad \text{...(A8.14)}$$

and

$$Z_0 = \sqrt{Z/Y}$$

where γ_0 is the propagation constant of the uncoupled helix and Z_0 is the characteristic impedance of the transmission line equivalent of the slow-wave circuit (helix).

From Eqn. A8.14

$$Z = -j\gamma_0 Z_0 \; ; \; Y = -j\gamma_0/Z_0 \qquad \text{...(A8.15)}$$

The expressions for voltage and current per unit length for the network shown in Fig. A8.1 are, respectively

$$\frac{\partial V_1}{\partial z} = -jZI_1 \qquad \text{...(A8.16)}$$

$$\frac{\partial I_1}{\partial z} = -jYV_1 + J \qquad \text{...(A8.17)}$$

where J is the impressed current per unit length, due to coupling between electron beam and circuit, and is given by

$$J = -\frac{\partial i_1}{\partial z} \qquad \text{...(A8.18)}$$

where $i_1 = (\text{beam area } A) \times (\rho_0 v + \rho_1 v_0)$ (See Eqn. A8.7)

Assuming the various ac quantities as $e^{j(\omega t - \gamma z)}$, from Eqns. A8.16 and A8.17, we get

$$-\gamma I_1 = -jYV_1 + \gamma i_1 \quad \text{...(A8.19)}$$

$$-\gamma V_1 = -jZI_1 \quad \text{...(A8.20)}$$

Substituting I_1 from Eq. A8.20 in Eq. A8.10, and re-arranging we obtain

$$\left(\gamma^2 + ZY\right)V_1 = \left(\gamma^2 - \gamma_0^2\right)V_1 = -j\gamma Z i_1 \quad \text{...(A8.21)}$$

$$\therefore \qquad V_1 = \frac{-\gamma\gamma_0 Z_0}{\left(\gamma^2 - \gamma_0^2\right)} i_1 \quad \text{...(A8.22)}$$

This is the **circuit equation** relating the r.f. circuit voltage with the r.f. beam current i_1.

(c) The gain parameter C

Substituting Eqn. A8.22 into electronic Eqn. A8.13 we obtain, the dispersion equation describing the TWT interaction of an electron beam with helix-the SWS as

$$\frac{jI_0\beta_0\gamma}{2V_0\left(j\beta_e - \gamma\right)^2} \frac{-\gamma\gamma_0 Z_0}{\left(\gamma^2 - \gamma_0^2\right)} = 1$$

$$-\frac{j\left(Z_0 I_0/2V_0\right)\gamma_0\beta_e\gamma^2}{\left(j\beta_e - \gamma\right)^2\left(\gamma^2 - \gamma_0^2\right)} = 1 \quad \text{...(A8.23)}$$

In this equations

Z_0 = the helix to beam coupling impedance (characteristic impedance of the line)
I_0 = the beam current
V_0 = d.c. voltage equivalent to the beam velocity v_0.
$\gamma_0 = j\sqrt{ZY}$ propagation constant of circuit in absence of beam
$\beta_e = \omega/v_0$ where v_0 is the dc beam velocity $\left(v_0 = \sqrt{2\eta V_0}\right)$
γ = the complex propagation constant in the presence of the beam.

Rearranging A8.23, we can write

$$\frac{jI_0 Z_0\gamma_0\gamma^2\beta_e}{2V_0\left(j\beta_e - \gamma\right)^2\left(\gamma_0^2 - \gamma^2\right)} = 1 \quad \text{...(A8.24)}$$

or

$$\left(\gamma_0^2 - \gamma^2\right)\left(j\beta_e - \gamma\right)^2 = \frac{j\gamma^2\gamma_0\beta_e Z_0 I_0}{2V_0} \quad \text{...(A8.25)}$$

Equation 8.25 is of fourth order in γ and thus has four roots. It means that, four waves are set up on the circuit in the presence of the beam. The exact solutions of Eqn. A8.25 can be obtained by numerical methods and a digital computer. However, the approximate solutions are given below.

For the travelling wave tube, we are particularly interested in a wave that propagates in the direction of electron flow and is nearly of the electron speed, which will account for the observed gain. Making a reasonable assumption that the electron speed is equal to that of circuit wave in the absence of beam interaction, we have

$$-\gamma_0 = -j\beta_e$$

If we further assume that γ differs from β_e by a small quantity ξ, we have

$$-\gamma = -j\beta_e + \xi \qquad \text{...(A8.26)}$$

Substitution of Eqn. A8.26 in Eqn. A8.24, results in an expression

$$\frac{-Z_0 I_0 \beta_e^2 \left(-\beta_e^2 - 2j\beta_e\xi + \xi^2\right)}{2V_0\left(2j\beta_e\xi - \xi^2\right)\xi^2} = 1 \qquad \text{...(A8.27)}$$

In accordance with the assumption $\xi < \beta_e$ we may neglect the terms $2j\,\beta_e\,\xi$ and ξ^2 in numerator and ξ^4 in denominator, we obtain

$$\xi^3 = -\frac{jZ_0 I_0 \beta_e^3}{4V_0}$$

$$= -jC^3\beta_e^3 \qquad \text{...(A8.28)}$$

$$\therefore \qquad \xi = [-j]^{1/3}\,\beta_e C$$

where $C = \left(\dfrac{I_0 Z_0}{4V_0}\right)^{1/3}$ is the travelling wave gain parameter first suggested by Pierce.*

---

* Pierce, J.R., **Theory of the Beam-type Travelling Wave Tube,** Proc. I.R.E., 35, pp. 111–123, 1947.

CHAPTER 9

Microwave Tubes: M-type Tubes (Magnetrons)

9.1 INTRODUCTION

In the last two chapters O-type devices, resonant and non-resonant, were discussed. As mentioned in Section 7.2, another class of microwave tubes is M-type tubes. These are widely used as high power (pulse as well as CW) sources of microwave power with limited electronic tuning and modulation capabilities. In these tubes electron motion takes place in a region where static electric and magnetic fields are present at right angles to each other. It is in contrast with O-type in which magnetic field (along the length of the tube) is used merely to focus the electron beam *i.e.*, preventing the space-charge spreading as beam travels the length of the tube. On the other hand, in M-type tubes beam dynamics requires the presence of crossed magnetic fields which strongly affects the radio frequency behaviour of the electrons under the action of radio frequency field. There are variety of M-type tubes that are tabulated in Table 7.3. These devices include oscillators like magnetron (the earliest source of significant

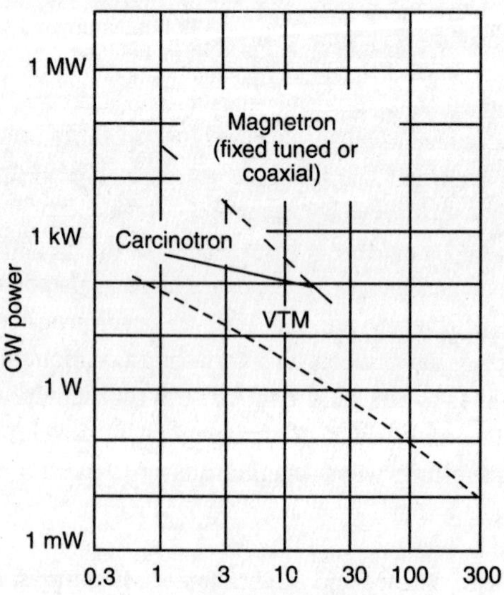

Fig. 9.1 Power versus frequency performance of magnetron and carcinotron (BWCFO)

amount of microwave power), M-type backward wave oscillators and amplifiers with different format, namely linear and circular, with various configurations such as forward/backward wave, injected-beam/distributed emission type and with/without re-entrant beam.

Performance

These devices are characterized by high power, high efficiency, adequate bandwidth and comparatively high noise output and low gain as compared to O-type tubes. Figures 9.1 and 9.2 summarize the power versus frequency performance of M-type oscillators and amplifiers.

The gyrotrons are relatively new type of microwave, millimetre are sub-millimetre crossed field devices.

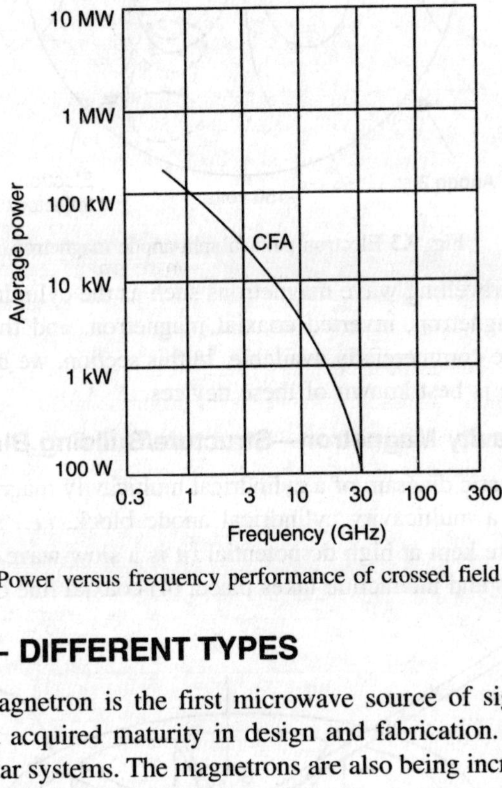

Fig. 9.2 Power versus frequency performance of crossed field amplifiers

9.2 MAGNETRONS – DIFFERENT TYPES

As mentioned earlier, a magnetron is the first microwave source of significant amounts of power. During the World War II it acquired maturity in design and fabrication. Magnetrons have been used extensively in variety of radar systems. The magnetrons are also being increasingly and widely used for cooking, industrial heating, diathermy equipments, etc.

The magnetrons are grouped into three types, viz. (*i*) the negative resistance magnetron (split-anode magnetron); this type of magnetron uses a negative resistance between two anode segments (Fig. 9.3) and Power output, efficiency are low; (*ii*) the cyclotron-frequency magnetron; it operates under the influence of synchronism between an alternating component of electric field and a periodic oscillations in a direction parallel to the field and (*iii*) the multicavity travelling wave magnetron. The first two were developed in the early days of the magnetron development. The reader is referred to literature*. The multicavity travelling wave magnetrons are superior in performance and technically highly developed devices.

---

* 1. Brownwell, A.B. and Beam, R.E., **Theory and Application of Microwaves,** McGraw-Hill, Inc. New York, 1947.
2. Collins, G.B., **Microwave Magnetrons,** MIT Radiation Laboratory Series, Vol. 6, McGraw-Hill, Inc., New York, 1948.

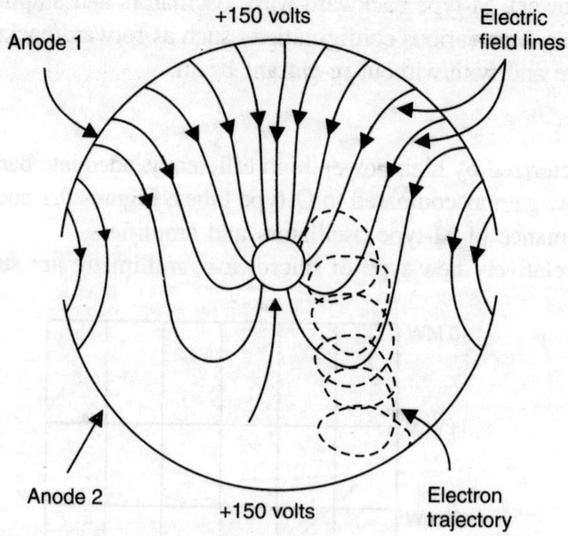

Fig. 9.3 Electron path in split-anode magnetron

Variety of multicavity travelling wave magnetrons such as the cylindrical magnetron, coaxial magnetron, voltage tunable magnetron, inverted coaxial magnetron, and the frequency-agile magnetron etc., were developed and are commercially available. In this section, we discuss cylindrical multicavity magnetron oscillator which is best known of these devices.

9.2.1 Cylindrical Multicavity Magnetron—Structure/Building Blocks

Figure 9.4 shows the schematic diagram of a cylindrical multicavity magnetron oscillator. It consists of (1) a coaxial cathode; (2) a multicavity cylindrical anode block, *i.e.,* several reentrant cavities are connected to the gap and are kept at high dc potential (it is a slow-wave structure); (3) cathode anode space where electron moves and interaction takes place; (4) coaxial line output (means of coupling the

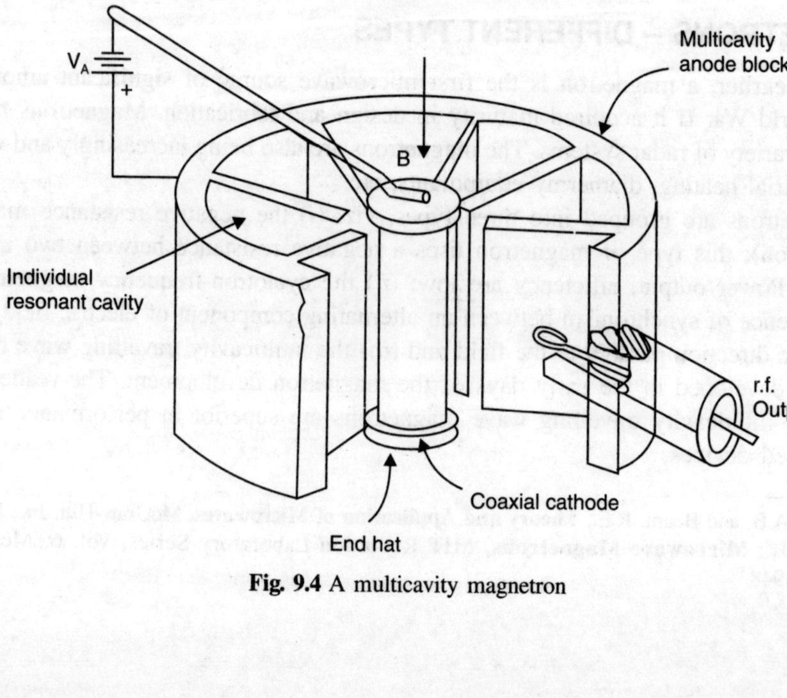

Fig. 9.4 A multicavity magnetron

generated microwave power to the outside); and (5) a permanent magnet (not shown in Fig. 9.4) to provide magnetic field along the axial length of the cathode and perpendicular to radial dc electric field.

Figure 9.5 shows various types of magnetron anode blocks, viz. (*a*) hole and slot; (*b*) vane; (*c*) slot; (*d*) rising sun; and (*e*) strap blocks.

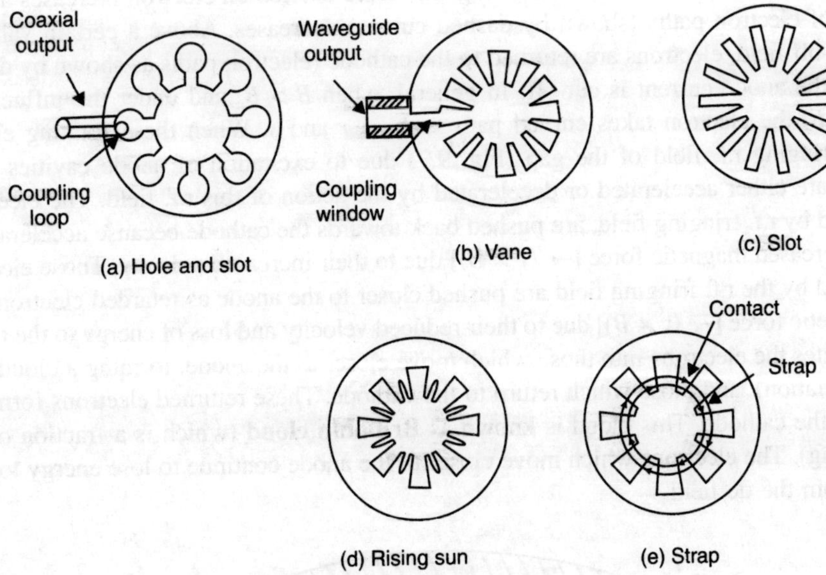

Fig. 9.5 Different types of anode blocks

9.2.2 Electron Trajectories and Operation Mechanism

In the absence of magnetic field, the electron emitted (assuming emission with zero velocity) from the cathode moves along the radial direction as shown by '*p*' in the Fig. 9.6 on account of the radial anode-

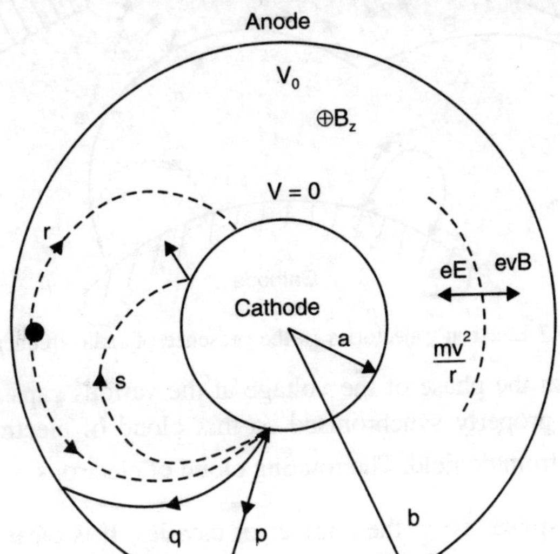

Fig. 9.6 Electron trajectories at various magnetic fields

to-cathode electric field. When a small magnetic field B is applied at right angle to radial electric field (axial to cathode cylinder) the electron trajectory bends and follows a curved path as shown by curve q in the Fig. 9.6. The force exerted on the electron by the magnetic field is given by

$$F = -e\,(v \times B)$$

As, the magnetic field is increased the magnetic force exerted on electron increases and the radius of curvature of electron paths (shown by dashed curves) decreases. Above a certain value of the B_c, called the cut-off field, electrons are returned to the cathode (electron paths as shown by dashed curves r and s) and the anode current is cut-off. In general, when $B > B_c$ and under the influence of radial electric field E_r the electron takes curved path such as r and s. When these rotating electrons pass through the fringing r.f. field of the gap (Fig. 9.7) due to excitation of anode cavities by r.f. noise voltage, they are either accelerated or decelerated by the action of this r.f. field. The electrons, which are accelerated by r.f. fringing field, are pushed back towards the cathode because accelerated electrons experience increased magnetic force [$-e\,(v \times B)$] due to their increased velocity. Those electrons which are decelerated by the r.f. fringing field are pushed closer to the anode as retarded electrons experience reduced magnetic force [$-e\,(v \times B)$] due to their reduced velocity and loss of energy to the r.f. field. This process separates the electrons into those which move closer to the anode, forming a cloud of electrons (density modulation), and those which return to the cathode. These returned electrons form an electron cloud around the cathode. This cloud is known as **Brillouin cloud** (which is a fraction of the anode-cathode spacing). The electrons which move closer to the anode continue to lose energy to the r.f. field and gain it from the dc field.

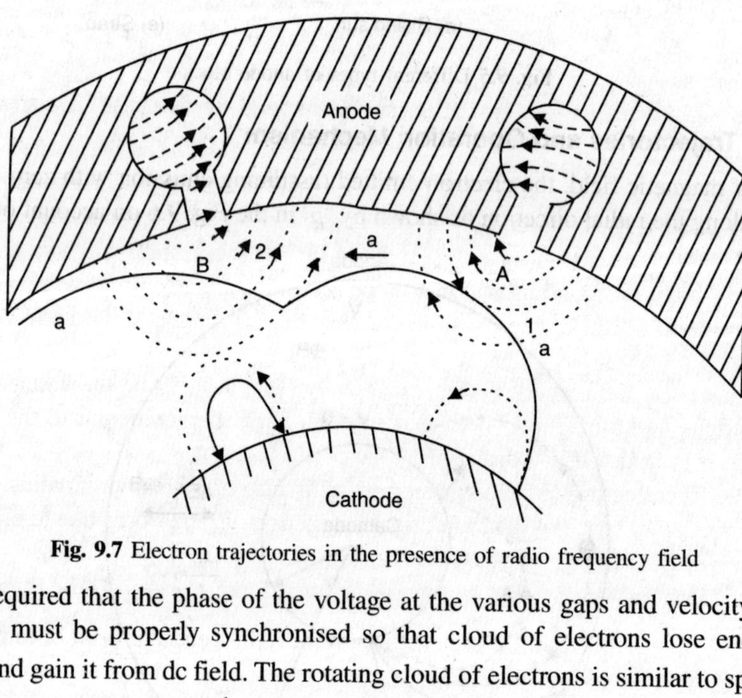

Fig. 9.7 Electron trajectories in the presence of radio frequency field

Thus, it is required that the phase of the voltage at the various gaps and velocity of the electrons from gap to gap must be properly synchronised so that cloud of electrons lose energy to the radio frequency field and gain it from dc field. The rotating cloud of electrons is similar to spokes of electrons (Fig. 9.8). The number of spokes is $\frac{1}{2}$ the number of cavities. It is clear that the electronic processes in magnetron are really the same as of travelling wave tube. The sorting of electrons takes place due to static crossed electric and magnetic fields, and transverse radio frequency field plays an important role.

The sorting of electrons (confined to spokes) is called **phase focussing.** The magnetron could also be operated as an amplifier in which radio frequency field grows like that in travelling wave tube.

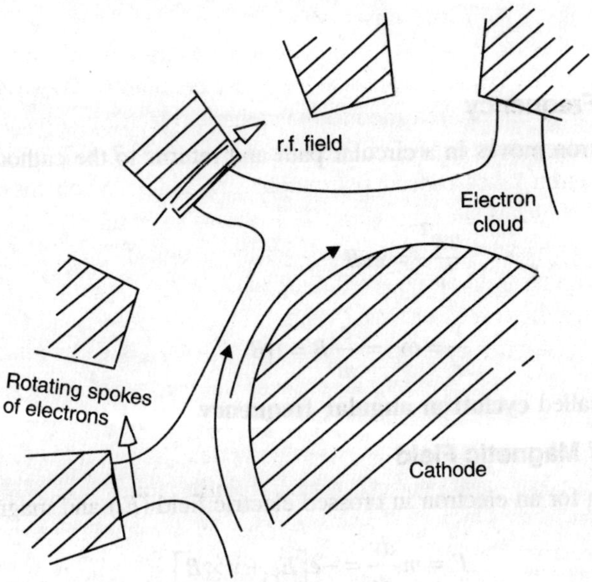

Fig. 9.8 Electron spokes

9.2.3 Hull Cut-off Magnetic Field and Cut-off Voltage in Cylindrical Magnetron

From the discussion in Section 9.2.2, it can be concluded that for a specific applied voltage on anode there is corresponding level of magnetic field B necessary for electrons emitted from the cathode to just graze the anode block and travel in interaction space between the cathode and anode. The anode current will be zero. This magnetic field is known as **Hull cut-off magnetic field** and conversely for a specific magnetic field there is a corresponding anode voltage for which anode current will be zero. This voltage is known as **Hull cut-off voltage.** The relations for Hull cut-off magnetic field and voltage are derived below.

Let b and a be anode and cathode radii, respectively (Fig. 9.6), V_0 be the voltage applied between the anode and cathode and a magnetic field B is into the page (perpendicular to the direction of v) as shown in Fig. 9.6. It is in positive z-direction.

There are three forces acting on an electron moving in curved path having radius of curvature r and velocity v in the interaction space of the cylindrical magnetron, *viz.*, (*i*) force due to electric field $(-eE_r)$, force due to magnetic field $[-e\,(v \times B)]$ and the centrifugal force $\left(\dfrac{mv^2}{r}\right)$. These three forces are radial as shown in Fig. 9.6. In equilibrium, for the electron to move in a circular orbit of radius r, the equation of motion is given by

$$\frac{mv^2}{r} + eE_r = evB \qquad ...(9.1)$$

where $E_r = \dfrac{V_0}{-r\ln\dfrac{b}{a}}$; here, b and a are anode and cathode radii respectively and V_0 is applied voltage on anode.

9.2.3 (a) Cyclotron Frequency

When $E_r = 0$, the electron moves in a circular path and returns to the cathode.

From Eqn. 9.1.

$$\frac{mv^2}{r} = evB \qquad \text{...(9.2)}$$

or

$$\frac{v}{r} = \omega_c = \frac{e}{m}B = \eta B \qquad \text{...(9.3)}$$

In this equation ω_c is called **cyclotron angular frequency.**

9.2.3 (b) Hull Cut-off Magnetic Field

The equation of motion for an electron in crossed electric field (E_r) and magnetic field (B) is given by

$$F = m\frac{dv}{dt} = -e\left[E_r + v \times B\right] \qquad \text{...(9.4)}$$

where v is the electron velocity.

As we are discussing the motion of an electron in cylindrical multicavity magnetron, the equation of motion (Eqn. 9.4) in cylindrical coordinates are

$$\frac{d^2r}{dt^2} - r\left(\frac{d\theta}{dt}\right)^2 = \eta E_r - \eta r\frac{d\theta}{dt}B_z \qquad \text{...(9.5}a\text{)}$$

and

$$\frac{1}{r}\frac{d}{dt}\left(r^2\frac{d\theta}{dt}\right) = \eta B_z\frac{dr}{dt} \qquad \text{...(9.5}b\text{)}$$

Where $B = B_z$ is assumed in the $+z$-direction

Equation 9.5b can be written as

$$\frac{d}{dt}\left(r^2\frac{d\theta}{dt}\right) = \frac{1}{2}\omega_c\frac{d(r^2)}{dt} \qquad \text{...(9.6)}$$

where $\omega_c = \eta B_z$ is the cyclotron frequency. Integration of Eq. (9.6), gives

$$r^2\frac{d\theta}{dt} = \frac{1}{2}\omega_c r^2 + K \qquad \text{...(9.7)}$$

where K is an integration constant. Applying boundary condition $r = a$, $\dfrac{d\theta}{dt} = 0$ in Eqn. 9.7, we get

$$K = -\frac{1}{2}\omega_c a^2$$

∴ Equation 9.7 becomes

$$r^2 \frac{d\theta}{dt} = \frac{1}{2}\omega_c r^2 - \frac{1}{2}\omega_c a^2$$

∴ The angular velocity is expressed by

$$\frac{d\theta}{dt} = \frac{1}{2}\omega_c \left(1 - \frac{a^2}{r^2}\right) \qquad ...(9.8)$$

The kinetic energy of the electrons is given by

$$eV = \frac{1}{2} m \left[\left(\frac{dr}{dt}\right)^2 + \left(r\frac{d\theta}{dt}\right)^2\right] \qquad ...(9.9)$$

as electron has a tangential as well as a radial velocity.

For grazing the anode $r = b$, $V = V_0$, and $\frac{dr}{dt} = 0$; Eqn. 9.9 becomes

$$b^2 \left(\frac{d\theta}{dt}\right)^2 = \frac{2e}{m} V_0 = 2\eta V_0 \qquad ...(9.10)$$

Substitution of Eqn. 9.8 into Eqn. 9.10 results in

$$b^2 \left[\frac{1}{2}\omega_c \left(1 - \frac{a^2}{b^2}\right)\right]^2 = 2\eta V_0 \qquad ...(9.11)$$

Substituting $\omega_c = \eta B_c$ at grazing and solving Eqn. 9.11 for B_c, we obtain

$$B_c = \frac{(8V_0/\eta)^{1/2}}{b\left(1 - \frac{a^2}{b^2}\right)} \qquad ...(9.12a)$$

The relation 9.12 is known as **Hull cut-off** magnetic field. Substituting the value of $\eta = 1.76 \times 10^{11}$ C/kg the Eqn. 9.12*a* becomes

$$B_c = \frac{6.75 \times 10^{-6} b}{(b^2 - a^2)} \sqrt{V_0} \text{ Wb/m}^2 \qquad ...(9.12b)$$

When $b >> a$

$$B_c = \frac{6.75 \times 10^{-6}}{b} \sqrt{V_0} \text{ Wb/m}^2 \qquad ...(9.12c)$$

Here V_0 is in volts and b is in metre.

This means that if $B > B_c$ for a given V_0 the electron will not reach the anode, *i.e.* anode current is zero. Conversely, the cut-off voltage for a given B is given by

$$V_c = \frac{\eta}{8} B^2 b^2 \left(1 - \frac{a^2}{b^2}\right) \qquad ...(9.13)$$

It is also called **Hull cut-off voltage** equation. It means that when $V_0 < V_c$ for a given B_z, the electron will not reach the anode. Figure 9.9 shows the typical trajectories of electron at various magnetic fields.

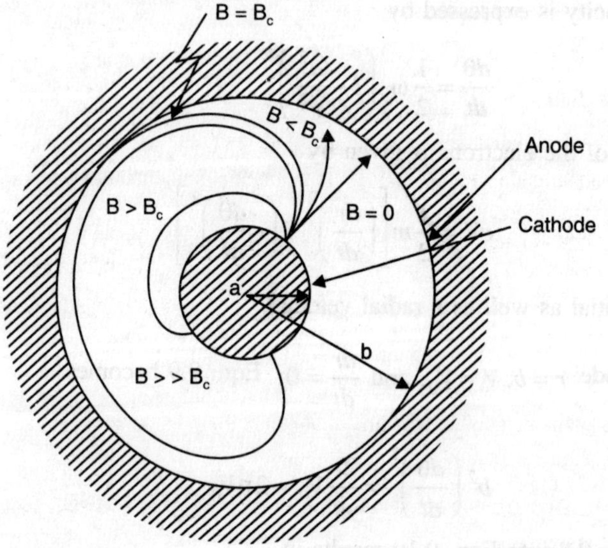

Fig. 9.9 Typical trajectories of electron at various magnetic fields

9.2.3 (c) Anode Current

Equation 9.12 predicts that anode current should fall abruptly to zero above cut-off magnetic field B_c for a given anode voltage. But it is not observed in actual measurement. The anode current decreases gradually as shown in Fig. 9.10. The gradual decrease of anode current is attributed to the space charge effects and in reality the electrons are emitted from cathode with the Maxwellian distribution of velocities (electron emitted with velocity $\neq 0$). In a practical magnetron, a magnetic field $B >> B_c$ is used which reduces the outer radius of Brillouin cloud around the cathode.

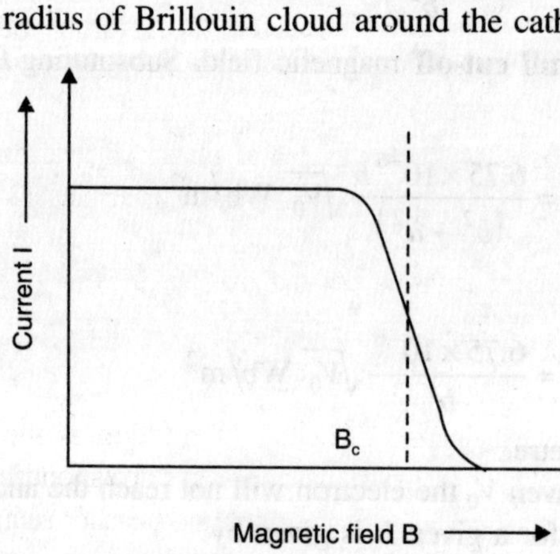

Fig. 9.10 Variation of anode current with magnetic field in magnetron

9.2.4 Hartree Condition/Voltage

Substituting the value of $E_r = -\dfrac{V_0}{r \ln \dfrac{b}{a}}$ in Eqn. 9.1 the anode potential V_0 is given by

$$V_0 = \left(\omega_e r^2 \ln \frac{b}{a}\right)\left(B - \frac{\omega_e}{\eta}\right) \quad ...(9.14)$$

where $\omega_e = v/r$ is electron angular velocity which is moving in circular path of radius r.

Taking, $\ln \dfrac{b}{a} \approx \dfrac{2(b-a)}{(b+a)}$ and $r = \dfrac{b+a}{2}$, Eqn. 9.14 becomes,

$$V_0 = \omega_e \frac{(b^2 - a^2)}{2}\left[B - \frac{\omega_e}{\eta}\right] \quad ...(9.15)$$

It is necessary to synchronise electron angular velocity to the velocity of travelling waves on the multicavity structure of magnetron.

9.2.5 Modes (Relationship between the Frequency of Oscillation ω_m and the Angular Velocity of Electron ω_e)

The angular velocity of the electron $\omega_e = d\theta/dt$ and it travels around the interaction space in a time

$$T = \frac{2\pi}{\omega_e} \quad ...(9.16)$$

The time taken by the electron to move from the centre A of one anode gap to B (Fig. 9.7) for next adjacent anode gap is given by

$$T_1 = \frac{2\pi}{N\omega_e} \quad ...(9.17)$$

where N = number of resonant coupled cavities in the anode block (slow wave structure).

Due to N resonant coupled cavities, there exists N resonate frequencies or modes (see equivalent resonant circuit in Fig. 9.13*a*).

As the slow-wave structure is closed on itself, the total phase shift around a closed path is $2\pi m$ radians for possible oscillations, where m is an integer = 0, ±1, ±2, ±3...±N/2 representing the number of cycles of phase shift around the closed path.

The phase shift between two adjacent cavities (A to B in Fig. 9.7) is given by

$$\phi_m = \frac{2\pi m}{N} \quad ...(9.18)$$

If we assume that the electron has the same direction of rotation as the a.c. field, the travelling electron sees a phase difference of $(\omega_m T_1 - \phi_m)$ radians between corresponding points A and B of two adjacent cavities (Fig. 9.7). Mechanism of oscillations as explained earlier requires that the electron has to experience a maximum retarding force in both positions and lose energy to the resonator (cavity), this phase shift seen by the electron must be an integral multiple of 2π radian *i.e.*,

$$\omega_m T_1 - \phi_m = 2\pi n \quad ...(9.19)$$

where $n = 0, \pm 1, \pm 2, \pm 3......$ Substituting Eqn. 9.17 and Eqn. 9.18 into Eqn. 9.19 and on arranging we obtain

$$\omega_e = \frac{\omega_m}{(m + nN)} \quad ...(9.20)$$

or

$$\omega_m = (m + nN)\,\omega_e = K\omega_e$$

where $K = (m + nN)$ may be negative or positive. Equation 9.20 expresses a relation between angular velocity of electron in terms of frequency of oscillation. Also different values of K represent different modes of oscillations.

The phase velocity in the azimuthal direction ϕ for the nth spatial harmonic for the mth resonant mode is

$$v_{pmn} = \frac{\omega_m r}{(m + nN)}$$

at the radius r; that is, angular phase velocity is ω_m/β_{mn}.

Substituting Eqn. 9.20 into Eqn. 9.15, we obtain

$$V_0 = V_H = \frac{\omega_m}{(m + nN)} \frac{(b^2 - a^2)}{2}\left(B - \frac{\omega_m}{\eta(m + nN)}\right) \quad ...(9.21)$$

π-Mode

This relation is known as **Hartree condition for oscillations.** When $m = N/2$, it means $\phi_m = \pi$ that is opposite phase shift in successive cavities and excitation is maximum in cavities. This mode is known as **π-mode** and is commonly used for magnetron oscillator. Substituting $m = N/2$ in Eqn. 9.21, we obtain Hartree voltage for π-mode

$$V_{H_\pi} = \frac{\omega_{N/2}(b^2 - a^2)}{N(1 + 2n)}\left[B - \frac{2\omega_{N/2}}{\eta N(1 + 2n)}\right] \quad ...(9.22)$$

For this mode (π), the resulting electric field pattern for an eight cavity circuit is shown in Fig. 9.11. A plot of Hull cut-off voltage and Hartree voltage V_H with magnetic field B is shown in Fig. 9.12.

The space between the Hull cut-off curve (Eq. 9.12) and the Hartree curve is the (Eq. 9.21) region of oscillations.

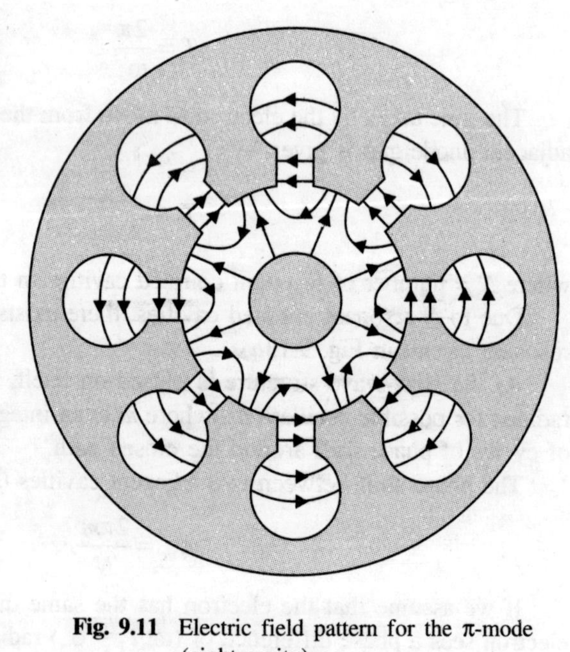

Fig. 9.11 Electric field pattern for the π-mode (eight-cavity)

From the above discussion and analysis, it becomes clear that it is necessary to adjust the magnetic field B and d.c. anode voltage V_0 to obtain the proper value of ω_e so that Eqn. 9.19 for the desired mode is satisfied. From Eqn. 9.20, there can be a large number of modes possible in a magnetron. However only few of these modes have practical significance, as most of them yield low power output and low efficiency.

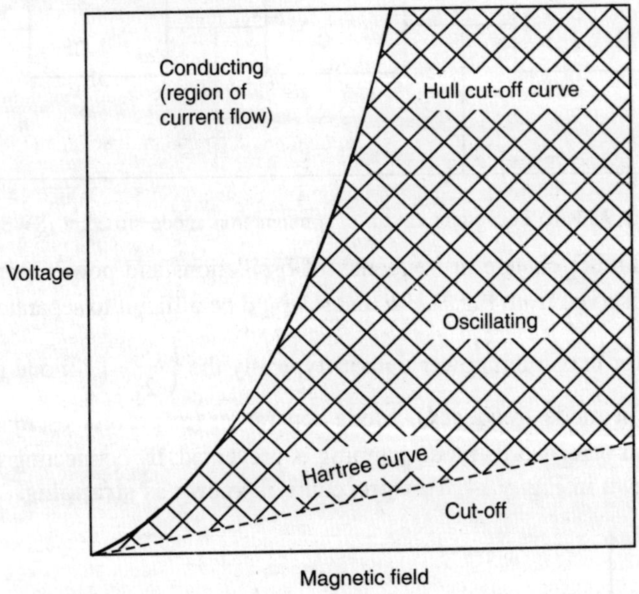

Fig. 9.12 A plot of Hull cut-off voltage and Hartree voltage with magnetic field (magnetron operation region)

9.2.6 Mode Jumping and Mode Separation-Strapping

From Eqn. 9.21 it is inferred that each mode corresponds to a different frequency and different operating condition. It has been shown by analysing* equivalent circuit of magnetron shown in Fig. 9.13*a*, which acts as a transmission line filter with lowpass characteristic, that the various modes differ very little in frequency from each other (Fig. 9.13*b*). It is a dispersion curve for a magnetron anode structure (SWS) with *N* cavities. Allowable mode numbers are represented by vertical lines. Due to this, magnetron oscillators have a tendency to jump from one mode to another mode under slight variation in operating conditions.

---

* From the low pass equivalent circuit shown in Fig. 9.13*a*, where each of the identical cavities is replaced by its L_1C_1 equivalent circuit and C_2 is the coupling capacitance between the cavity vane and the cathode due to which mutual coupling between various cavities take place, it can be shown that $\left(\frac{\omega_m}{\omega}\right)^2 \approx \frac{1}{1+\frac{C_2}{C_1\left(1-\cos\frac{2\pi m}{N}\right)}}$ where ω_m is mode frequency, and ω is the fundamental resonant frequency of the cavities given by $\omega = \frac{1}{\sqrt{L_1C_1}}$. In obtaining the above relation we used the relation $\cos\phi_m = 1+\frac{Z_1}{2Z_2}$, where $\phi_m = \frac{2\pi m}{N}$ the phase shift (Eq. 9.18); Z_1 = total series impedance and Z_2 = total shunt impedance per section.

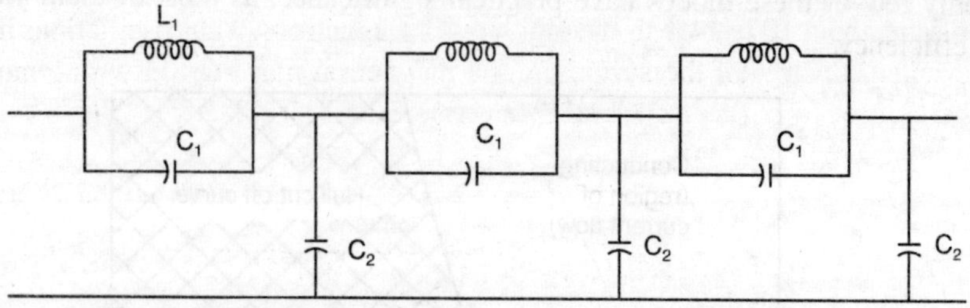

Fig. 9.13 (*a*) Equivalent circuit of a magnetron anode structure (SWS)

This results in an abrupt change in frequency of oscillations and power output. This is known as **mode jumping**. It is observed from Fig. 9.13*b* that it would be difficult to separate the π-mode from the next immediate mode for maximum power output (typically the $\left(\frac{N}{2}-1\right)$ mode frequency is only 1 to 3% different from the π mode frequency). Mode separation (to prevent oscillation in the undesired mode) is achieved, or in other words mode jumping is prevented, by connecting two metallic straps to anode segments as shown in Fig. 9.14. This procedure is known as **strapping.**

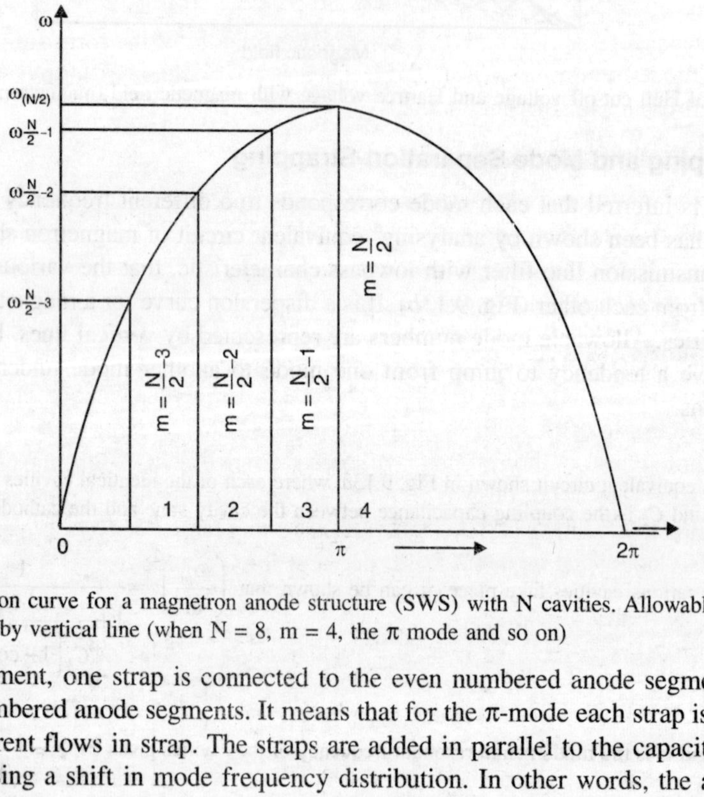

Fig. 9.13 (*b*) Dispersion curve for a magnetron anode structure (SWS) with N cavities. Allowable mode numbers are represented by vertical line (when N = 8, m = 4, the π mode and so on)

In this arrangement, one strap is connected to the even numbered anode segments and the other strap to the odd numbered anode segments. It means that for the π-mode each strap is at same potential and no π-mode current flows in strap. The straps are added in parallel to the capacitor (Fig. 9.13*a*) of each slot, thus causing a shift in mode frequency distribution. In other words, the added capacitance depresses the π-mode resonance frequency greatly. Gilmour indicated that strapping causes the frequency separation between the π-mode and the $\left(\frac{N}{2}-1\right)$ mode to the extent of about 25 to 30% in the lower power magnetrons and about 10 to 14% in the high power magnetrons. Thus oscillations in the undesired mode can be prevented. In spite of the strapping, the magnetron may start in wrong mode due to lack of adequate mode control in conventional magnetrons.

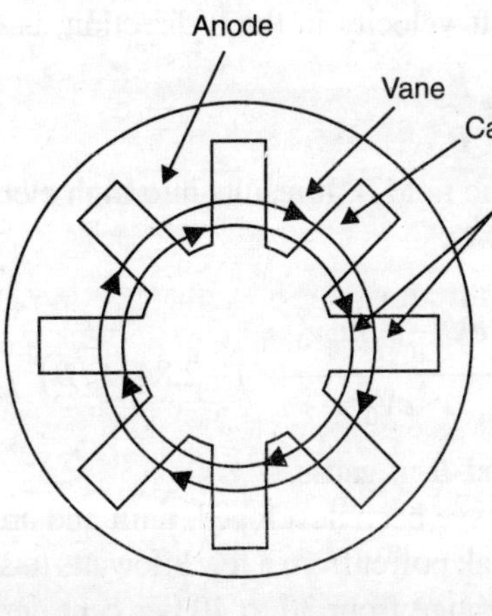

Fig. 9.14 Metallic straps to anode segments of a magnetron

At higher frequencies (≥ 10 GHz), rising-sun anode structure of the form shown in Fig. 9.18 can be used to separate mode frequencies. At higher frequencies, fabrication of straps become very difficult because of the small dimensions of the anode.

9.2.7 Pulling and Pushing of Frequency

The variation in frequency due to variation in load impedance of a magnetron is called **Frequency Pulling** and the variation in frequency with variation in operating voltages and currents is called **Frequency Pushing.** Depending on the extent of these two, the frequency stability is affected and sometimes the magnetron may fail to produce a pulse.

9.2.8 Tuning of Magnetron

It is achieved by varying the effective inductance or the effective capacitance of the resonators. The arrangement is shown in Fig. 9.15. In case of arrangement 9.15(*a*) an inductance is varied by sliding a conducting plunger into or out of the resonator. In case of arrangement 9.15(*b*) the capacitance is varied by moving annular strap into or out of groves in the anode. Tuner motion is produced by a mechanical connection through flexible bellows in the vacuum wall. Tuning ranges of 10 to 12 per cent of bandwidth are possible for pulse tubes, and as much as 20 per cent for CW tubes.

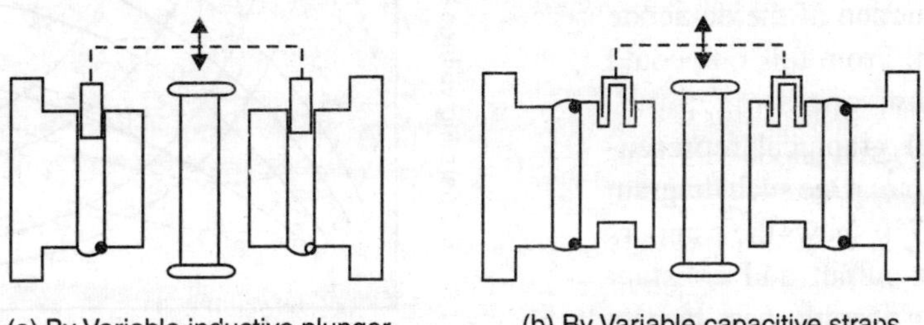

(a) By Variable inductive plunger (b) By Variable capacitive straps

Fig. 9.15 Tuning of magnetron

9.2.9 Power, Efficiency and Applications of Magnetron

We have seen that r.f. voltage grows because of the energy supplied by the electrons moving towards the anode. Thus potential energy of electron is transferred to the r.f. field while kinetic energy remains more or less constant. The average drift velocity in the θ direction, is constant for a given

$$v_\theta = \frac{E}{B}$$

applied dc voltage V_0 and axial mangetic field B. It results into high electronic efficiency (η) of conversion of d.c. power into microwave power.

$$\eta = \frac{eV_0 - \frac{1}{2}m_e v_0^2}{eV_0} \approx 1 - 2.84(E/B)^2 / V_0 \qquad ...(9.22)$$

where V_0 is applied voltage in volts and B in gauss.

Both pulsed and CW magnetrons have been developed, built and are commercially available. Pulse magnetrons are capable of producing peak power from a few kilowatts to several megawatts at frequencies up to 100 GHz. Efficiencies typically range from 30 to 40 per cent depending on the power level and operating frequency. Continuous wave (CW) magnetrons are also capable of producing continuous output power from a few hundred watts in a tunable tube to up to 25 kW in a fixed frequency device. Efficiencies typically range between 30 to 70 per cent or more. Most CW magnetrons oscillate at a frequency of 2450 or 915 MHz, but CW magnetrons have been developed at other frequencies.

Magnetrons are relatively small in size and weight, low-cost devices and operate at relatively lower voltages. High power pulsed magnetrons are extensively used in radar system, while the low-power pulsed devices find application in beacons. Tunable continous wave magnetrons are used in electronic countermeasures (ECM).

Fixed frequency magnetrons are used as microwave heating sources such as consumer oven magnetrons (output power of under 1kW), commercial oven magnetrons (1 to 2 kW output power), industrial magnetrons (2 to 10 kW output power or more), etc.

9.2.10 Operating Characteristics

Commercially available magnetrons are supplied with performance charts. One typical chart is shown in Fig. 9.16. It gives contours of constant power output, constant efficiency and constant magnetic field strength as function of the dc anode voltage and current. From this one could select the optimum operating point. Another method of graphical representation is Rieke diagram. One such diagram is shown in Fig. 9.17. It shows the contours of constant power output and constant frequency plotted on a Smith chart. Its data are experimentally obtained by varying the load impedance and observing the power output and frequency. Rieke diagram is

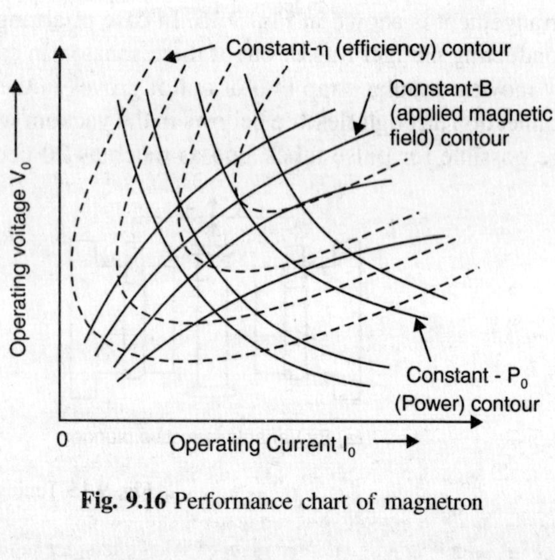

Fig. 9.16 Performance chart of magnetron

generally supplied by the manufacturer along with data sheet. From such performance chart one could select the optimum load impedance.

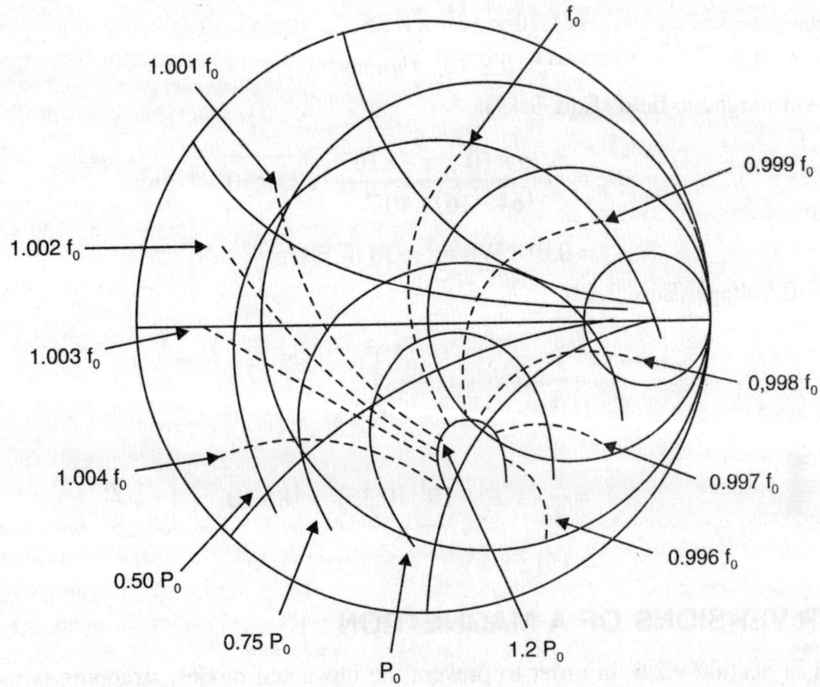

Fig. 9.17 Rieke diagram

Example 9.1. A 250 kW pulsed cylindrical magnetron is operated with the following parameters—

Anode voltage = 25 kV
Peak anode current = 25 A
Magnetic field $B = 0.35$ Wb/m^2
Radius of cathode cylinder = 4 cm
Radius of cylinder = 8 cm

Calculate:

(*i*) The efficiency η of the magnetron
(*ii*) The cyclotron frequency ω_c
(*iii*) The cut-off magnetic field
(*iv*) The cut-off voltage.

Solution.

(*i*) Efficiency $\eta = \dfrac{P_{out}}{P_{in}} \times 100$

$$= \frac{250 \times 10^3 \times 10^2}{25 \times 10^3 \times 25}$$

= 40 per cent

(*ii*) The cyclotron frequency $\omega_c = \dfrac{e}{m}B$

$$= 1.76 \times 10^{11} \times 0.35$$
$$= 0.616 \times 10^{11} \text{ radians.}$$

(*iii*) The cut-off magnetic field (Eqn. 9.15),

$$B_c = \frac{6.75 \times 10^{-6} \times 8 \times 10^{-2}}{(64-16) \times 10^{-4}} \sqrt{25 \times 10^3}$$

$$= 0.018 \text{ Wb/m}^2 = 18 \text{ m Wb/m}^2$$

(*iv*) The cut-off voltage (Eqn. 9.13),

$$= \frac{1}{8}\eta B^2 b^2 \left(1 - \frac{a^2}{b^2}\right)^2$$

$$= \frac{1}{8} \times 1.76 \times 10^{11} (0.35)^2 \times (8 \times 10^{-2})^2 \left(1 - \frac{4^2}{8^2}\right)^2$$

$$= 151.58 \text{ kV}$$

9.3 OTHER VERSIONS OF A MAGNETRON

As mentioned in Section 9.2.6, in order to prevent the unwanted modes, strapping is used so that the π-mode (the most efficient mode) is present. However, as the frequency is increased (≥ 10 GHz) the resonator size decreases to such as extent that straps become very difficult to fabricate; or in other words strapping becomes physically impossible in the millimetre and submillimetre wavelength region. Rising-Sun anode structure shown in Fig. 9.18 was used to separate mode frequencies. Other versions, those separate mode frequencies, are (*i*) co-axial magnetron; and (*ii*) inverted coaxial magnetron.

9.3.1 Rising-Sun Magnetron

The transverse cross-section of rising-sun magnetron is shown in Fig. 9.18. In this, there are two sets of resonators. The resonate frequency of the system lies somewhere between the two resonant frequencies. All the resonators are coupled together. The operation of rising-sun magnetron is exactly the same as discussed in Section 9.2. The number of modes is the same as that in the strapped magnetron, with a difference that half the number of modes are resonant at frequencies below that of the π-mode and the other half resonant at frequencies above that of the π-mode. In strapped system, all the modes are resonant at frequencies below that of the π-mode (Fig. 9.13*b*). Thus, rising-sun structure increases the frequency separation between the π-mode and the two other adjacent modes (one above π-mode and one below π-mode) in comparison to an unstrapped device. This characteristic of rising-sun structure has facilitated the design and fabrication of millimetre wavelength rising-sun magnetrons. The number of resonators is limited by the practical mode separation requirements. The size of the cathode is limited by the number of resonators. As the frequency of operation is increased with a fixed number of resonantors, the cathode size also decreases. This limits the input current and hence the output power. Higher the frequency *i.e.*, lower the wavelength, it is quite difficult to fabricate the rising-sun structure. For an example, the vane thickness for a 5 mm wavelength rising-sun magnetron is about 1 mil (0.001 inch), and the cathode diameter is $\frac{1}{2}$ mil (0.0005 inch). Such dimensions are difficult to fabricate. With these small sizes, heat dissipation, voltage gradient etc., become limiting factors.

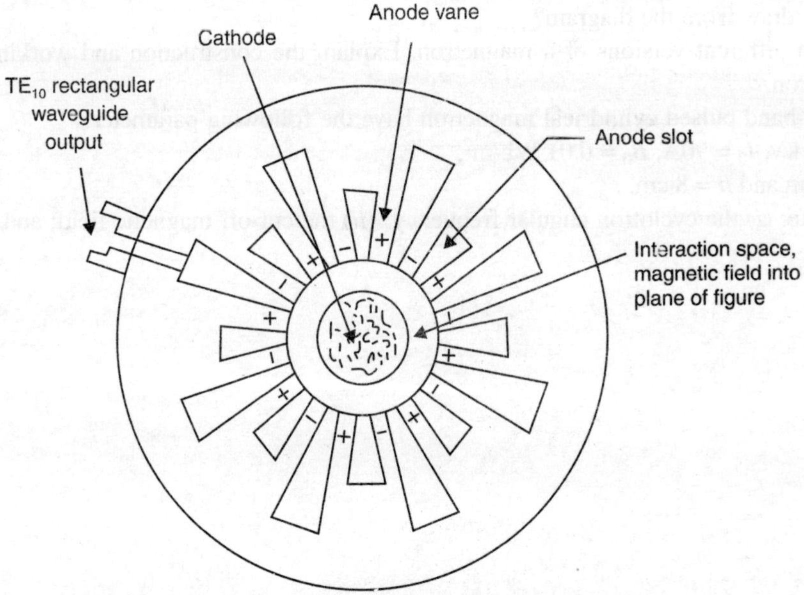

Fig. 9.18 Cross-section of a rising-sun magnetron

The other two versions namely coaxial and inverted coaxial magnetrons have been designed and fabricated to overcome the limitation of rising-sun structure.

REVIEW QUESTIONS AND PROBLEMS

1. Give construction of a magnetron. Derive Hull cut-off voltage equation and explain the process of generation of microwaves by a cavity magnetron operated in π-mode. Describe in brief applications of magnetrons.
2. Giving necessary theory explain the mechanism of oscillations in a multicavity magnetron. Discuss various modes of oscillations and their separation by means of straps. Obtain an expression for the efficiency of this oscillator. How would you represent its operating characteristics graphically?
3. Describe the construction and working of a magnetron as oscillator. Show that in a magnetron the angular velocity of electron is related to the frequency of oscillation by the relation:

$$\omega_e = \frac{2\pi f}{(m + nN)}$$

where N is the number of resonators in magnetron and m and n are integers.
4. Describe the principle of operation for a normal cylindrical magnetron and its characteristics.
5. Derive an expression for the cut-off voltage for a fixed B_z the axis magnetic field in case of cylindrical magnetron.
6. Derive an expression for the cut-off magnetic field for a given anode voltage V_0 in case of linear magnetron.

7. What is meant by π-mode operation in a magnetron? Describe how strapping separates the π-mode from other possible modes.
8. Show by a diagram the paths of electrons in crossed electric and magnetic field. What inference you can draw from the diagram?
9. Mention different versions of a magnetron. Explain the construction and working of rising-sun magnetron.
10. Does X-band pulsed cylindrical magnetron have the following parameters:
    $V_0 = 30$ kV, $I_0 = 90$A, $B_0 = 0.01$ Wb/m^2,
    $a = 4$ cm and $b = 8$ cm.
    Calculate: (*i*) the cyclotron angular frequency; (*ii*) the cut-off magnetic field; and (*iii*) the cut-off voltage.

CHAPTER 10

Microwave Solid State Devices: Transfer Electron Devices (TEDs – Gunn Diode) and Avalanche Transit Time Devices

10.1 INTRODUCTION, CLASSIFICATION, APPLICATION

Microwave solid-state devices are increasingly utilised for several applications (where low power and low frequencies) that include modern communication systems, radars, navigation, radiometer and radio astronomy receiver systems, etc. Solid state devices have the advantages of small size, ruggedness, low cost and compatibility with microwave integrated circuits. Microwave solid state devices are broadly classified into two groups, namely:

(1) Two terminal devices listed in Fig. 10.1 and (2) Three terminal devices listed in Fig. 10.2.

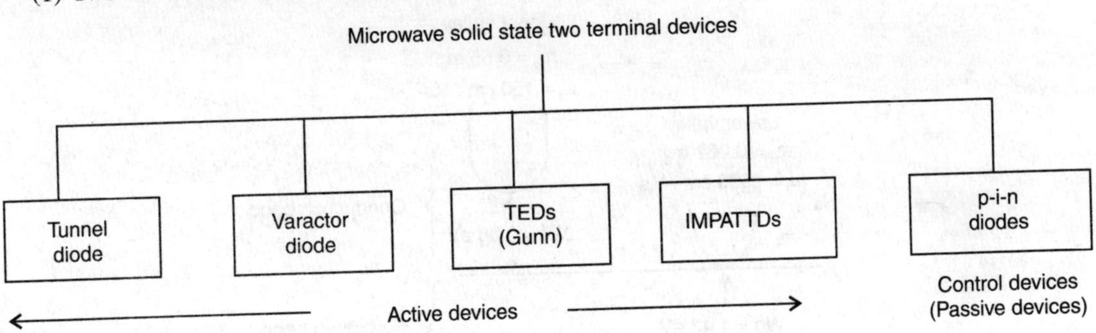

Fig. 10.1 Microwave solid state two terminal devices

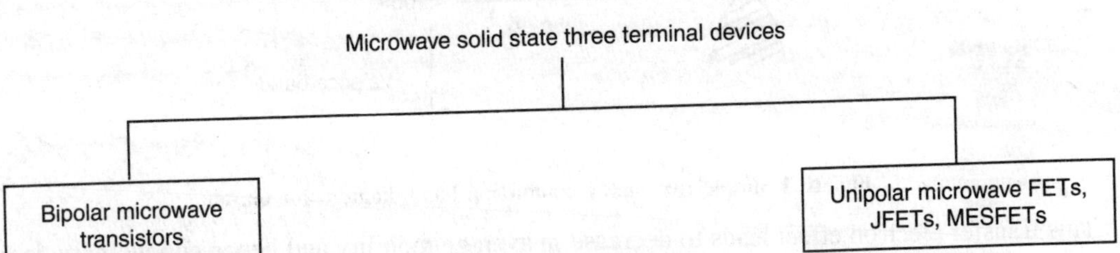

Fig. 10.2 Microwave solid state three terminal devices

10.2 INTRODUCTION TO TRANSFER ELECTRON DEVICES (TED'S)

Most of the devices mentioned in Figs. 10.1 and 10.2 are junction devices except Transfer Electron Devices (TEDs). The TEDs uses bulk semiconductors, rather than junctions. Thus these would be, naturally, simpler in fabrication and operating parameters and cheaper when compared to junction devices. Ridley, Watkins and Hilsum theoretically predicted the possibility of the reduction of electron drift velocity with increase in electric field (Fig. 10.3) above a threshold value due to gradual transfer of a fraction of the electrons from low-energy high-mobility state to high-energy low-mobility state in conduction band in semiconductors like **GaAs** (Fig. 10.4).

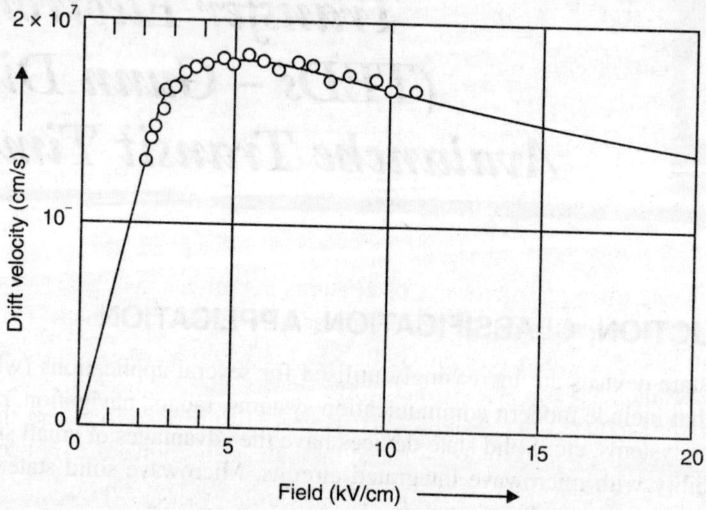

Fig. 10.3 Electron drift velocity versus electric field

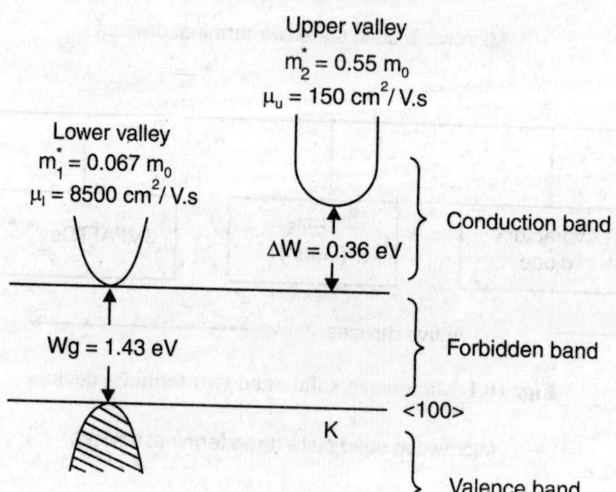

Fig. 10.4 Simple two-valley conduction band diagram for GaAs

This transfer-electron effect leads to decrease in average mobility and hence conductivity decreases with an increase in electric field above threshold value. Thus, there is a range of applied voltage over which the current decreases with increasing voltage (Fig. 10.5) and exhibits negative-conductance characteristics that can be exploited for the generation and amplification of microwave frequencies.

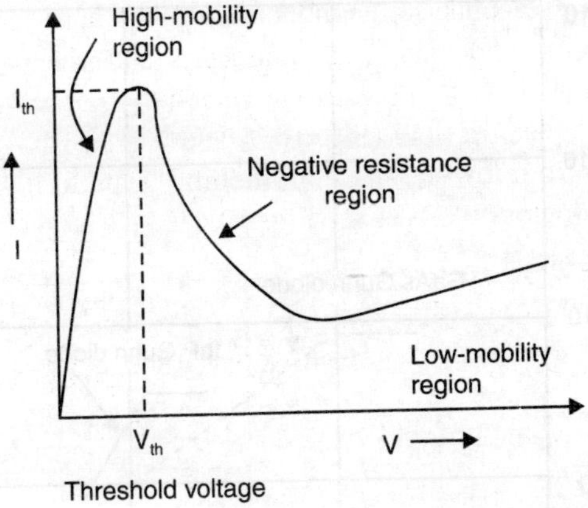

Fig. 10.5 Current-voltage characteristic for a Gunn diode

Gunn was the first to discover microwave oscillations of currents in bulk GaAs samples with ohmic contacts (Fig. 10.6), which were being subjected (pulsed) to high electric field. This effect was named as **Gunn effect** and devices which rely on Gunn effect are known as Gunn devices/diodes. Kromer showed that the experimental Gunn effect (microwave oscillation of current in bulk semiconductor) and transfer-electron effect predicted by Ridley, Watkins and Hilsum were one and the same. The Gunn devices rely on **transfer-electron effect;** as such, they are also known as Transfer Electron Devices (TEDs).

The transfer electron devices use mostly III-V compounds GaAs and InP. The Gunn effect has been observed in CdTe, ZnSe and InAs. More other compound semiconductors are almost certain to be added to the Gunn family in future.

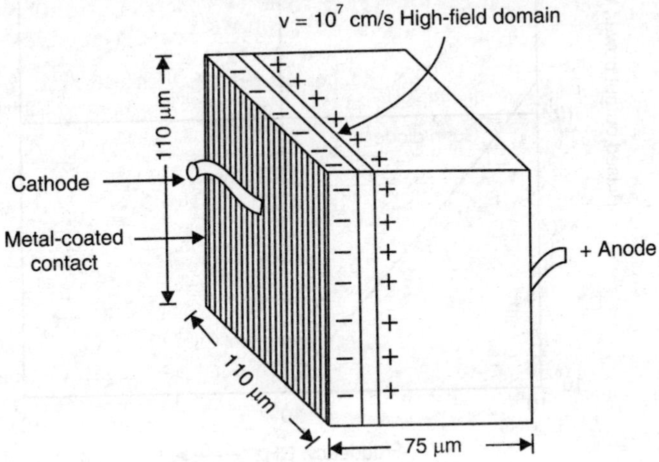

Fig. 10.6 Schematic structure of *n*-type GaAs Gunn diode

The Gunn diodes have been operated over a frequency range from below 1 GHz to above 140 GHz. Figure 10.7 shows the typical maximum CW power output from Gunn devices versus frequency. The pulse performance is shown in Fig. 10.8. Figure 10.9 shows the typical maximum efficiency of CW Gunn oscillator versus frequency.

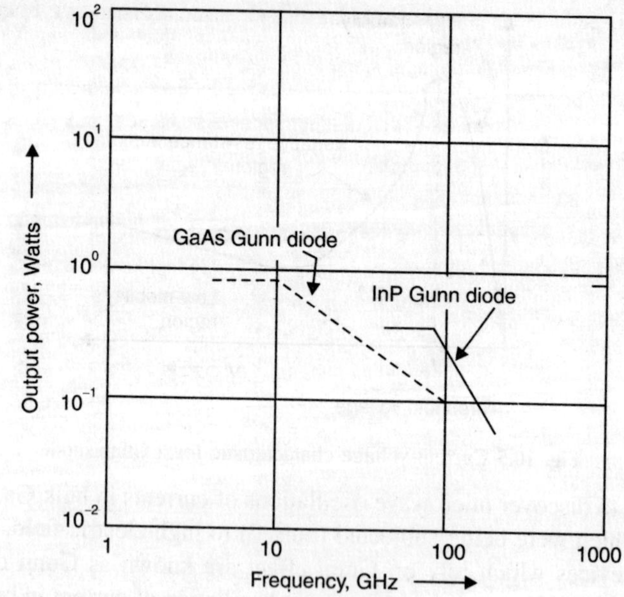

Fig. 10.7 Typical maximum CW power output from Gunn devices versus frequency

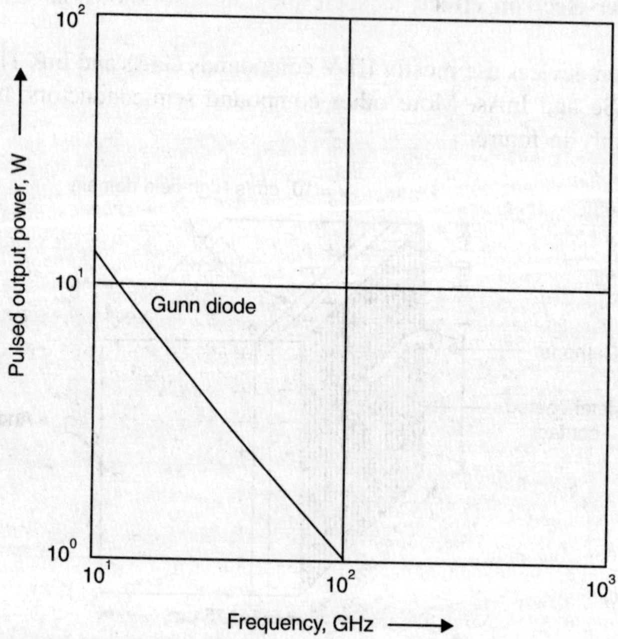

Fig. 10.8 Typical maximum pulse power output from Gunn oscillators versus frequency

These devices are tunable over a wide frequency range, have low noise characteristics, simple to fabricate and are inexpensive. They find application as driver stages, power stages and local oscillators in radar and microwave communication systems. Gunn devices have also been used as logic element. These devices are extensively used in millimetre-wave system design due to their low noise properties as compared to IMPATT devices. Gunn diodes can also be used as negative-resistance reflection-type amplifiers.

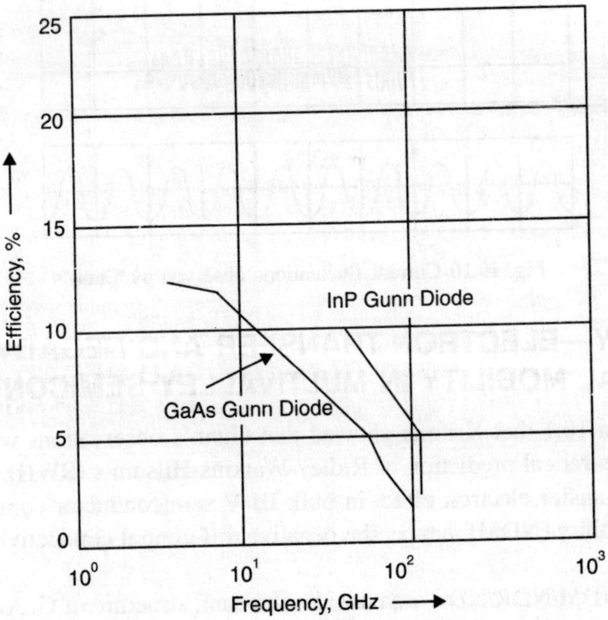

Fig. 10.9 Typical maximum efficiency of CW Gunn oscillators versus frequency

In this section, we will briefly discuss Gunn effect, electron transfer and negative differential mobility, electronic modes of operation, and InP TEDs.

10.3 GUNN EFFECT

In 1963 Gunn observed that, when a dc voltage V is applied to the ohmic contacts on the ends of a rectangular bar of *n*-type GaAS or InP (Fig. 10.6), the current first rises linearly as V is increased from zero and then begins to oscillate (Fig. 10.10) when the average electric field exceeds a threshold value of several kilovolts per centimetre (typical 3000 V/cm). The current fluctuations are shown in Fig. 10.10. The current wave form was produced by a voltage pulse of 16 V amplitude and 10 nanosec duration applied to a specimen of *n*-type GaAs 2.5×10^{-3} cm in length. The frequency of the oscillating component is 4.5 GHz. Gunn found that the period of oscillation observed was very close to the transit time of electrons calculated from the threshold current I_{th}. Gunn also discovered that the threshold electric field $E_{th} = \left(\frac{V_{th}}{l}\right)$ varied with the length (l) and type of material. For an example $V_{th} = 59$ V and $l = 210$ μm, then $E_{th} = 59/210 \times 10^{-6} \times 10^2 = 2810$ V/cm.

Gunn also found that the magnitude of E_{th} in both GaAs and InP and the nature of the instabilities in GaAs were unaffected by the nature of the contacts, specimen surface condition, light irradiation or application of a moderate static magnetic field. On the other hand in InP, the nature of instability was partly dependent on contact material. However, in heavily doped n^+-contact the instability closely resembles that in GaAs.

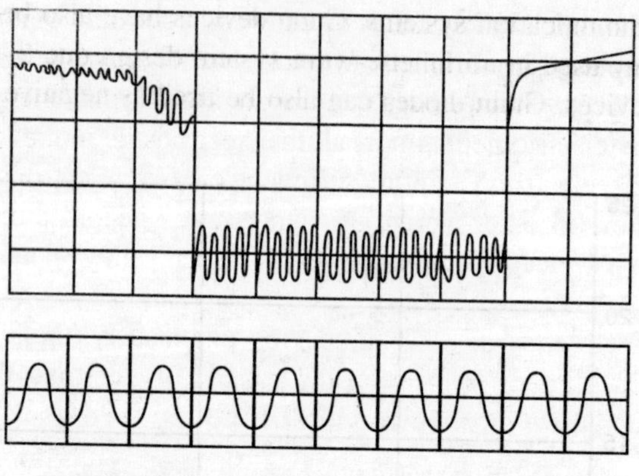

Fig. 10.10 Current fluctuations observed by Gunn

10.4 RWH THEORY—ELECTRON TRANSFER AND NEGATIVE DIFFERENTIAL MOBILITY IN MULTIVALLEY SEMICONDUCTORS

As mentioned in Section 10.1 that Kromar showed that Gunn's observations were in complete agreement with the earlier theoretical prediction of Ridley-Watkins-Hilsum's (RWH). Now, we shall discuss mechanism of RWH's transfer electron effect in bulk III-V semiconductor compounds which leads to negative differential mobility (NDM), hence, the negative differential conductivity (NDC) or resistivity (NDR).

For explanation of NDM/NDR/NDC we consider the band structure of GaAs, which is mostly used to fabricate Gunn diodes or TEDs as mentioned in Section 10.2. GaAs has two closely spaced energy bands (valleys)—lower and upper, in the conduction band as depicted in the energy band (W-k) diagram of Fig. 10.4, with their pertinent parameters. The electron effective mass is larger in the upper valley by

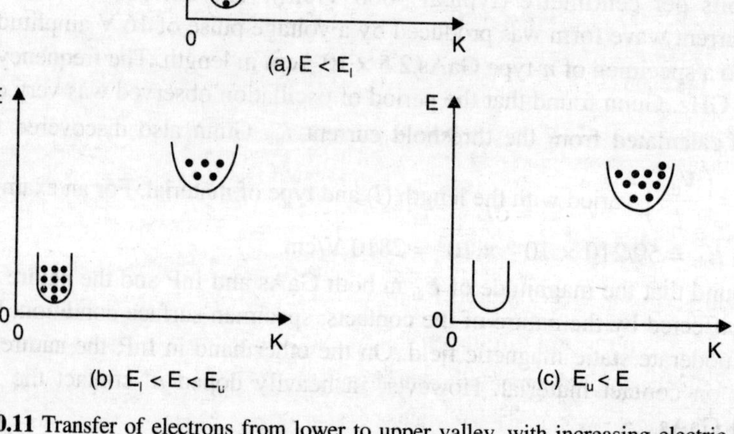

Fig. 10.11 Transfer of electrons from lower to upper valley, with increasing electric field

a factor of about 8.2, while the mobility is smaller by an even larger ratio of about 53. Energy difference $\Delta W \approx 0.36$ eV between minima of two bands is much smaller than the gap energy $W_g = 1.43$ eV. However, energy difference ΔW is much larger as compared to thermal energy $k_B T$ (0.025 eV at room temperature). As such, at low electric field almost all the electrons lie in the lower valley (Fig. 10.11*a*). As the field is increased the electrons acquire sufficient energy (electrons become hotter) and get transferred to the upper valley (in other words, the upper valley becomes more and more populated). The dependence of transfer of electrons from lower to upper valley in electric field is depicted in Fig. 10.11.

This redistribution of electrons takes place almost instantaneously (the internal scattering times are of the order of 10^{-11}s or even shorter). Because of the transfer of electrons, the fraction of (in terms of effective mass) electrons increases very much with the electric field above the threshold value, the average velocity decreases as shown in Fig. 10.3. As a consequence mobility decreases $\left(\mu = \frac{v}{E}\right)$ and hence conductivity decreases ($\sigma = en\mu$). The transfer of electron effect leads to NDM/NDC in GaAs and other bulk semiconductors like InP and CdTe. The mechanism of Gunn effect is represented in block diagram Fig. 10.12. Figure 10.3 also indicates that at fields much higher than the threshold field, the electron velocity in the upper band does saturate.

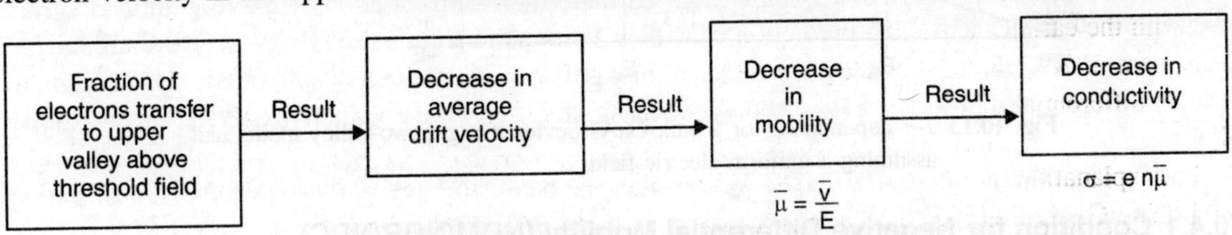

Fig. 10.12 Block diagram showing mechanism of Gunn effect (events are instantaneous)

Thus, it is clear that there is a range of applied voltage above the threshold value over which the current decreases (Fig. 10.5) with increasing voltage and exhibits negative differential resistivity/ conductivity.

Let n_l and n_u be, respectively, the electron concentrations in the lower and upper valleys. The conductivity of the *n*-type GaAs sample is

$$\sigma = e(\mu_l n_l + \mu_u n_u) \qquad ...(10.1a)$$

$$= en\bar{\mu} \qquad ...(10.1b)$$

$$= en\frac{\bar{v}}{E} \qquad ...(10.1c)$$

The current density is given by

$$J = \sigma E = en\bar{\mu}E = en\bar{v} \qquad ...(10.2)$$

where e = the electronic charge

μ_l = the electron mobility in lower valley

μ_u = the electron mobility in upper valley

$n = n_1 + n_u$, the total electron concentration

$\overline{\mu} = \dfrac{\mu_l n_l + \mu_u n_u}{n}$; the average mobility

$\overline{v} = \overline{\mu}E$, the magnitude of the average drift velocity.

J-E dependence is shown in Fig. 10.13, at very low and high fields *J* versus *E* are straight lines (dotted in the Figure 10.13) and at intermediate fields *J* decreasing.

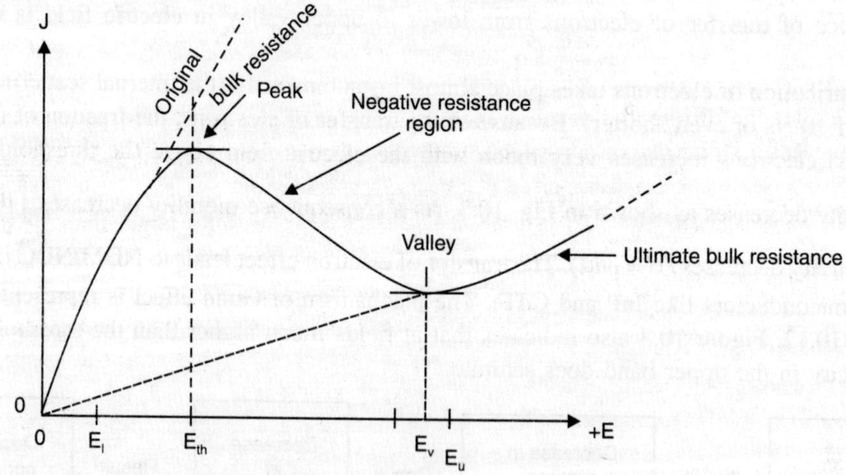

Fig. 10.13 *J-E* dependence for a bulk GaAs device using a two-valley model and assuming a uniform electric field

10.4.1 Condition for Negative Differential Mobility (NDM/NDR/NDC)

The condition for negative differential conductance can be derived in a simple manner. The incremental variation of current density with field is obtained by differentiating Eqn. 10.2 with respect to *E*.

$$\frac{dJ}{dE} = \sigma\left(1 + \frac{E}{\sigma}\frac{d\sigma}{dE}\right) = en\frac{d\overline{v}}{dE} \quad \text{...(10.3)}$$

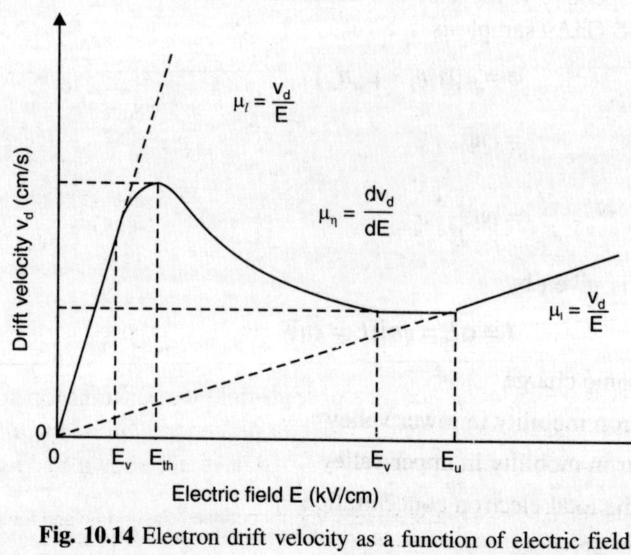

Fig. 10.14 Electron drift velocity as a function of electric field

The condition for negative differential conductance (or resistance) is

$$\frac{dJ}{dE} < 0$$

or

$$1 + \frac{E}{\sigma}\frac{d\sigma}{dE} < 0$$

or

$$\frac{d\bar{v}}{dE} = \mu_n < 0 \qquad \text{...(10.4)}$$

where μ_n denotes the differential negative mobility. This is shown in Fig. 10.14.

Thus, the **band structure of a semiconductor** must satisfy the following three criteria in order to exhibit negative resistance for transferred-electron effect to occur.

1. Separation energy between two valleys ΔW must be several times larger than the thermal energy k_BT (= 0.025 V) at room temperature. That is, $\Delta W >> k_BT$ or $\Delta W >> 0.025$ eV.
2. Energy separation between two valleys ΔW must be much smaller than the gap energy between the conductance and valence bands. That is $\Delta W << W_g$. Otherwise the semiconductor will breakdown and become highly conductive before the electrons begin to transfer to the upper valley due to hole-electron pair formation.
3. Electrons in the lower valley must have small effective mass, high mobility and a low density state and on the other hand, electrons in the upper valley must have large effective mass, low mobility and a high density state. This means the electron velocities (*dW/dk*) must be much smaller in the upper valleys than in the lower valleys. Table 10.1 lists the data for few two-valley semiconductors. From the data we find that GaAs, InP and CdTe do satisfy above mentioned characteristics of band structure to exhibit negative resistance. Other semiconductors listed in Table 10.1 do not have the necessary band structure characteristics to exhibit negative conductance.

Table 10.1 Properties of two-valley semiconductors

Semiconductor	*Gap energy W_g(eV) (at 300K)*	*Separation energy between two valleys ΔW (eV)*	*Threshold field E_{th} (kV/cm)*	*Peak velocity v_p (cm/s)*
Ge	0.80	0.18	2.3	1.4×10^7
GaAs	1.43	0.36	3.2	2.2×10^7
InP*	1.33	0.60	10.5	2.5×10^7
		0.80		
CdTe	1.44	0.51	13.0	1.5×10^7
InAs	0.33	1.28	1.60	3.6×10^7
InSb	0.16	0.41	0.6	5×10^7

Several workers carried out detailed analysis of high-field carrier transport to obtain a theoretical velocity field characteristics of two valley semiconductor, in general, and particularly for GaAs. Theoretical and experimental velocity-field characteristics of GaAs are shown in Fig. 10.15. We see that

InP is a three-valley semiconductor; 0.60 eV is the separation energy between the middle and lower valleys, 0.80 eV is the separation energy between the upper and lower valley.

there is a good agreement between measured and calculated velocity-field characteristics for GaAs. Thus, one can say that theoretical analysis is correct.

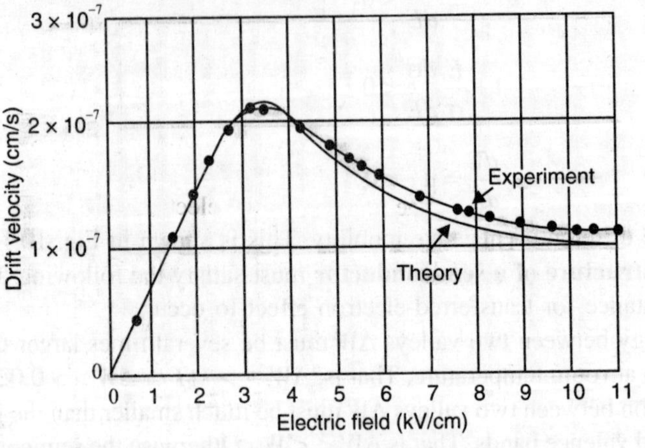

Fig. 10.15 Velocity-field characteristics of GaAs (experimental and theoretical)

10.4.2 High-Field Domains in NDM Semiconductors

In the last section, we discussed how negative differential resistance (or negative differential mobility) can occur when an electric field above threshold is applied to a bulk multi-valley semiconductor compound sample such as *n*-type GaAs. In this section, it will be shown that a decrease in drift velocity with increasing electric field (Fig. 10.17*a*) can lead to the formation of high field domains, as observed by Gunn and predicted by Ridley and Watkins, the propagation of which in turn gives rise to microwave generation.

Figure 10.16 shows the doping profile and a charge distribution when the $n^+ \, n \, n^+$ GaAs structure is biased by an external voltage. In this the majority carriers are electrons that are drifting through a background of fixed positive charge-donar atoms. At low voltage, the structure is ohmic, since the drift velocity of the electrons is proportional to the electric field, the conduction current density in the diode is given by

$$J = eN_d v_0$$
$$= en_0 v_0 \text{ [assuming } N_d = n_0] \qquad ...(10.5)$$

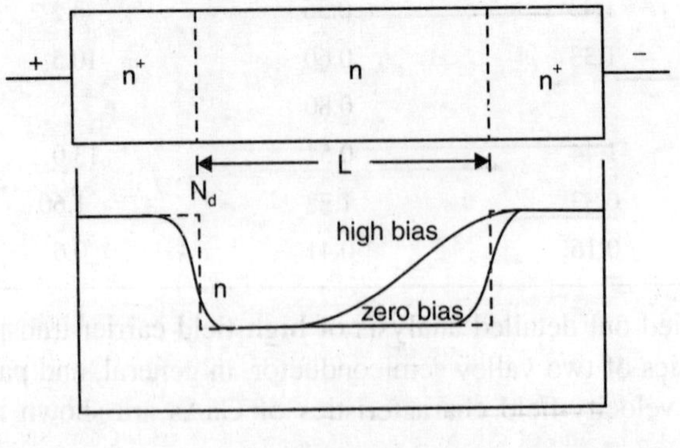

Fig. 10.16 Doping profile and charge distribution when $n^+ \, n \, n^+$ structure (TED) is biased by an external voltage

where the electron velocity v_0 should be determined from the $-E$ characteristic for a given bias field V/L. However, this is not the case in a region where dv/dE is negative, the charge distribution is not stable.

10.4.3 Formation of High Field Domain and Instability Condition

Let us suppose that there exists an excess (or accumulation) of negative space charge at *A* (Fig. 10.17*b*) either due to a random fluctuation or possibly by a permanent non-uniformity in doping of the *n*-GaAs sample as mentioned earlier. These excess charges (or accumulated charges) then create an electric field, as shown in Fig. 10.17*d*. The field on the left side of the accumulation layer (*A* point) is lower than that of the right. As a consequence of this the velocity of the electrons are lower to the right than to the left (see Fig. 10.17*a* *v-E* curve). This situation implies that charges are arriving at the accumulation layer faster from the left than they are leaving to the right. Thus, increasing the excess negative space charge in accumulation layer (or at point *A*) which further increases the field on the right and decreases on the left than the original one. This process continues to build the excess negative space charge in accumulation layer until the low and high fields reach values outside the differential negative mobility

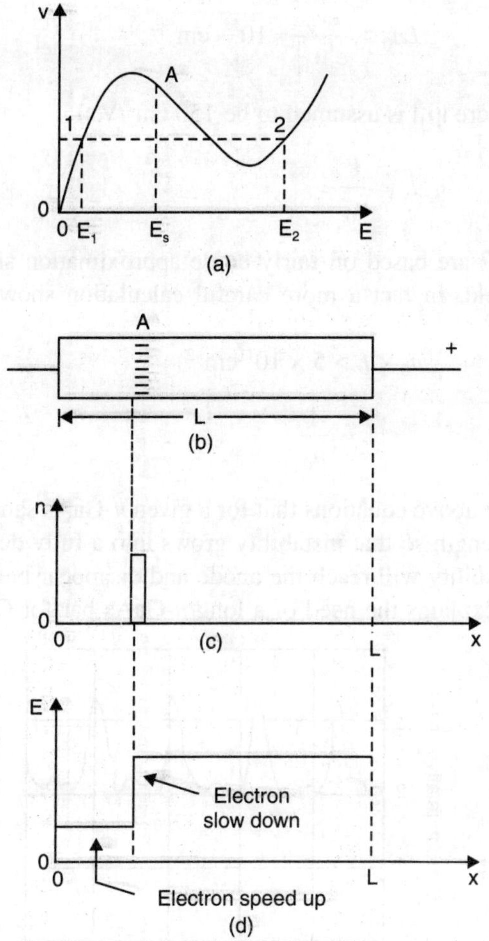

Fig. 10.17 Formation of electron accumulation layer in *n*-GaAs sample biased in the NDM region

(v-E curve or J-E curve) region and settles at points 1 and 2 (Fig. 10.17*a*) where the velocity of carrier (or charges) reaching accumulation layer equals the velocity of the carriers leaving the accumulation layer. It also implies that currents in the two field regions are equal. Because of the drift of the charges, the space-charge accumulation drift until it is collected out of the anode end and the process repeats itself from the nucleation site (site where initial disturbance developed) usually in the neighbourhood of the cathode (the negative contact). As a result of this process, a travelling space-charge layer is formed (which is a high field domain). We can see that for this instability to grow (accumulation layer to grow) into a fully developed domain before it drifts out of the sample at anode, it is necessary that transit time $\frac{L}{v_0}$ be larger than the dielectric relaxation time τ_d $\left(\tau_d = \frac{\varepsilon}{e|\mu_1|n_0}\right)$ *i.e.*

$$\frac{L}{v_0} > \frac{\varepsilon}{e|\mu_1|n_0} \quad ...(10.6)$$

or

$$Ln_0 > \frac{\varepsilon v_0}{e|\mu_1|} \approx 10^{12} \text{ cm}^{-2}$$

(for n-type GaAs sample where $|\mu_1|$ is assumed to be 150 cm^2/V.s)

$$L \propto \frac{1}{n_0} \quad ...(10.7)$$

Equations 10.6 and 10.7 are based on fairly crude approximation since $|\mu_n|$ is not a constant but varies considerably with field. In fact a more careful calculation shows that the critical product of doping and device length is

$$n_0 \times L > 5 \times 10^{11} \text{cm}^{-2} \quad ...(10.8)$$

$$L \propto \frac{1}{n_0}$$

Thus, it is clear from the above equations that for a given n-GaAs sample bar with n_0 concentration, it should have a minimum length so that instability grows into a fully developed domain. If the length is shorter than this, the instability will reach the anode and disappear before it has time to grow into a fully developed domain. It explains the need of a long n-GaAs bar for Gunn effect to occur. A pulsed

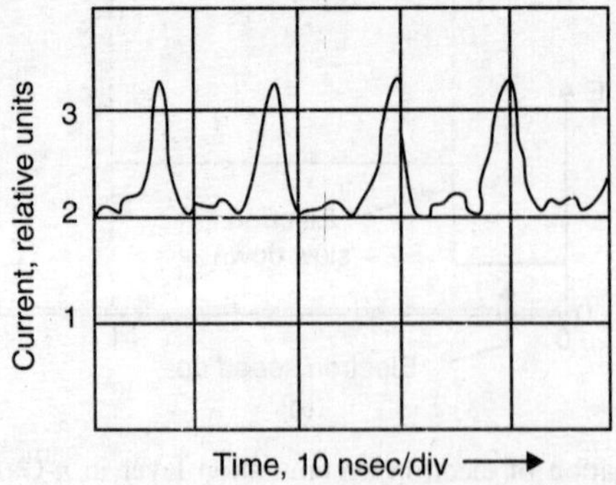

Fig. 10.18 Current waveform due to Gunn device

current output is obtained in the external circuit as shown in Fig. 10.18 (one pulse for every high field domain). The oscillation frequency of the pulsed current is roughly given by

$$f = \frac{v_{dom}}{L_{eff}} \quad ...(10.9)$$

where v_{dom} is the velocity of the domain or roughly the drift velocity of the electron in the domain and L_{eff} is the effective length that the domain travels from the time it is formed till the time that a new domain begins to form.

Further, devices with the same value for the product of doping and length will behave similarly with regard to the product of frequency and length or with the ratio of voltage to length or with efficiency.

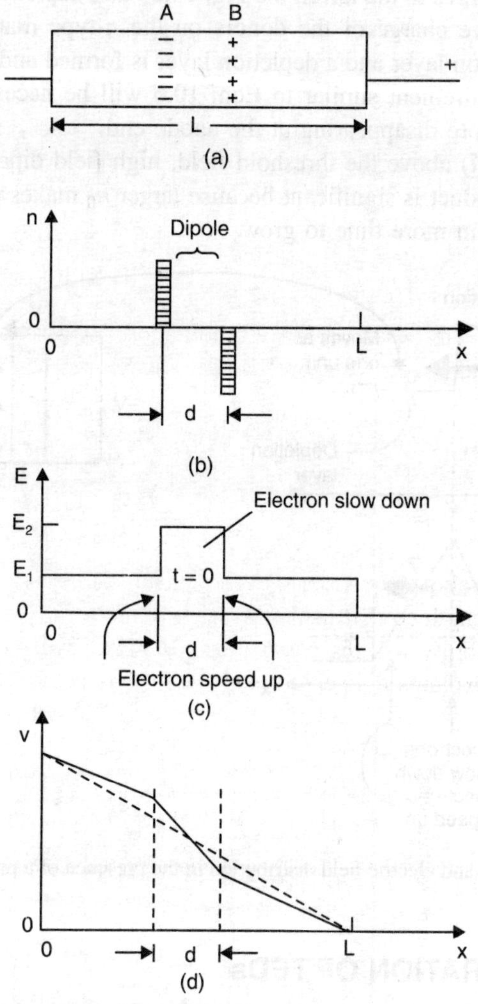

Fig. 10.19 Formation of an electron dipole layer in an *n*-GaAs sample bias in the NDM region

The pure accumulation layer instability described above is the simplest form of space charge instability. Another space charge instability subject to fairly similar consideration is a *dipole layer nucleate at inhomogeneties in the sample,* or in some samples a space charge layer starts at the cathode and then changes into dipole layer at an inhomogenety depicted in Fig. 10.19*a*. The electric field inside the dipole domain is high (Fig. 10.19*c*) and builds up as the domain drifts towards the anode. The cause of this built up is a higher velocity in the low field region on either side of the dipole layer. This results in the electrons at the edge *B* of the dipole layer moving faster than they are arriving at *B*. Similarly, the electrons arrive faster from the left at edge *A* than they move through the high field region towards the right. This growth in the charge densities at the edge of the dipole layer (domain) continues until the electric field within the dipole layer has built up, as the domain drifts towards the anode, to a value so that a balance between the rate of arrival and departure of electrons at edges *A* and *B* is restored (similar to points 1 and 2 of Fig. 10.16*a*). When the high field domain disappears at the anode a new dipole field starts forming at the cathode end and the process repeats.

With accumulation of charges to the left in the Fig. 10.20 and depletion of charges to the right. The depleted layer has the positive charge of the donors on the *n*-type material. Thus dipole layer *i.e.*, combination of an accumulation layer and a depletion layer is formed and drifts with constant velocity.

In this case also the requirement similar to Eqn. 10.6 will be necessary for the growth of fully developed dipole domain before disappearing at the anode end.

We can summarize that (*i*) above the threshold field, high field dipole domains will develop and propagate. (*ii*) The $n_0 \times L$ product is significant because larger n_0 makes the domain grow faster and a longer device gives the domain more time to grow.

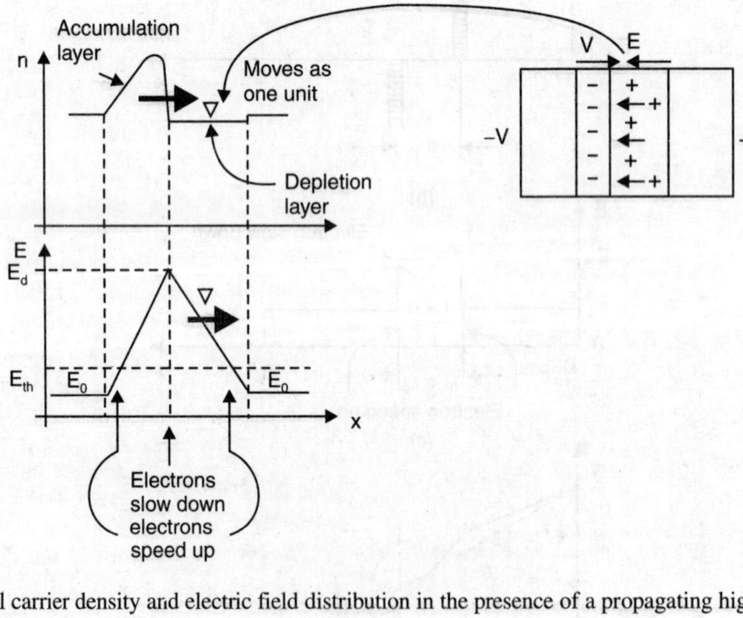

Fig. 10.20 Typical carrier density and electric field distribution in the presence of a propagating high field dipole domain in a bulk *n*-GaAs

10.5 MODES OF OPERATION OF TEDs

Since the discovery of Gunn effect in *n*-type GaAs and *n*-type InP, which is exploited for the generation or amplification of microwave power, several different modes of operation were discovered, namely, (*i*) Dipole domain modes (transit time mode, delayed domain mode, quenched domain mode, limited space-charge accumulation modes); (*ii*) Accumulation layer modes; (*iii*) Relaxation and hybrid modes and; (*iv*) Amplifier mode. Which mode will be operative is largely determined by the product of doping and *n*-region length as well as frequency doping length product. This can be visualized by mode diagram which summarises the relationship between the modes. The mode diagram is shown in Fig. 10.21 in which the product of doping length and doping ($n_0 \times L$ cm^{-2}) is plotted on one axis and that of product of frequency and length ($f L$ cms^{-1}) on the other axis.

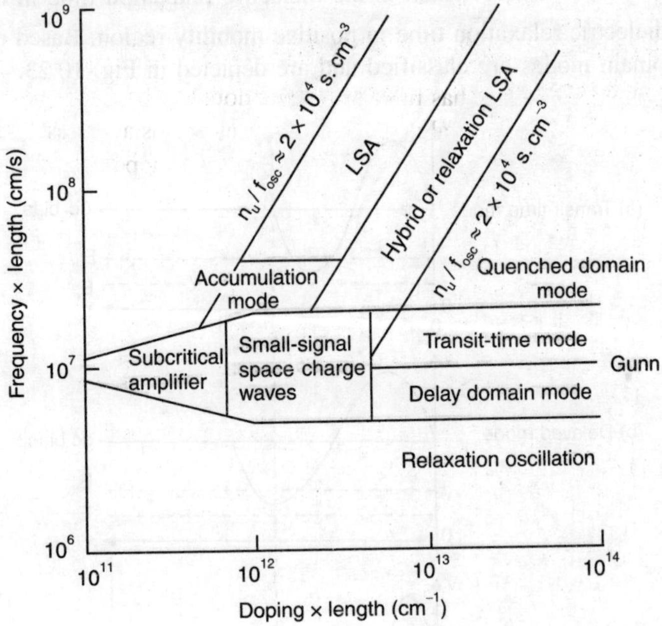

Fig. 10.21 Mode diagram for Gunn diodes (TEDs)

10.5.1 Dipole Domain Modes (Gunn Oscillation Modes)-time-parameters

The packaged Gunn diode is put in a microwave resonant circuit. Figure 10.22 shows a Gunn diode in a typical circuit. In this microwave resonant circuit is represented by *L*, *C* and G_L.

The functions of other components, indicated in the circuit are self-explanatory. The resonant frequency of the microwave circuit (without the Gunn diode) is f_r, and the corresponding period of

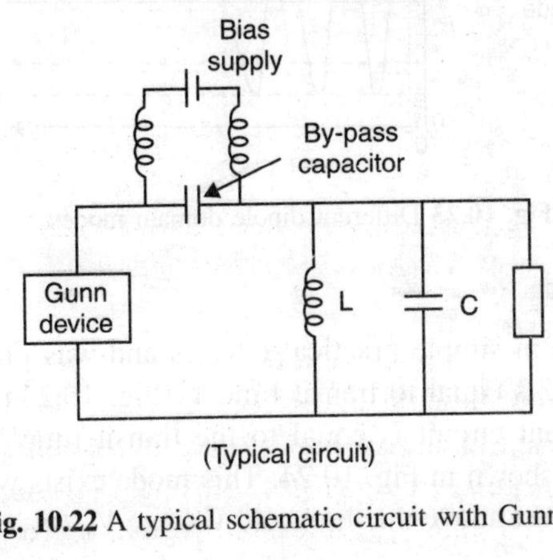

Fig. 10.22 A typical schematic circuit with Gunn diode

oscillation is $\tau_r = \frac{1}{f_r}$. The Gunn device will oscillate at a frequency f_{osc}. The transit time of matured domain across the full length of the device is $\tau_t = \frac{L_{eff}}{v_{dom}}$. Another important parameter is domain growth time constant τ_g. This time constant is the dielectric relaxation time in the negative mobility region and τ_d is the dielectric relaxation time in positive mobility region. Based on these various time parameters, dipole domain modes are classified and are depicted in Fig. 10.23.

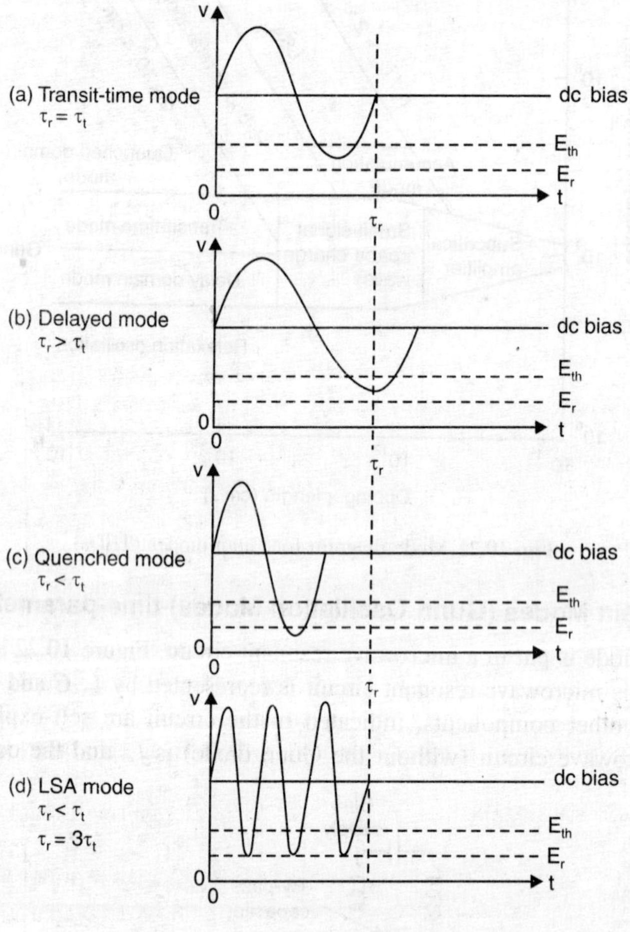

Fig. 10.23 Different dipole domain modes

10.5.1 (a) Transit Time Mode

It is most commonly used mode in simple practical devices and was first observed by Gunn. In this mode the period of oscillation τ_r is equal to transit time τ_t (Fig. 10.23*a*). It means that resonant frequency of the microwave resonant circuit is equal to the transit time frequency of this mode. The voltage and current waveform is shown in Fig. 10.24. This mode exists when the device is operated in a low Q circuit at a bias voltage just above the threshold voltage V_{th} (E_{th}). A domain is nucleated at the

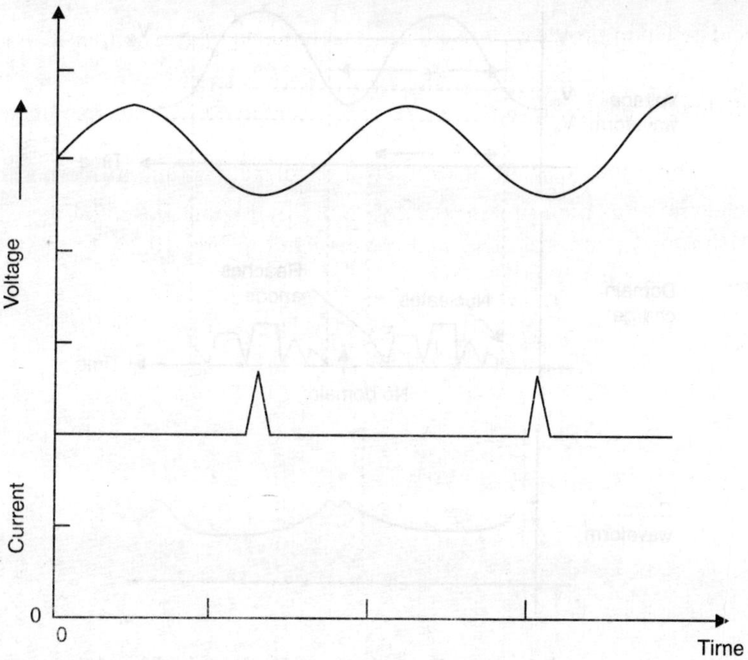

Fig. 10.24 Voltage and current waveforms in the transit time mode of operation of Gunn diode

cathode of the device when the voltage increases above the threshold voltage (field) V_{th} (E_{th}). A roughly constant current* in external circuit flows while the domain travels through the device. There is a short pulse of current when the domain is collected at the anode. As soon as the domain is collected at the anode, the voltage swings over the threshold value and the cycle repeats. This predicated an efficiency of about 10% for $n_0 \times L = 10^{12}$ cm^{-2}. It is quite low as compared to maximum expected efficiency of 32% because the current pulses are typically quite short.

For this mode, $fL \approx 10^7$ cms^{-1}.

10.5.1 (b) Delayed Domain Mode

In this mode the period of oscillation τ_r is more than $\tau_t (\tau_r > \tau_t)$. It means that the frequency is lower than the inverse of the domain transit time $\left(\frac{1}{\tau_t}\right)$. In this case the domain will reach the anode before or at the time the voltage goes below the critical value necessary to sustain a domain (Fig. 10.23*b*). Since the voltage keeps going down, the formation of new domain will be delayed until the voltage rises again. Thus, it is called the **delayed-domain mode.** Figure 10.25 shows the voltage waveform, the moving domain charge and the current waveform for this mode. The current pulse is somewhat broader than in case of transit time mode (Fig. 10.25), as such the efficiency is higher and which could reach to 20%. For this mode fL lies between 10^6 and 10^7 cms^{-1}. The frequency of oscillation is lower than $\frac{1}{\tau_t}$, as mentioned above, because of the delay information of new domain. The frequency of oscillation can

* No major displacement current occurs in the contacts since the domain contains equal amount of positive and negative charges.

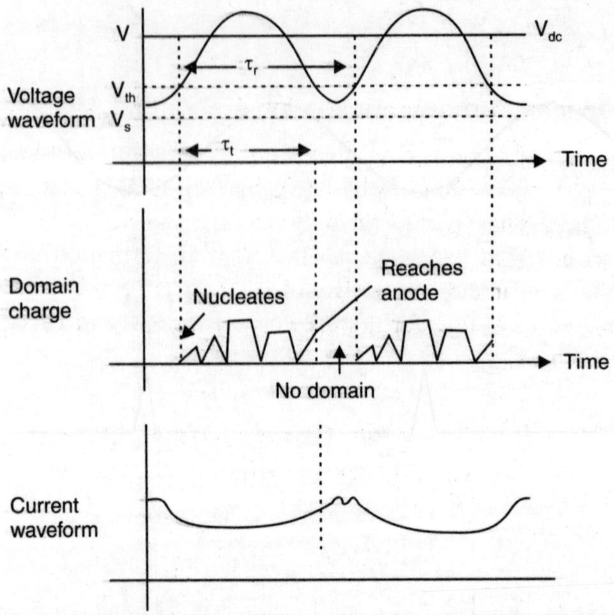

Fig. 10.25 Different waveforms (voltage, charge and current) in delayed domain mode

be varied by the resonant circuit and requires more careful tuning of the device than the transit time mode.

10.5.1 (c) Quenched Domain Mode

In this mode, when the voltage (field) swings below the sustaining voltage V_s (Field E_s Fig. 10.23c) the domain collapses (quenches) before it reaches the anode. The new domain is nucleated only when the bias voltage goes above V_{th} (field E_{th}), then the process is repeated. This mode is called the **quenched domain mode.** The waveform for this mode is shown in Fig. 10.26. The condition for the occurrence of this mode is that

$$\tau_d > \tau_r > \tau_s$$

and $\qquad f_{osc} > 1/\tau_t$

i.e., frequency of oscillation can be several times the transit time frequency.

Here τ_s is the time which it takes for domain to be quenched. It is roughly equal to the dielectric relaxation time τ_d with positive differential mobility which we mentioned earlier. For this mode $fL > 2 \times 10^7$ cm s^{-1} and the efficiency of the device

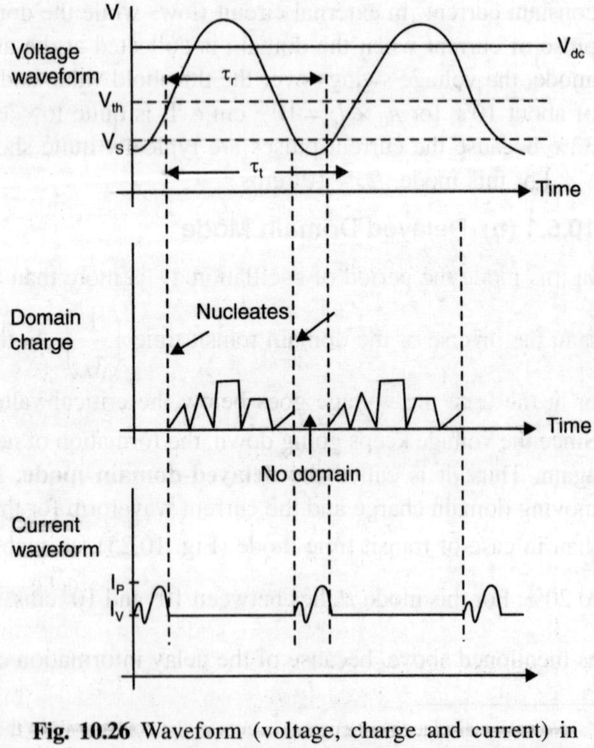

Fig. 10.26 Waveform (voltage, charge and current) in quenched domain mode

for this mode can be as high as 13%. The device can be made longer since the transit time condition no longer applies.

10.5.1 (d) Limited Space-charge Accumulation Mode (LSA Mode)

Limited space-charge accumulation (LSA) mode can occur when very high frequency (*i.e.* $\tau_r < \tau_t$) oscillating signal is superimposed on a steady state value (Fig. 10.27). Here domain formation is largely prevented because for the greatest part of the cycle the average electric field is in the range where the differential mobility is negative but small, so that domain formation goes slowly. Therefore, any accumulation of electrons near the cathode has enough time to collapse while the signal is below threshold (Fig. 10.10). The condition for the full LSA mode (domain never develops) requires that:

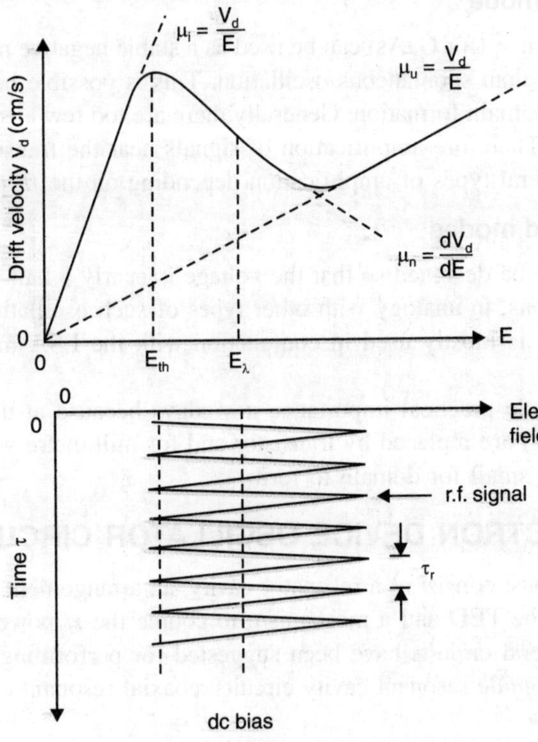

Fig. 10.27 LSA mode operation

(*i*) the excess-charge must not grow to a full domain in one period; and (*ii*) the quenching (dielectric relaxation) time must be less than the period. Using Eqn. 10.9*b*, we then obtain

$$\frac{\varepsilon}{n_0 e|\mu_n|} > \tau_r >> \frac{\varepsilon}{ne\mu_0} \quad ...(10.10)$$

The range of acceptable value of n_0/f_{osc} is about 10^4 (upper frequency limit) to 10^5 (low frequency limit) for GaAs. Both upper and lower boundaries of the LSA mode are indicated in Fig. 10.21. Predicted theoretical efficiency is around 25%, however in practice only about 20% has been reached. The LSA mode of oscillation has the advantage that the operating frequency is not limited by transit time effect as in case of quenched domain mode. Therefore, the sample length L, can be made large and the device can sustain higher voltage and higher power dissipation. Therefore, this mode is preferable when appreciable power is necessary at very high frequencies. However, the device fabrication and circuit adjustment (tuning) are much more difficult than for other modes of operation. This mode is not likely to occur in device for frequencies higher than 20 GHz.

10.6 OTHER MODES

(a) Accumulation layer modes

The accumulation layers will be formed instead of dipole domains when $n_0L \approx 10^{12}$ cm^{-2} (Eqn. 10.3) in *n*-GaAs device. These modes can also be delayed or quenched as for the dipole domain modes (See Fig. 10.19).

(b) Stable amplification mode

A device with $n_0 \times L < 10^{12}$ cm^{-2} (for GaAs) can be used as a stable negative resistance amplifier at the transit time frequency rather than spontaneous oscillation. This is possible because the negative conductance is utilized without domain formation. Generally there are too few electrons for domain formation within the transit time. Therefore amplification of signals near the transit time frequency can be accomplished. There are several types of amplification depending on the *fL* product of the device.

(c) Relaxation and hybrid modes

In this mode, the circuit may be designed so that the voltage is nearly a half sinusoidal. This mode is called as relaxation oscillations, in analogy with other types of such oscillations. The efficiencies are somewhat higher. The mode is mostly used in conjunction with the LSA mode and thus termed as hybrid mode (Fig. 10.21).

All these modes have little practical importance nowadays because at the lower microwave frequencies Gunn diodes (TEDs) are replaced by transistor and for millimetre wave, where they are still used, the dimensions are too small for domain to form.

10.7 TRANSFER-ELECTRON DEVICE OSCILLATOR CIRCUIT

Any TED oscillator circuit must consist of a resonator cavity, an arrangement for coupling TED to the cavity, a circuit for biasing the TED and a mechanism to couple the *rf* power from the cavity to the external circuit or load. Several circuits have been suggested for performing these functions. In this section we will discuss waveguide resonant cavity circuits, coaxial resonant cavity structure and YIG-tuned stripline configurations.

10.7.1 Waveguide Resonant Cavity Circuits

The simplest waveguide resonant cavity circuit is shown in Fig. 10.28 in which TED is mounted under a cylindrical metal post centred in the broad wall of a rectangular waveguide. The cavity is resonated at the desired frequency by adjusting the short circuit position $\left[f = \dfrac{v_d\,(\text{drift velocity})}{L_{\text{eff}}\,(\text{effective length})}\right]$. An inductive diaphragm separates the external waveguide from the cavity (Fig. 10.28). It is used to adjust the coupling between the output waveguide and cavity by changing the window opening. The top of the post is insulated from the waveguide to form *rf* bypass capacitor and thus *rf* currents do not flow through the bias voltage supply (Fig. 10.28). A tuning screw is inserted perpendicularly at the centre of the broadwall for fine tuning of the cavity. The resonator waveguide cavity has an impedance transforming property in order to reduce the high impedance of the output waveguide to a lower impedance at the TED.

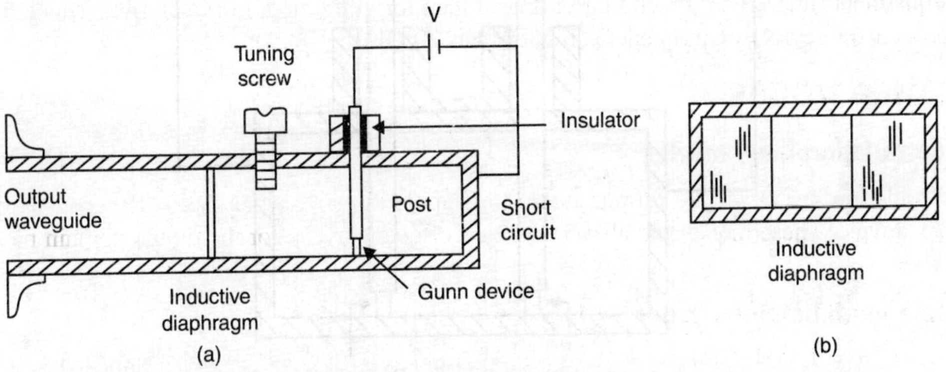

Fig. 10.28 Schematic diagram of waveguide resonant cavity TED oscillator circuit

Figure 10.29 is another arrangement for mounting TED in a resonant waveguide cavity. It used a two section quarter wave transformer to transform the high impedance of the waveguide to a low impedance at the TED (Gunn device). The cavity resonant frequency can be adjusted by changing the location of the short circuit. A tuning screw is provided (Fig. 10.29) for fine tuning of the cavity.

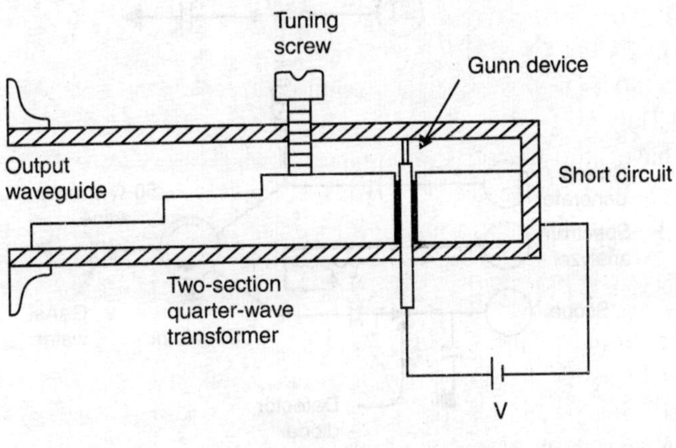

Fig. 10.29 Schematic arrangement of Gunn oscillator cavity which uses a two-section quarter-wave transformer

Another resonant waveguide-cavity structure is shown in Fig. 10.30 which has two posts, one for mounting the Gunn device and other for mounting a varactor diode for cavity tuning. A sawtooth sweep voltage applied to the varactor diode will produce frequency modulation of the oscillator.

10.8 TED AMPLIFIER CIRCUIT CONFIGURATION

A device with product of doping and length n_0L of a bulk GaAs, to be about sub-critical *i.e.,* it is about 10^{12} cm^{-2} or less, can be used as a stable amplifier mentioned in Section 10.5 where we have discussed various modes of operation. A typical TED amplifier circuit, used in conjunction with circulator-coupled network is shown in Fig. 10.31. It has a voltage gain-bandwidth product excess of 10 dB and frequencies from 4 to about 16 GHz. Linear gains of 6 to 12 dB per stage and saturated output power levels in excess of 0.5 W have been realized.

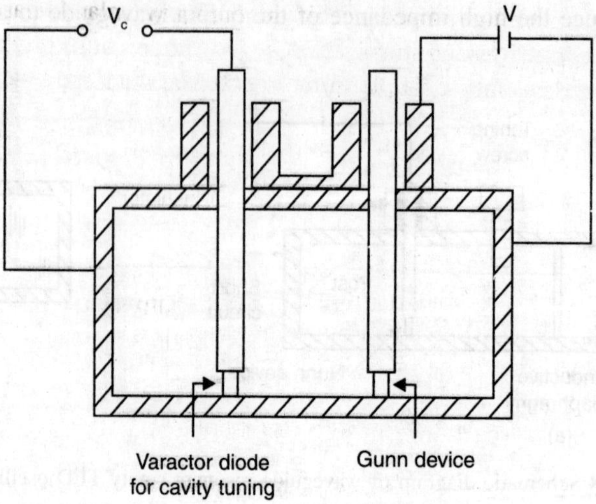

Fig. 10.30 A TED oscillator which has post-mounted varactor diode

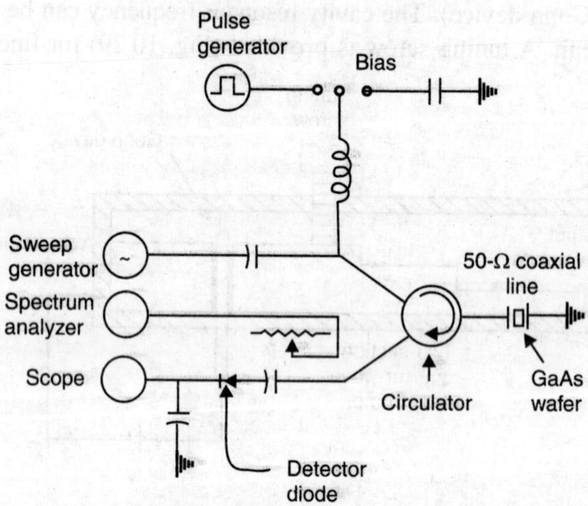

Fig. 10.31 TED amplifier circuit layout

10.9 INTRODUCTION–IMPATT AND TRAPATT DIODES

Impact ionization avalanche transit-time (IMPATT) devices are relatively high power (Figs. 10.32 and 10.33) and high efficiency (Fig. 10.34) solid state two-terminal generators or amplifiers for microwave and millimetre wave frequencies up to 300 GHz and beyond. Efficiencies as high as 25% have recently been obtained with GaAs devices, but the traditional IMPATT is a silicon device with up to 15% efficiency. Their basic physical mechanism is the interaction of the impact ionization avalanche and the transit time of charge carriers. This interaction results in a 180° phase-delay of the current with respect to voltage which clearly makes the device to exhibit a differential negative resistance characteristic. This negative resistance has been exploited to generate microwave and millimetre wave power and to amplify microwave and milimetre wave signals. This origin lies in a classic paper of 1958 by Read* in which he theoretically predicted that an idealized n^{+}-p-i-p^{+} diode could produce negative resistance effect at microwave frequencies. Only a couple of months before the Read type operation was realized

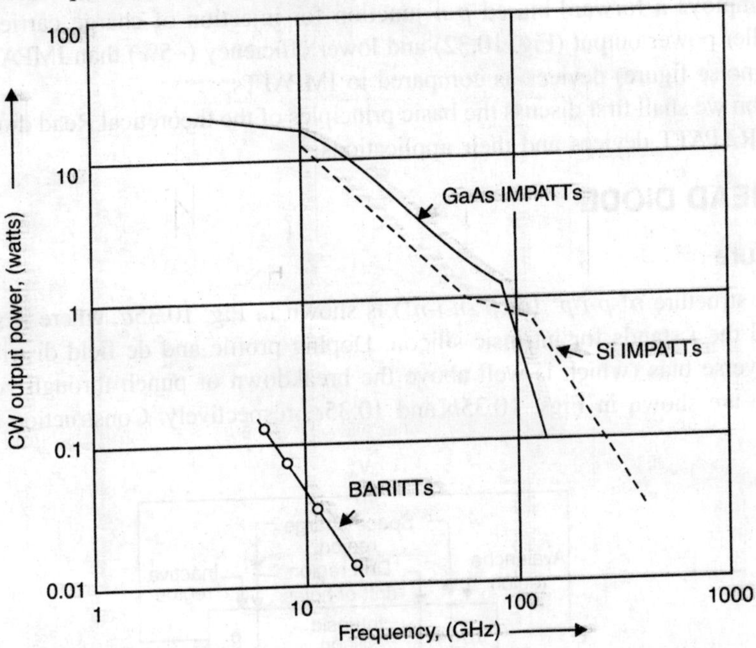

Fig. 10.32 CW output power versus frequency (IMPATTs and BARITTs)

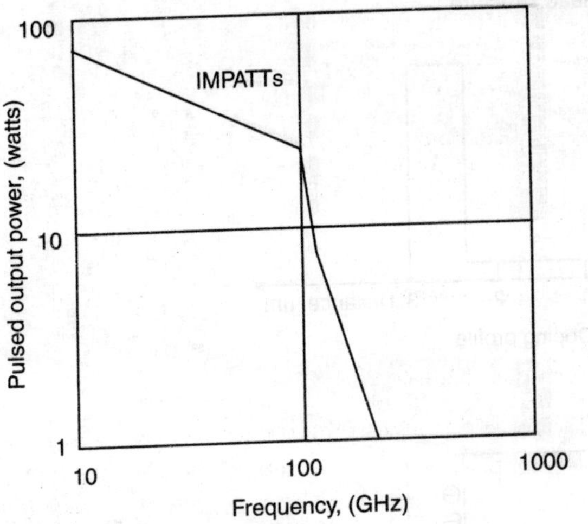

Fig. 10.33 Pulsed output power versus frequency (IMPATTs)

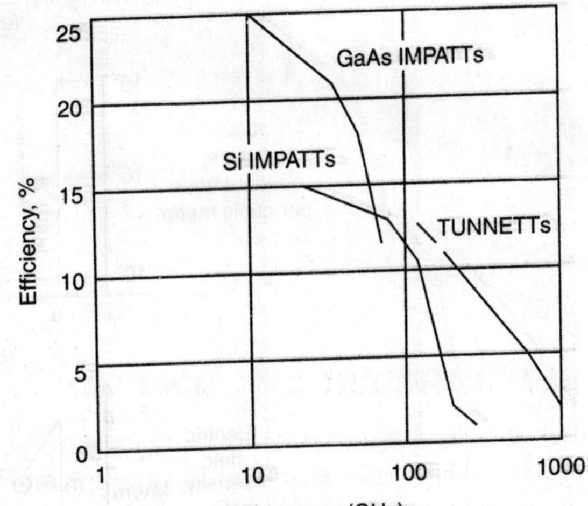

Fig. 10.34 Efficiency versus frequency (IMPATTs, TENNETTs)

* In 1954, Shockley, who first time proposed the use of transit time to obtain negative resistance from semiconductor diodes (Shockley, W., **Negative Resistance Arising from Transit Time in Semiconductor Diodes,** Bell Sys. Tech. J., 33, 4, pp. 799-826. 1954.)

by Lee *et al.*, 1965, Johnston *et al.*, 1965 reported IMPATT (Impact Ionization Avalanche and Transit-time) type microwave oscillator which could give 80 mW of pulsed power at 12 GHz using simple silicon *p-n* junction. In 1971, Coleman *et al.*, discovered BARITT (Barrier Injected Transit-time) diode. The BARITT employs a forward-biased *p-n* junction for injection of charge carriers. These devices have much smaller power output (Fig. 10.32) and lower efficiency (~5%) than IMPATTs, but these are less noisy (low-noise figure) devices as compared to IMPATTs.

In this section we shall first discuss the basic principles of the theoretical Read diode and then study IMPATT, and TRAPATT devices and their applications.

10.10 THE READ DIODE

10.10.1 Structure

The Read diode structure n^+-p-i-p^+ (or p^+-n-i-n^+) is shown in Fig. 10.35*a*, where + sign denotes very high doping and the *i* stands for intrinsic silicon. Doping profile and dc field distribution that exists when a large reverse bias (which is well above the breakdown or punch through voltage) is applied across the diode are shown in Figs. 10.35*b* and 10.35*c*, respectively. Construction details of silicon

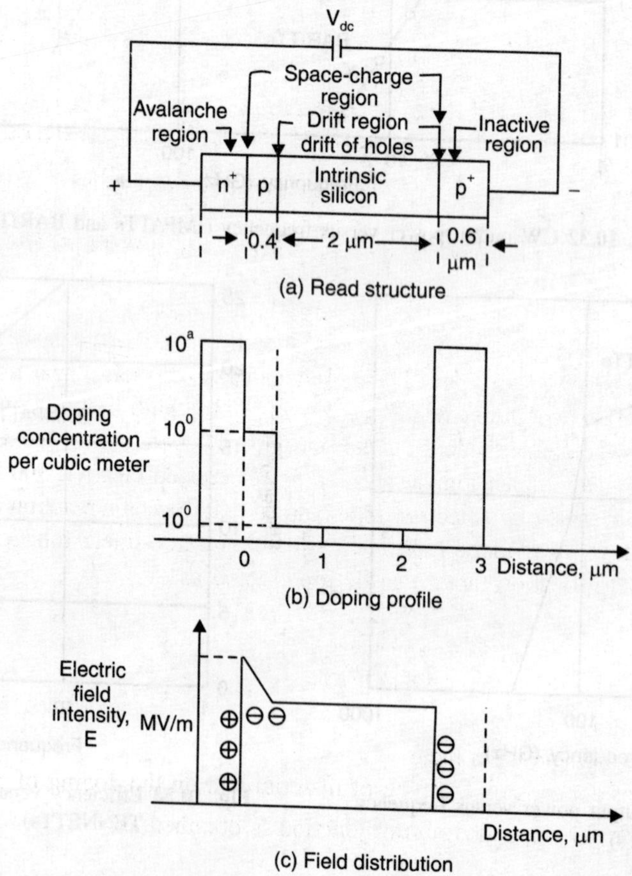

Fig. 10.35 (*a*) Read diode structure showing four regions; (*b*) doping profile; and (*c*) dc field distribution under high reverse bias

n^+-p-i-p^+ Read diode is shown in Fig. 10.36. As seen from Fig. 10.35 the Read diode has four regions which together form an n^+-p-i-p^+ structure. The four regions are:

1. The avalanche or high field region, which is in the vicinity of the n^+-p junction, where holes and electrons are generated.
2. The space between n^+-p junction and i-p^+ junction is called the space charge region.
3. The i or v region through which the generated holes must drift in moving to the p^+ contact. The region is called the intrinsic region or drift region.
4. The second drift region or inactive region which is associated with the heavily doped p^+ section.

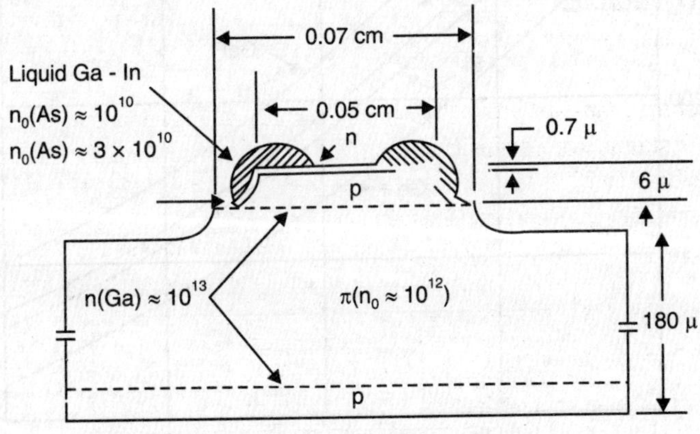

Fig. 10.36 Lee's Read diode construction details

10.10.2 Operation—Physical Processes Leading to the Negative Resistance

(a) Avalanching

When the reverse bias voltage is well above the breakdown voltage, one can see from Fig. 10.35(*c*), the electric fields have a relatively sharp peak (several mega volts per metre) near n^+-p junction (depending upon the doping profile). Those holes which are moving in the high field region near the n^+-p junction acquire enough energy to knock valence electrons into the conduction band thus producing hole-electron pair. This is called Impact ionization. The newly created electron and hole are accelerated in the field and generate, in turn, extra electron-hole pair, so that the chain reaction occurs. The rate of pair production is referred to as avalanche multiplication and is a non-linear function of the field. Empirically the avalanche multiplication factor can be represented as

$$M = \frac{1}{1-(V/V_b)^n} \quad ...(10.11)$$

where V = applied voltage

V_b = avalanche breakdown voltage

n = 3 to 6, that depends on the type of material and on the doping of p^+-n or n^+-p junction.

The breakdown voltage for one-sided abrupt junction is obtained as

$$|V_b| = \frac{\varepsilon E_m^2}{2eN_B} \quad ...(10.12)$$

where E_m = Critical field for breakdown (maximum field at breakdown)

ε = Semiconductor permittivity

N_B = Impurity concentration

As seen from Eqn. 10.12, the avalanche breakdown voltage is inversely proportional to the impurity concentration.

A generalized avalanche condition, *i.e.*, condition for self-sustained avalanching (the avalanche keeps itself growing independent of charge carrier entering in avalanche region) is obtained as where

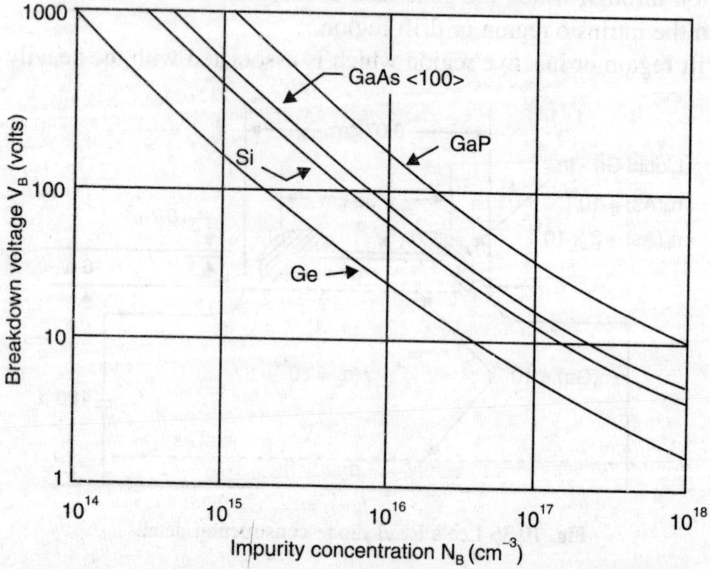

Fig. 10.37 Avalanche breakdown voltage versus impurity concentration

$\bar{\alpha}$ is the average ionisation rate* for electrons and holes and l_a is the length of the avalanche region.

$$\int_0^{l_a} \bar{\alpha}\, dx = 1 \qquad \text{...(10.13)}$$

Due to the strong dependence of α on electric field, as such avalanche multiplication will be confined very closely to the peak of the electric field (Fig. 10.35(*c*)), narrow region near the n^+-p junction), this region of the diode is termed as the avalanche region (Fig. 10.35(*a*)).

(b) Drifting of charge carriers

Because of the reverse bias, the electrons generated due to avalanching move immediately into the n^+-region. Holes, on the other hand, drift through the space charge (drift) region without multiplying ($\alpha = 0$) with the saturation velocity** (drift velocity) v_d which is of the order of 10^5 m/s for silicon. Thus

---

* α, the ionisation rate is the probability of impact ionisation occurring in unit length of travel of an accelerated carrier. The units are usually cm^{-1}. The ionisation rates for electrons and holes are not the same. Also, the ionisation rates depend very strongly on the electric field. A typical dependence is $\alpha(E) = \alpha_0 E^6$.

** The electric field intensity in the space-charge (drift) region is about 0.5 MV/m and is considerably smaller than E_m at the junction. However, it is still in excess of the E field needed for velocity saturation scattering limited velocity for the device material. Figure below shows the electron velocity as function of dc electric field for the two commonly used materials Si and GaAs. We note that E in excess of 0.5-1.5 MV/m is consequently needed for Si and GaAs, and saturation velocity of the order of 10^5 m/s is obtained for the drift region.

we see that there is a narrow avalanche region on the p-side of the n^+-p junction that supplies the holes and a wide space charge region in which the holes drift at a constant velocity but do not multiply ($\alpha = 0$). If the length of the drift region (space charge region or intrinsic region) is l_d, the drift time through this region is

$$\tau_d = \frac{l_d}{v_d} \qquad \text{...(10.14)}$$

For $\quad l_d = 2 \times 10^{-6}$ m and $v_d = 10^5$ m/s

$$\therefore \quad \tau_d = \frac{2\times10^{-6}}{10^5} = 2\times10^{-5}\ \mu\text{s}.$$

(c) Avalanche current (carrier current) and external current

The hole current $I_0(t)$ generated at n^+-p junction will induce a current $I_e(t)$ in the external circuit. Since the drift velocity of the holes in the space-charge region is constant, the induced current $I_e(t)$ in the external circuit is simply $\frac{1}{\tau_d}$ times the total charge of moving holes. Here l_d is the length of the drift region and τ_d is the associated drift time. If a pulse of holes of charge ΔQ is suddenly generated at the n^+-p junction a constant current $I_e(t) = \frac{\Delta Q\, v_d}{l_d}\left(\text{or } \frac{\Delta Q}{\tau_d}\right)$ will flow immediately in the external circuit and continue to flow during the time τ_d in which the holes are moving across the drift region and collected by the p^+-region. On the average, therefore, the circuit current $I_e(t)$ is delayed by $\tau_d/2$ sec. relative to hole current $I_0(t)$ generated near the n^+-p junction.

In this case avalanche multiplication is confined to very narrow high-field region at the n^+-p junction as mentioned earlier, as such one can regard the avalanche multiplication rate as a function of the peak field E_0. At a maximum field (critical field) E_m breakdown will occur and multiplication will be self-sustaining, *i.e.*, self-sustaining current. When the field is above E_m, $I_0(t)$ will be more than self-sustaining and will build up and when the field is below E_m the current is less than self-sustaining. Here, E_m is the maximum field a breakdown voltage.

(d) Subjected to ac voltage (microwave voltage)

In addition to dc reverse bias, which is just below the critical field E_m for breakdown, the Read diode (n^+-p-i-p^+ structure) is also subjected to an alternating voltage (Fig. 10.38). Thus the total field across the diode is the sum of the dc and ac fields. The peak field is above the E_m during the positive half and below E_m during the negative half. Therefore, $I_o(t)$ builds up during the positive half and dies down during the negative half (Fig. 10.38(*d*)). In other words $I_0(t)$ reaches its maximum in the middle of the ac voltage cycle (Fig. 10.38(*d*)) or one-quarter of a cycle later than the voltage. It means $I_0(t)$ at the n^+-p edge of the space-charge (depletion) layer is delayed from the voltage by 90°. This current in turn gives rise to an $I_e(t)$ in the external circuit. The finite transit time of the holes (carriers) results in a current pulse of a duration equal to the transit time τ_d. If the length of the drift region is such that the transit time (drift time) of carriers through drift region is $\frac{T}{2}$ (where T is the period of the ac voltage) then $I_e(t)$ is delayed by 90° with respect to $I_0(t)$. Since the holes (carriers) $I_0(t)$ is delayed by 90° relative to the voltage, the external current is then delayed by 180° relative to the ac (microwave) voltage as shown in Fig. 10.38(*d*). This implies that the resistance of the diode (device) is negative.

The device exhibits differential negative resistance when length of the space charge (*i*-layer) region l_d is such that the drift time

$$\tau_d = \frac{l_d}{v_d} = \frac{T}{2} = \frac{1}{2f} \qquad \text{...(10.15)}$$

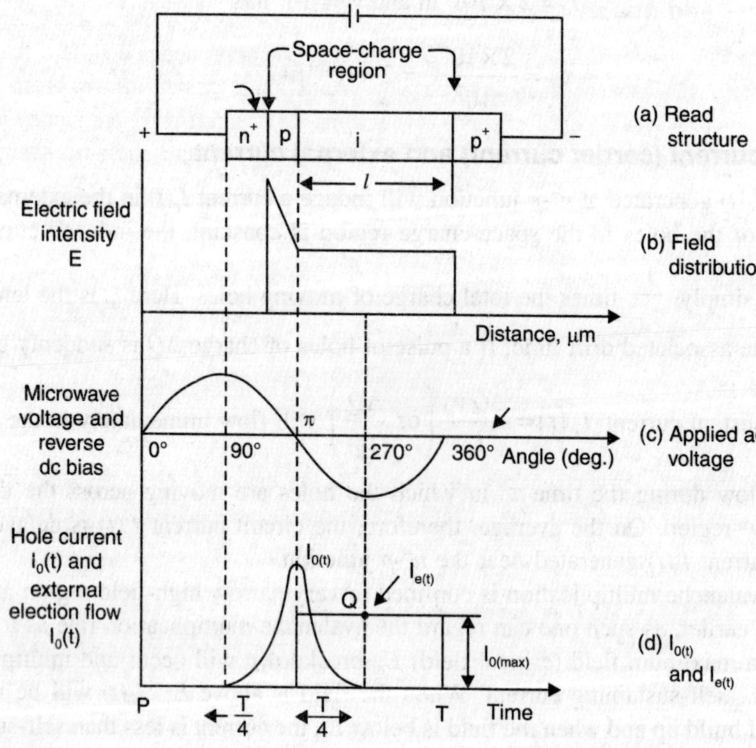

Fig. 10.38 Phase relationship between ac voltage and the external current associated with read diode

$$\therefore \qquad l_d \approx \frac{v_d}{2f} \qquad \text{...(10.16)}$$

or

$$f = \frac{v_d}{2l_d} \qquad \text{...(10.17)}$$

For example, when

$v_d = 10^5$ m/s, $l_d = 2 \times 10^{-6}$ m, the diode can operate at frequency

$$f = \frac{10^5}{2 \times 2 \times 10^{-6}} = 2.5 \times 10^{10}$$

$$= 25 \text{ GHz}$$

Read* pointed out that one could design a high frequency negative resistance diode that can be used for microwave oscillations and amplifications. When such a diode, which is reverse biased just below the critical field E_m for avalanche breakdown, is mounted in a resonator cavity and tuned to the frequency (Eqn. 10.17) then the negative resistance of the diode allows a continuous oscillation to be sustained by converting a portion of the dc bias into microwave power output from the cavity. The impedance of the cavity is mainly inductive and is matched to the capacitive impedance of the diode to form a resonant circuit.

(e) Output power and quality factor

One can note from Fig. 10.38(*d*) that the external current $I_e(t)$ approaches a square wave with top and bottom levels at $I_{e(\max)}$ and nearly zero, respectively. The average current drawn from the reverse bias supply is $I_{e(\max)}/2$ and the fundamental current component at the microwave frequency has a peak value $2I_{e(\max)}/\pi$. If V_a is the amplitude (peak value) of the microwave voltage, the ac power delivered to the cavity is [15, 18]

$$P = \frac{V_a}{\sqrt{2}} \times \frac{2I_{e(\max)}}{\pi} = \frac{\sqrt{2}}{\pi} V_a I_{e(\max)} \text{ watts} \qquad ...(10.18)$$

The quality factor Q of a circuit is defined as

$$2\pi \times \frac{\text{peak power stored during a cycle}}{\text{average power dissipated over the cycle}} \qquad ...(10.19)$$

Since the Read diode delivers energy to the cavity, its effective Q is negative in contrast to the positive Q of the cavity. The negative Q of the diode must balance with the cavity's positive Q to maintain a continuous oscillation.

(f) Analytical results

Real part of the Read diode impedance is plotted against transit angle $\left(\theta_d = \omega\tau_d = \dfrac{\omega l_d}{v_d}\right)$ in Fig. 10.39.

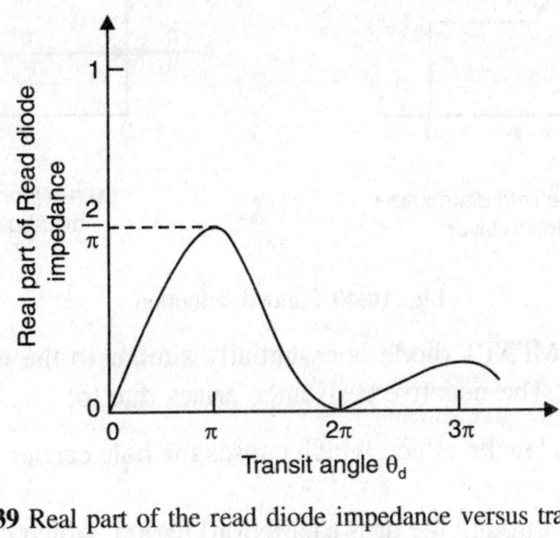

Fig. 10.39 Real part of the read diode impedance versus transit angle

* One may note that there is further drift of the charge carriers in p^+ layer, so that the Read diode has a double drift region which results in the most efficient form of avalanche (IMPATT) diodes.

It shows strong negative resistance effect for relatively small transit angle when $\omega > \omega_r$ (avalanche resonant frequency). The peak value of the negative resistance occurs near $\theta_d = \pi$ (Fig. 10.40). As such, Read diode work well only in a frequency range around the π transit angle. That is

$$f = \frac{1}{2\tau_d} = \frac{v_d}{2l_d} \quad ...(10.20)$$

It is the same as Eqn. 10.17.

10.11 IMPATT DIODES

10.11.1 Introduction

As mentioned in Section 10.9 that Johnston *et al.* fabricated a Si *p-n* junction diode which exhibited negative resistance characteristic associated with its avalanche transit time properties at microwave frequencies. This led to the construction of avalanche *p-n* microwave oscillators and amplifiers. Figure 10.40(*a*) shows the field distribution in Read diode (n^+-*p*-*i*-p^+) and Fig. 10.40(*b*) shows the field distribution in *n-p* diode. One can observe that in both cases, there is a region within which the field varies linearly and there is a constant field region in the Read diode and not in the *n-p* (*p-n*) junction. This difference is of no real consequence as long as the avalanche region is much smaller as compared to the length of the depletion region (drift region), because carriers once generated drift with the saturation velocity characteristics of the material without avalanching (as the field is low enough). The principle

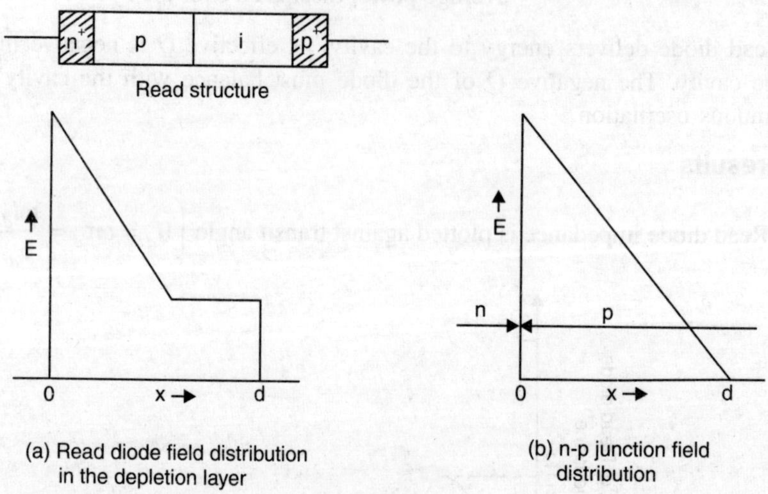

Fig. 10.40 Field distribution

of operation of *p-n* junction IMPATT diode is essentially similar to the mechanism described for the Read diode (Section 10.10.2). The negative resistance arises due to:

1. The impact ionisation avalanche effect, which causes the hole carrier current $I_0(t)$ to lag the microwave voltage by $\pi/2$.
2. The effect of *transit time* through the drift (depletion) region, which causes the exteral current $I_e(t)$ lagging the microwave voltage by a further $\pi/2$.

Thus the Read diode is the basic type in the IMPATT diode family.

A great variety of doping profiles have been used for IMPATT diodes such as p^+-n-n^+ (single drift region), p^+-p-n-n^+ (double drift) and p^+-i-n^+.

10.12 PERFORMANCE OF IMPATT DIODES

The IMPATT diodes have been fabricated from germanium, silicon, gallium arsenide or indium phosphide and can probably be constructed from other semiconductors as well. Schottky barriers rather than *p-n* junction can be used for IMPATTs. Figure 10.41 shows the physical layout of a Schottky barrier IMPATT diode and Fig. 10.42 shows a typical package for an IMPATT diode.

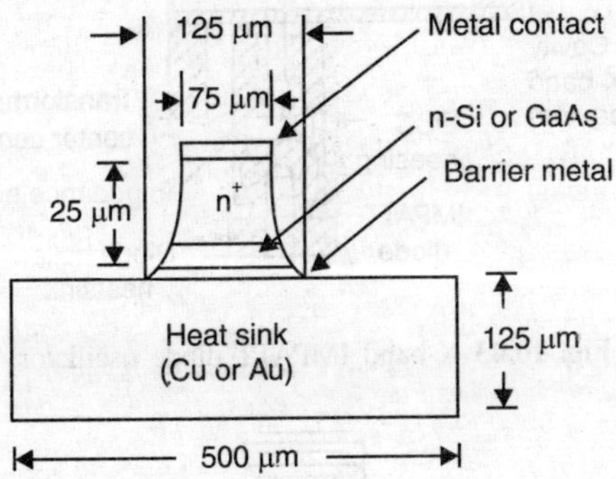

Fig. 10.41 Physical layout of a Schottky barrier IMPATT diode

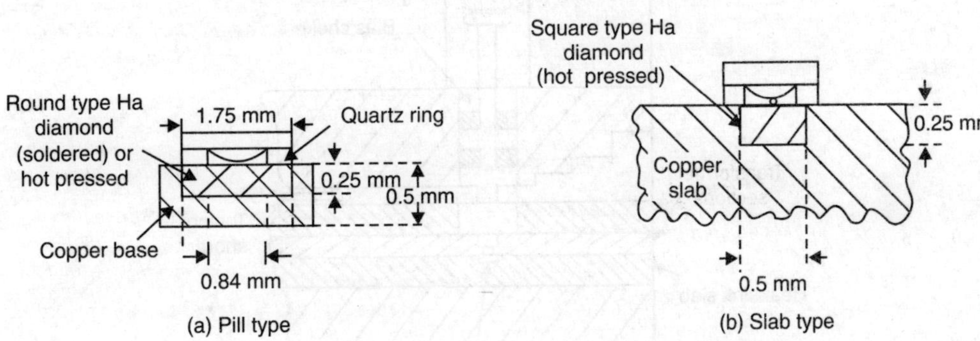

Fig. 10.42 A typical package for an IMPATT diode

Silicon IMPATTs have achieved oscillations up to 423 GHz in harmonic mode and in a fundamental mode up to 341 GHz. GaAs devices have gone up to 230 GHz. The performance of the IMPATTs depends on the various factors such as doping density, the thickness of the *n*-epitaxial layer, breakdown voltage, the ratio of the width of the avalanche region to that of the total width of the drift region.

10.13 IMPATT OSCILLATORS AND AMPLIFIERS

IMPATT diodes are used in driver stages, power stages and as a local oscillator in communication systems as well as in Radar. A number of oscillator circuits have been developed for use with IMPATT microwave and millimeter devices. In general, a circuit which has a low impedance, and also a convenient way for introducing the bias, is required. A physical layout of IMPATT diode oscillator (X-band 8-12 GHz) is shown in Fig. 10.43. It is a combination of a coaxial line and a waveguide circuit. This circuit makes it easy to combine the power of several IMPATT diodes by placing coaxial mounts at a

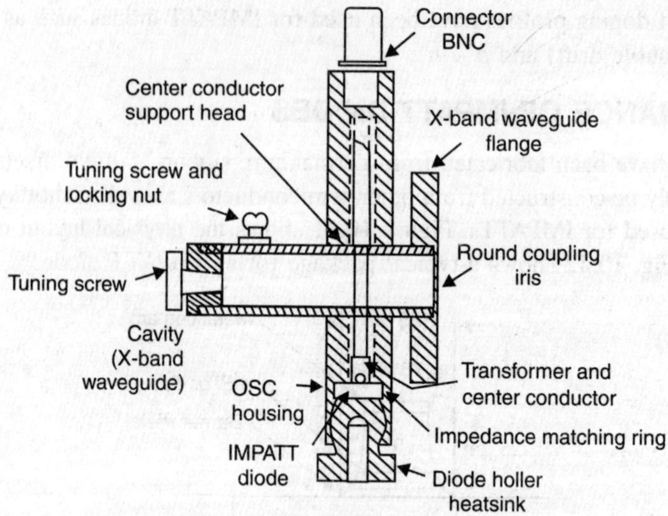

Fig. 10.43 X-band IMPATT diode oscillator

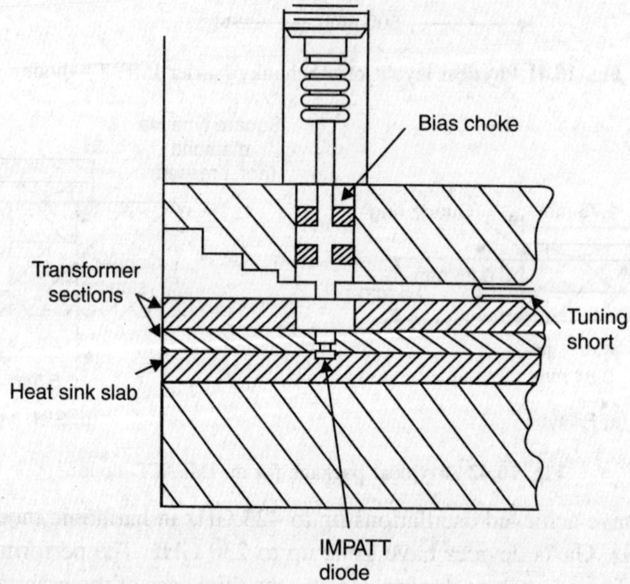

Fig. 10.44 Q-band IMPATT diode oscillator

distance of one wavelength along the waveguide. A layout of Q band (30-50 GHz) oscillator is shown in Fig. 10.44. In which the waveguide height is reduced in steps to provide impedance matching. Two microwave strip circuits are shown in Fig. 10.45.

A single stage amplifier connected via a circulator is shown in Fig. 10.46(*a*) and three stage configuration of IMPATT amplifiers is shown in Fig. 10.46(*b*). The circulator is used to separate incoming and reflected signals. **IMPATT amplifiers are used where high power is necessary and low noise is less important.**

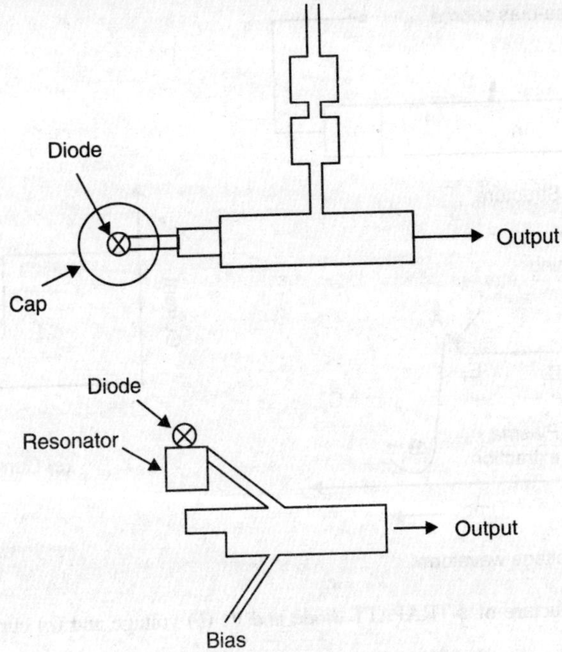

Fig. 10.45 Two microstrip circuits

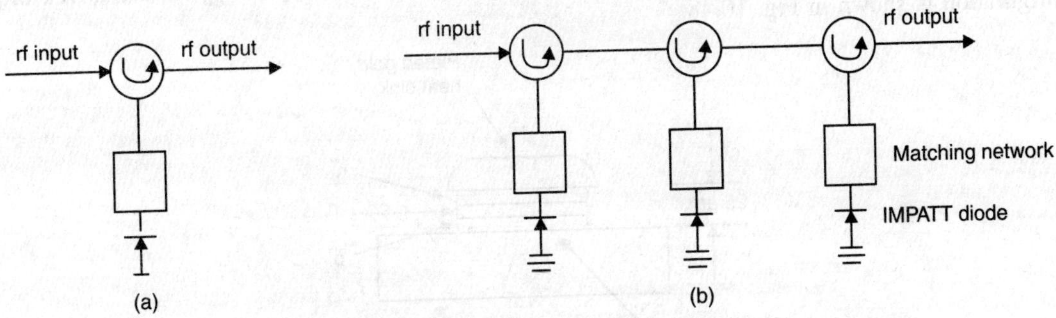

Fig. 10.46 (*a*) Single and (*b*) three stage IMPATT amplifier

10.14 TRAPPED PLASMA AVALANCHE TRIGGERED TRANSIT-TIME (TRAPATT) DEVICES

10.14.1 Structure and Operation

(a) Structure

The other transit-time device is trapped plasma avalanche triggered transit-time (TRAPATT) device. It is a high-efficiency device as compared to IMPATT diode and is capable of operating from several MHz to several GHz and requires a large repetitive pulse of current for its operation as an oscillator or as an amplifier. The typical TRAPATT is p^+-n-n^+ (or n^+-p-p^+) structure (Fig. 10.47(*a*) made from silicon or germanium with the *n*-type depletion region width varying from 2.5 μm to 12.5 μm. The device's p^+ region is kept as thin as possible at 2.5 μ to 7.5 μm. The doping of the depletion region is such that the

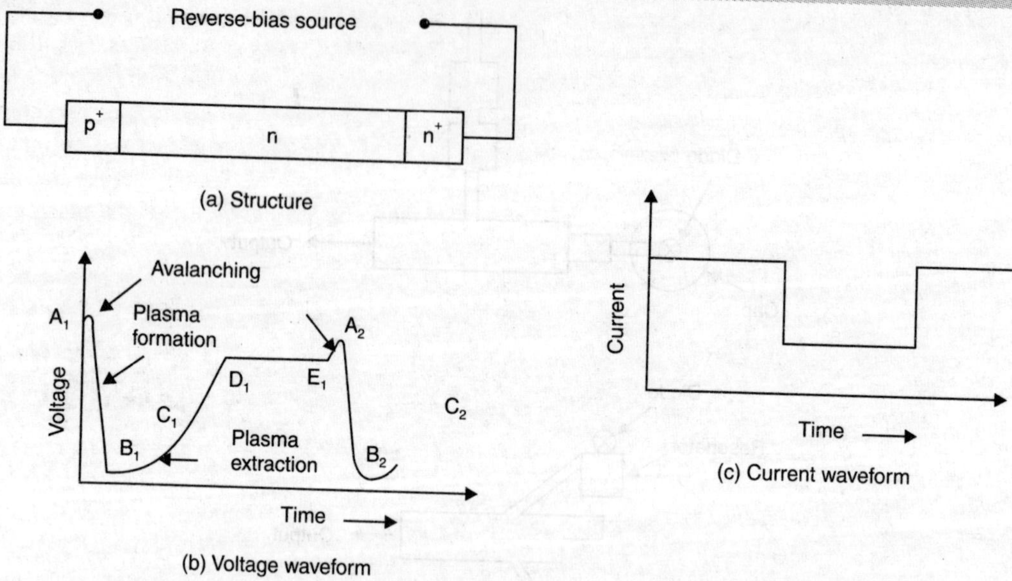

Fig. 10.47 (*a*) Structure of a TRAPATT diode and its (*b*) voltage and (*c*) current waveforms

diode is fully punched through when breakdown occurs, in other words, the depletion layer's electric field intensity just prior to breakdown is well above the saturation velocity. A typical TRAPATT diode configuration is shown in Fig. 10.48.

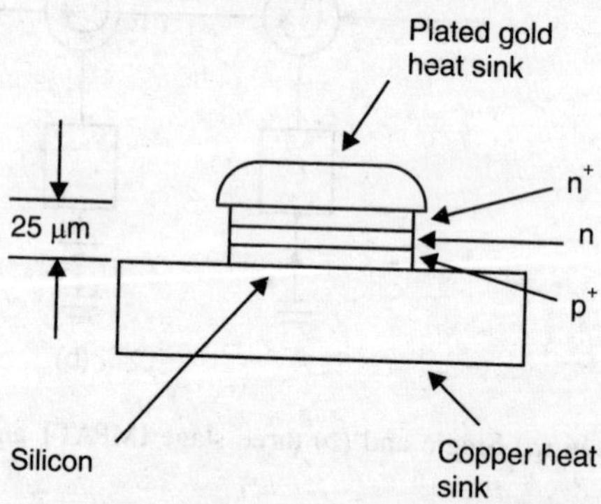

Fig. 10.48 An L-band TRAPATT diode configuration with double heat sink (*n*-layer the thickness is about 3 µm)

(b) Operation

Several workers developed the approximate analytical solutions for the TRAPATT mode in p^+-n-n^+ diodes. They have shown that a high field avalanche zone propagates through the diode and fills the depletion layer with a dense plasma of electrons and holes that become trapped in the low field region behind the zone. A typical voltage and current waveforms* across a TRAPATT mode of avalanche p^+-n-n^+ diode is shown in Fig. 10.47(*b*) and (*c*), respectively. The TRAPATT operation can be explained

* These waveforms are closest to an ideal 100 per cent efficient negative resistance diode which would produce square-wave voltages and currents that are exactly 180° out of phase with one another.

by referring these figures. During the operation of the device as an oscillator, the reverse biased TRAPATT diode is mounted in a high-Q-resonator at a point where the *rf* voltage swing is maximum. Oscillations start with the diode operating as an IMPATT device. When the amplitude of oscillation becomes large enough, the total voltage (dc bias together with *r.f.* swing) swings beyond the threshold of breakdown, a plasma of holes and electrons are generated. The presence of the plasma creates across the junction a high potential difference which opposes the reverse bias voltage and the voltage across the diode drops to a low value (points B_1, B_2, etc.). At this reduced voltage (it means low-field) the plasma gets trapped. But since external current is flowing so that voltage rises, the plasma trapped in low-electric field region is released from the diode (over periods B_1C_1, B_2C_2, ... etc.) producing current pulse across the drift region. Once the plasma is released (or extracted) completely, the cycle will repeat.

The electric field can be expressed as

$$E(x, t) = E_m - \frac{eN_a}{\varepsilon}x + \frac{Jt}{\varepsilon} \quad \text{...(10.21)}$$

where

N_a = Doping concentration of *n*-region

ε = Semiconductor permittivity

E_m = The value of electric field at a given distance *x* into the depletion region and is obtained by setting

$E(x, t) = E_m$. We get from Eqn. 10.21.

$$t = \frac{eN_a}{J}x \quad \text{...(10.22)}$$

Differentiating Eqn. 10.22 with respect to *t*, we get v_{az} the avalanche zone velocity

$$v_{az} = \frac{dx}{dt} = \frac{J}{eN_a}$$

Thus the avalanche zone will quickly sweep across almost entire diode, and the diode will be filled by highly conducting electron hole plasma whose space charge depresses the voltage to low values. The drift velocity depends on particles' mobilities, hence the transit time of the carrier becomes much longer than the transit time τ_d

$$\tau_d = \frac{l}{v_d} \quad \text{...(10.23)}$$

(c) Performance

Because of the large extraction time of the electron hole plasma, the total transit time of TRAPATT is much larger as compared with the IMPATT diode, so that its operating frequency is only of the order of a few gigahertz (around 10 GHz). However, since the current pulse is associated with the relatively low voltage, the amount of power dissipation is low and the efficiency and output power are correspondingly high. These oscillators are mostly used for pulse operation. The maximum efficiency achievable in this mode is about 80%. A typical silicon TRAPATT diode when used for the pulsed operation in 1-2 GHz range gave output power of 1 kW and efficiency of about 60 per cent. Table 10.2 lists some of the parameters of TRAPATT oscillators.

Table 10.2 Some parameters of TRAPATT oscillators

Parameters	*Frequency (GHz)*				
	0.5	1	2	4	8
Peak power (W)	600	200	100	100	50
Average power (W)	3	1	1	1	1
Operating voltage (V)	150	110	80	80	60
Efficiency (%)	40	30	25	20	15

Because extremely large fluctuations occur in the startup of the current by avalanche multiplication, TRAPATT generally exhibits a considerably higher noise, which severely limits its use as an amplifier. Further, their bandwidth is limited to about 15 per cent. Because of these limitations TRAPATTs, in spite of their high-power output and efficiency are not used extensively or their use is decreasing.

(d) IMPATT diode power combining

The power output from a single device decreases rapidly with increasing frequency and is limited to milliwatts of power (Fig. 10.11 and Fig. 10.12). However, achievable power levels can be significantly increased by combining the output of several devices. Several techniques have been developed for power combining. These are:

1. Chip-level combining,
2. Circuit level combining,
3. Spatial (Quasi-optical) combining, and
4. Multi-level combining (combinations of above three).

Example 10.1. An N type GaAs Gunn diode has the following specification:

Threshold field E_{th} = 3 kV/m

Applied field E = 3.5 kV/m

Device length L = 10 μm

Doping constant n_o = 10^{14} cm^{-3}

Operating frequency f = 10 GHz

Calculate (*a*) Current density (*b*) (–ve) electron mobility

Solution.

(*a*) At the low frequency limit the electron drift velocity for *n*-type GaAs is

$$v_d = fL = 10 \times 10^9 \times 10 \times 10^{-6} = 10^5 \text{ m/sec} = 10^7 \text{ cm/sec}$$

(*b*) The current density

$$J = en_0 v_d = 1.6 \times 10^{-19} \times 10^{14} \times 10^7 = 160 \text{ A/cm}^2$$

(*c*) The negative electron mobility is

$$\mu_n = \frac{-v_d}{E} = -\frac{10^7}{3500} \approx -2857 \text{ cm}^2/\text{V.sec}$$

Example 10.2. For a transit time, the domain velocity is equal to the carrier drift velocity and is equal to 10^7 cm/s. Determine the drift length of the diode at a frequency of 10 GHz.

Solution. The drift length of the diode is

$$L = \frac{v_{dm}}{f} = \frac{10^7}{10 \times 10^9} = 10\ \mu\text{m}$$

Example 10.3. GaAs has the following specification

$$\mu_1 = 8500\ \text{cm}^2/\text{V.s.}$$
$$n_o = 10^{15}/\text{cm}^3$$
$$\varepsilon_r = 13$$

Calculate dielectric relaxation time for GaAs assuming $\varepsilon_0 = 8.85 \times 10^{-12}$ *f*/m and $e = 1.6 \times 10^{-19}$ C

Solution. Dielectric relaxation time $\tau_d = \dfrac{\varepsilon}{e n_0 \mu_1}$ as per Eqn. 10.9(*b*)

Substituting the values

$$\tau_d = \frac{8.85 \times 10^{-12} \times 13}{1.6 \times 10^{-19} \times 10^{15} \times 8500 \times 10^{-4} \times 10^6}$$

$$\approx 0.84 \times 10^{-12} = 0.84\ \text{ps.}$$

REVIEW QUESTIONS AND PROBLEMS

1. Discuss the Gunn diode in relation to its negative resistance property.
2. Derive the condition for negative resistivity/NDC in two valley semiconductors.
3. Discuss the difference between transfered electron devices and avalanche transit time device.
4. Describe the modes of operation for Gunn diodes.
5. Describe the Ridley-Watkins-Hilsum theory.
6. Discuss the formation of high field domain in TEDs.
7. Explain dielectric relaxation time concept and instability condition and formation of high field domains.
8. A Gunn diode has a drift length of 2 μm. Find the minimum voltage required to set in Gunn mode.
9. A Gunn diode oscillating in the stable domain mode has a sample length of 5 μm. Calculate the natural frequency of operation drift velocity is 2×10^7 cm/s.
10. Calculate the natural frequency of the Gunn diode if the drift length is 10 μm and drift velocity of electrons is 2×10^5 m/s.
11. Determine the drift length of the Gunn diode oscillating at a frequency of 8 GHz in a transit time mode. The domain velocity is equal to the carriers drift velocity and is equal to 10^5 m/s.
12. A Gunn-device using GaAs operated at 20 GHz in the transit time mode. Find the drift length of the device.
13. Explain the principle of operation of Read diode.
14. Draw Read structure and mark the four regions which together form Read structure.
15. Explain Read diode is the basic type in the IMPATT diode family.

16. Discuss the phase relationship between the various quantities in a Read diode.

17. An IMPATT diode has a drift length of 2 μm and drift velocity is of the order of 10^5 m/s. Determine (*a*) the drift time of the carriers and (*b*) the operating frequency of the IMPATT diode.

18. Explain the operation of TRAPATTs and discuss why operating frequency is only of the order of a few gigahertz.

19. Determine the avalanche-zone velocity in case of TRAPATT diode which has the following parameters:

(*i*) doping concentration $N_a = 2 \times 10^{21}$ m^{-3}

(*ii*) current density $J = 20$ kA/cm^2

CHAPTER 11

Microwave Measurements

11.1 INTRODUCTION

Microwave measurements are concerned with distributed circuit elements, which are acted upon by electromagnetic waves having interdependent E and H fields. The time varying magnetic fields generate the time varying electric fields and *vice versa.* The electromagnetic fields produced, propagate in the free space with a velocity which approaches the velocity of light. Once the electromagnetic waves leave their source of generation, these fields are completely independent of any changes occurring at the source. The motion of an electromagnetic wave is two fold (*i*) a translational velocity which is outward from the source and is dependent on the medium, and (*ii*) an oscillatory motion in a plane perpendicular to the direction of propagation.

Microwave system which include power generation, transmission and detection are in many ways different than low frequency radio wave systems. Hence, the measurement techniques and parameters measured are also different.

At microwave frequencies most important parameters to be measured are the power, impedance and the frequency. Other quantities which are measured are the phase-shift, attenuation, VSWR and the noise factor. In this chapter, we shall concentrate on the measurement of above mentioned parameters.

11.2 FREQUENCY MEASUREMENTS

11.2.1 Various Components of a Microwave Bench

The basic experimental arrangement used in the laboratories for these measurements is described below (Fig. 11.1). X-band microwave sources, detectors and components are easily available, hence X-band set up is generally used for teaching purposes. Figures 11.2(*a*), (*b*), (*c*), (*d*) show most commonly used measurement equipments.

(*i*) A microwave source used is either a Gunn diode oscillator or a reflex klystron, the output is of the order of few milliwatts. Signal output from these sources is generally square-wave modulated at an audio-frequency of 1 kHz. (see Fig. 11.7 in Section 11.3.3).

(*ii*) The second component is an attenuator which can provide 0-50 dB attenuation above its insertion loss (Fig. 11.2*a*) or an isolator to avoid loading of the klystron or microwave source.

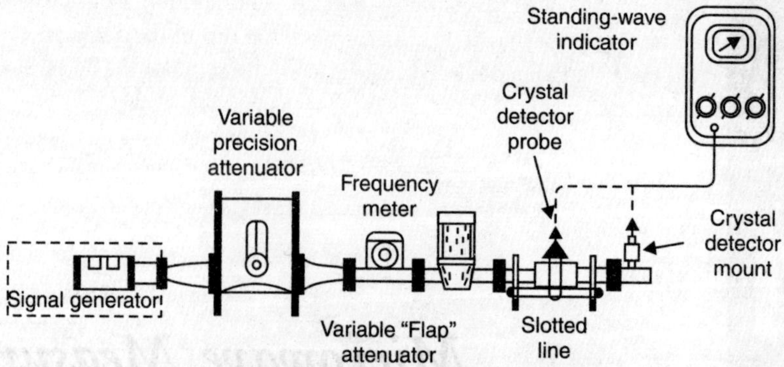

Fig. 11.1 Frequency measurement set up

Fig. 11.2 (*a*) Precision attenuator

(*iii*) A frequency meter in the form of a cylindrical cavity, which can be adjusted to resonante and is coupled to the waveguide through a slot (Fig. 11.2(*b*)).

(*iv*) A slotted line section, which has a cut along the centre of broad wall of the wave guide is connected. This unit is required primarily for SWR measurements (Fig. 11.2*a*).

(*v*) A SWR meter is connected for the direct measurement of SWR. This meter is basically a sensitive, tuned voltmeter (Fig. 11.2*d*). It consists of a high gain amplifier, which operates at a fixed audio frequency (1 kHz). The amplifier output is measured with a square-law calibrated voltmeter which can provide direct reading of the SWR. A gain control allows the reading to be adjusted to the required value.

11.2.2 Procedure for Frequency Measurement

The procedure for determining an unknown microwave frequency is as follows:

(*i*) The frequency meter cavity should be at off resonance at the unknown frequency. The attenuator should be adjusted to provide approximately 6-10 dB loss. The measurement set up is shown in Fig. 11.1.

(*ii*) Activate the microwave source and switch to square wave modulation. Vary the amplitude and the frequency (1 kHz) for a peak reading in the SWR meter. If necessary adjust the meter's gain control and the precision attenuator so that the needle is on the '0' degree mark on the right hand side of the scale.

(*iii*) Adjust the size of the cavity until resonance occurs. A small amount of microwave energy from the waveguide will be delivered to the cavity. An appreciable dip in the SWR meter will be indicated. The unknown frequency can then be read directly from the scale of the calibrated frequency meter. A typical Q value for the cavity is about 5000.

Fig. 11.2 (*b*) Resonant cavity frequency meter

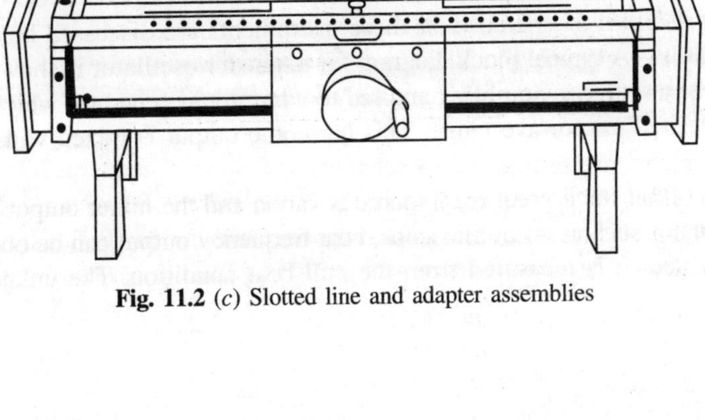

Fig. 11.2 (*c*) Slotted line and adapter assemblies

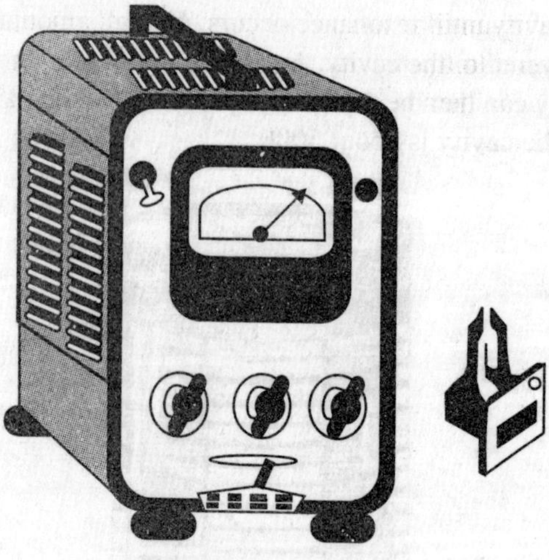

Fig. 11.2 (*d*) Microwave VSWR meter

The basic accuracy of the cavity wavemeter is ±1%. This value is subjected to changes in temperature and humidity. These effects can be overcome to some extent by sealing the cavity properly.

An alternative method of determining the frequency is to use a slotted line. The slotted line is terminated by a short circuit so that a standing wave pattern appears. Accurately locate the positions of two adjacent nulls by moving the probe along the slotted line, read the position of nulls in the vernier scale of the slotted line. These two positions are separated by half a guide wavelength, $\lambda_g/2$. The free space wavelength λ can be calculated by using the relationship:

$$\lambda = \lambda_g\sqrt{1+\left(\lambda_g/2a\right)^2} \qquad ...(11.1)$$

The frequency is calculated using f (GHz) = 30/λ cm. The accuracy of the result will depend partly on the measurement accuracy of the waveguides inner wide dimension a. The overall accuracy obtained is of the order of ±1%.

The methods of measuring frequency described so far depend on mechanical devices such as the cavity and the slotted line. However, for precision measurement of frequency, electronic devices such as frequency counters are used to compare the unknown microwave frequency with a high order harmonic of an accurately determined lower frequency.

11.2.3 Electronic Techniques for Frequency Measurement

These techniques are based on the comparison of the unknown frequency with a harmonic of the known standard frequency. Instruments that use null-beating technique for this comparison are known as transfer oscillators. There are automatic transfer oscillators which provide direct display of microwave frequencies. Basic principles of operation of these instruments are discussed below.

Transfer oscillator—A typical block diagram for a transfer oscillator is shown in Fig. 11.3. Signal from a stable frequency source is amplified and fed to a harmonic generator which provides a comb of frequencies in the desired microwave range. This harmonic output is mixed with the unknown signal frequency.

If the frequency of the stable frequency source is varied and the mixer output is connected through a detector to an indicator such as an oscilloscope, beat frequency output can be observed. Frequency of the stable frequency source is measured from the null beat condition. The unknown frequency is an integral multiple of this value. The value of the integral multiples can be obtained by noting the next higher frequency which produces null beat. If two frequencies are f_1 and f_2, the unknown frequency f may be found by the following equations

$$nf_1 = f \quad \text{...(11.2}a\text{)}$$

$$(n-1) f_2 = f \quad \text{...(11.2}b\text{)}$$

Eliminating n, we may write

$$f = \frac{f_1 f_2}{f_2 - f_1} \quad \text{...(11.3)}$$

Since f_1 and f_2 can be measured very accurately this method provides very precise frequency measurement. In fact, this method is generally used to calibrate resonant cavity type wavemeters.

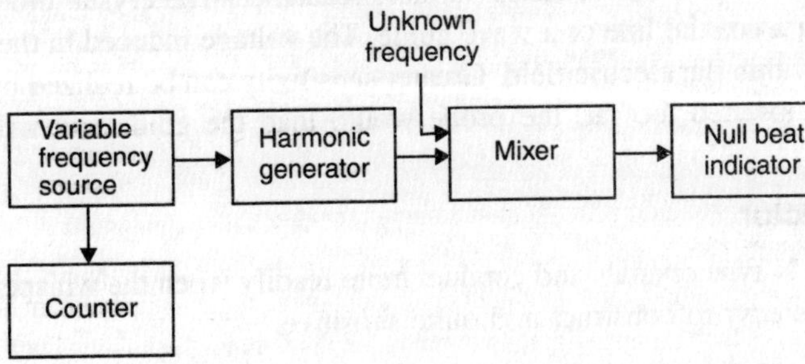

Fig. 11.3

11.3 SLOTTED LINE TECHNIQUES FOR VSWR MEASUREMENT

Standing wave patterns are usually measured by means of a travelling detector which probes the electric field intensity along the axis of propagation. The whole assembly comprises of a slotted section, an

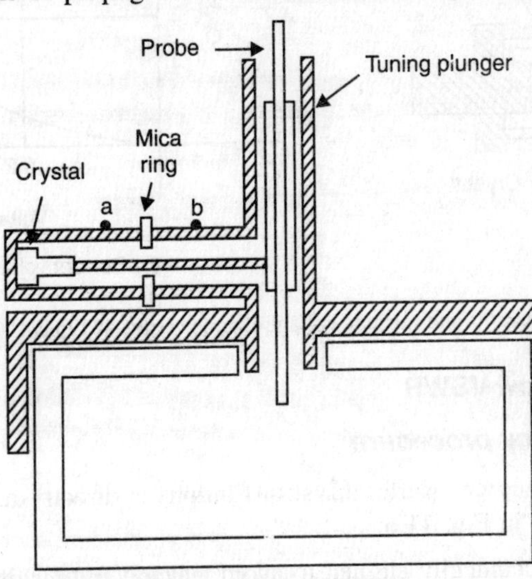

Fig. 11.4 Standing-wave crystal detector unit

electric probe and a detector. This assembly is often referred to as a standing wave detector. The slotted sections are both coaxial and waveguide type. It consists of a longitudinal slot milled along its surface. The slot is located at the place where surface current is minimum so that its presence do not change the field configuration considerably. The slot is tapered at both the ends so that a smooth impedance transfer is achieved and the discontinuity effect of the slot reduces to a minimum. The electric probe is inserted within the holder which rides above the slot as shown in Fig. 11.4.

The probe length inside the guide is adjusted by a screw mechanism to provide loose coupling to the electric field. The voltage is detected by any of the devices, *viz.* crystal, bolometer or thermocouple. A crystal diode is simplest of all and is generally used in the teaching laboratories. The crystal current-voltage characteristic is square low in nature *i.e.*, the crystal current varies nearly square of the voltage detected, *i.e.*, $i = KE^2$. Where i is the current read by microammeter, E is the voltage on the line at the probe and K is constant introduced due to the detector and probe coupling.

To match the probe and crystal assembly for maximum power transfer, a tuning plunger composed of a sliding short circuit is used to introduce variable reactance. The crystal diode may be mounted in a short length of either a coaxial line or a wave guide. The voltage induced in the probe is proportional to its exposed length within the electric field. Greater sensitivity can be realized by increasing the probe insertion. However if inserted too far, the probe would load the guide and will change the original pattern.

11.3.1 Crystal Detector

Most crystals used are N-type crystals and conduct more readily when the whisper is negative as shown in Fig. 11.5. These are easy to construct and quite sensitive.

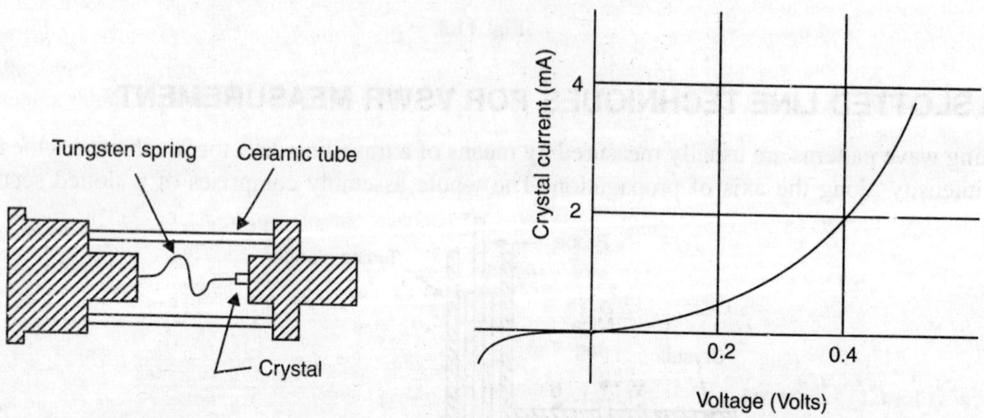

(a) Structure (b) Current voltage characteristic

Fig. 11.5 Structure and current voltage characteristic of typical crystal diode

11.3.2 Measurement of Low VSWR

(a) Unmodulated frequency procedure

In this case, the microwave source (Reflex klystron) output is directly read by a microammeter, the experimental set up is shown in Fig. 11.6.

1. All components are mechanically aligned to avoid leakage and residual standing waves.

2. Set the microammeter shunt so that sensitivity is minimum. Even if large signal is passed through the meter it will not be damaged.
3. Switch on the reflex klystron oscillator. Vary the reflector voltage until oscillations are available on galvanometer.

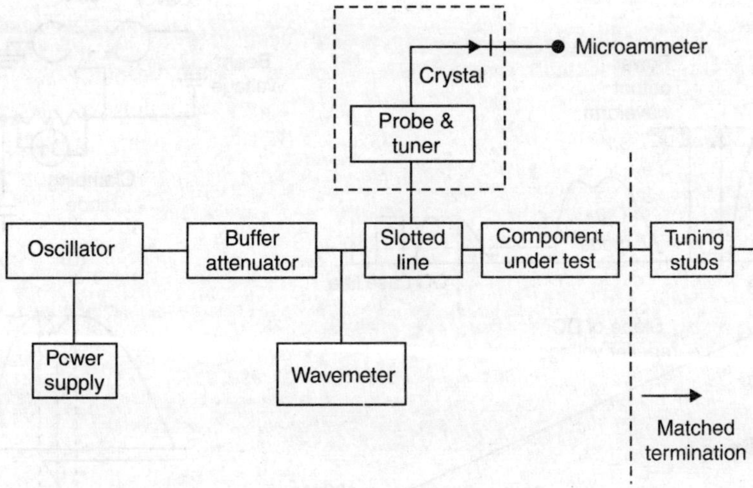

Fig. 11.6 Slotted line section measurement of VSWR-method unmodulated frequency method

4. When oscillations are obtained, tune the system. Vary reflector voltage and antenna stub till maximum power is indicated by the galvanometer, probe in the slotted line sections at the maximum of standing wave pattern. Now tune the probe for maximum sensitivity by adjusting the tuning plunger on the slotted section. Adjust the probe depth to a minimum, increase the galvanometer sensitivity.
5. Measure output frequency with a calibrated frequency meter.
6. Remove component under test and measure standing wave pattern for matched condition. When the probe is located at the voltage minimum $i_{min} = KE_{min}^2$ and probe is located at the voltage maximum,

$$i_{max} = KE_{max}^2$$

then
$$\text{VSWR} = \frac{E_{max}}{E_{min}} = \sqrt{i_{max}/i_{min}} = \sqrt{\frac{\text{Maximum meter reading}}{\text{Minimum meter reading}}}$$

It should be less than ≤ 1.02.

7. Insert the component again and measure standing wave pattern and note the maximum and minimum value.

11.3.3 Modulated Frequency Procedure

If a greater sensitivity is desired, the source can be audio frequency modulated and the crystal output amplified by a tuned audio frequency amplifier. The amplifier output scale is calibrated to read VSWR directly when the amplifier is fed by a "square law" r.f. detector such as a crystal.

To prevent frequency modulation of the r.f. output, generally a square wave modulation is used at 1000 Hz per second. If the reflector voltage is square wave modulated as shown in Fig. 11.7, the envelope frequency of the output corresponds to the modulation frequency, which therefore appears as the output frequency of the crystal detector. The scale of the meter is calibrated in such a way that if the pointer is set to full scale when the probe rests on a voltage maximum, the VSWR is read directly when the probe is moved to a voltage minimum.

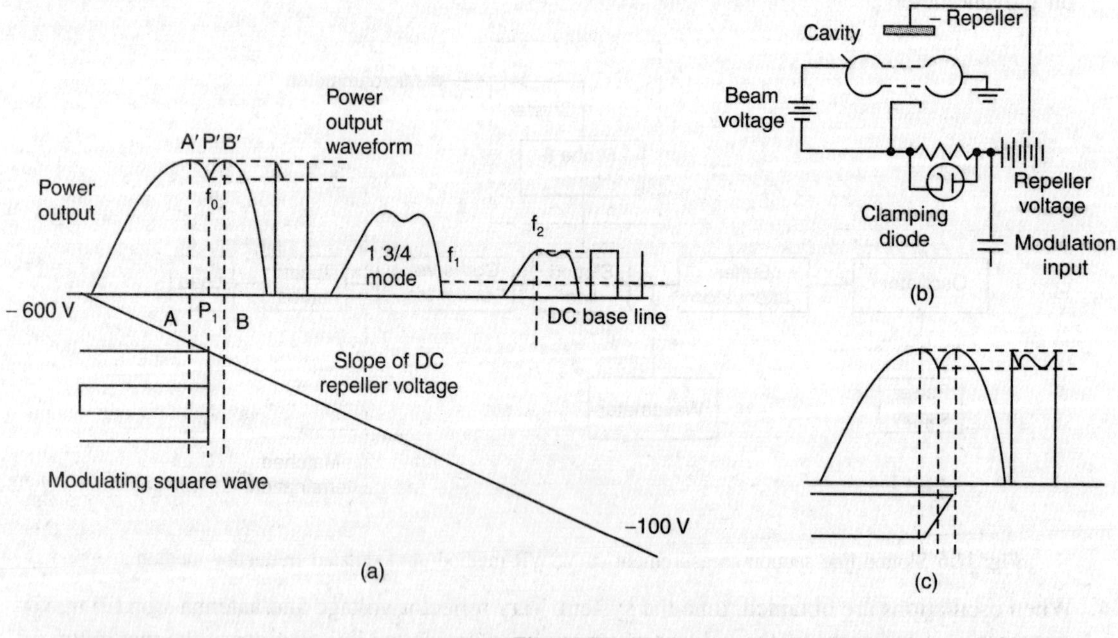

Fig. 11.7

The set up for measuring VSWR by means of a modulated source is shown in Fig. 11.8.

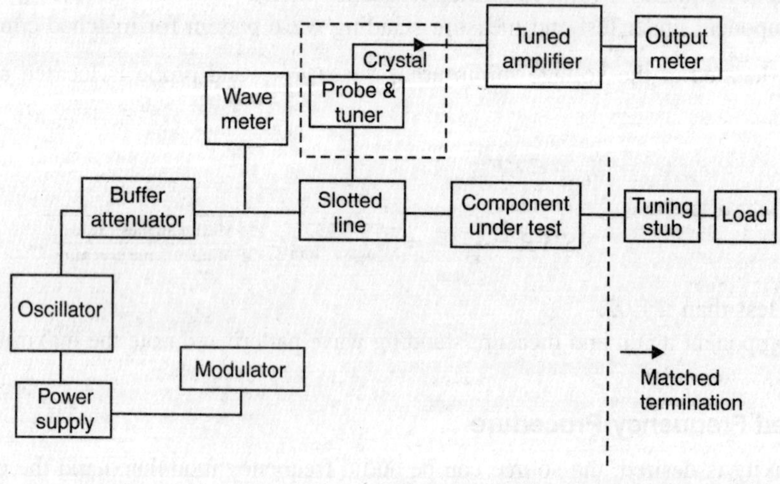

Fig. 11.8 Set up for measuring VSWR by means of a modulated source

Procedure to measure VSWR is described in the earlier section. The averaging technique of locating a VSWR is recommended whenever the VSWR is less than 2.

1. In order to locate the minimum x_{min} the positions x_1 and x_2 corresponding to equal meter reading on either side of x_{min} as shown in Fig. 11.9, are found and their mean is taken. That is,
$$x_{min} = \frac{(x_1 + x_2)}{2}.$$

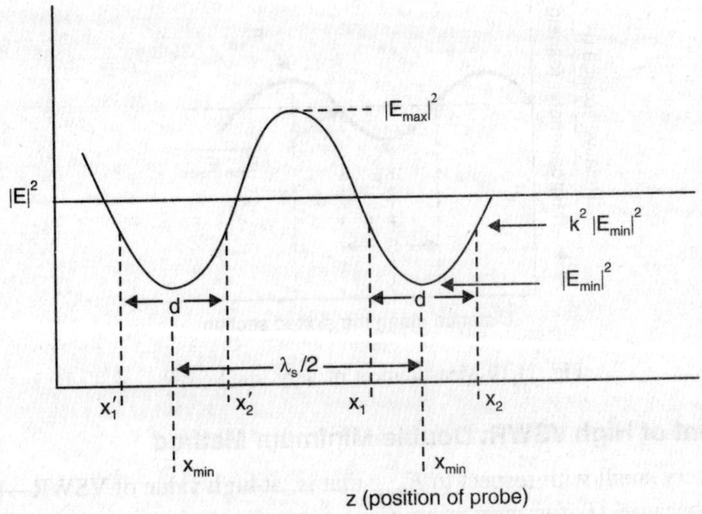

Fig. 11.9 Accurate location of V_{min}

2. The guide wavelength is obtained by finding the distance between two successive minima.
$$\lambda_g = 2\left[(x_1 + x_2) - (x_1' + x_2')\right] \quad ...(11.4)$$
3. The free space wavelength λ_0 is determined by measuring the source frequency. The frequency meter reading corresponding to the maximum dip of the meter will give the frequency of the source from which λ_0 is determined. The cut-off wavelength λ_c is calculated from the relation.
$$\lambda_c = \lambda_0 \left[1 - \left(\frac{\lambda_g}{\lambda_0}\right)^2\right]^{-1/2} \quad ...(11.5)$$
and is verified to be $\lambda_c = 2a$.

Precautions

During VSWR measurement the following conditions should be considered:

1. The amplifier should be run at almost full gain.
2. The probe depth should be small and fixed.
3. The tuning stub should be adjusted for maximum power transfer.
4. The noise level of the amplifier when the source is not oscillating, should not be more than a very small fraction of the full scale reading.

11.3.4 Measurement of Very Low VSWR

When the VSWR is very low, the minimum will be very broad and it may be difficult to locate its position accurately. In this case, better results can be obtained by measuring the position of the points, on either side of the voltage minimum, at which the voltage is roughly the mean of the minimum and maximum voltage as shown in Fig. 11.10. The minimum is located midway between these points. Similarly, the maximum can also be located. (Either the geometric or the arithmetic mean can be taken.)

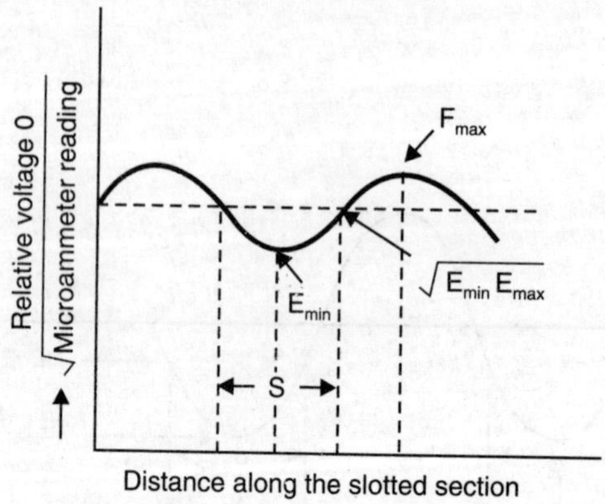

Fig. 11.10 Measurement of very low VSWR

11.3.5 Measurement of High VSWR: Double Minimum Method

When E_{min} becomes very small with respect to E_{max}, that is, at high value of VSWR—the direct method becomes inaccurate because (*i*) for measuring E_{min}, a greater probe penetration is required which produces distortion in VSWR pattern and (*ii*) the voltage range becomes too broad to permit operation of crystal rectifier entirely in square law region.

Accurate measurements of VSWRs greater than 10, can be made using the 'double minimum' method. The variation of square law detector output, *i.e.*, amplitude square of the standing wave along the slotted section, is shown in Fig. 11.11(*a*).

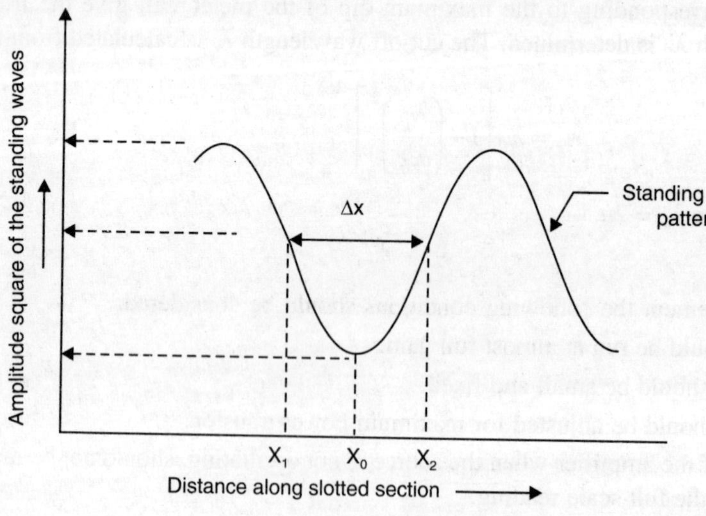

Fig. 11.11 (*a*) Typical standing wave pattern for the measurement of high VSWR

The equation for E_x^2, the square of the amplitude of standing wave, at any point x which lies between the minimum and maximum amplitude is given by

$$E_x^2 = E_{max}^2 \sin^2 \theta + E_{min}^2 \cos^2 \theta \qquad ...(11.6)$$

where $\theta = \dfrac{\pi \Delta x}{\lambda}$; x being the distance from the minimum to some point x (Fig. 11.11*b*). Thus,

$$\text{VSWR} = \frac{E_{max}}{E_{min}} = \left\{ \left(E_x / E_{min} \right)^2 - \cos^2 \theta \right\}^{1/2} \Big/ \sin \theta \qquad ...(11.7)$$

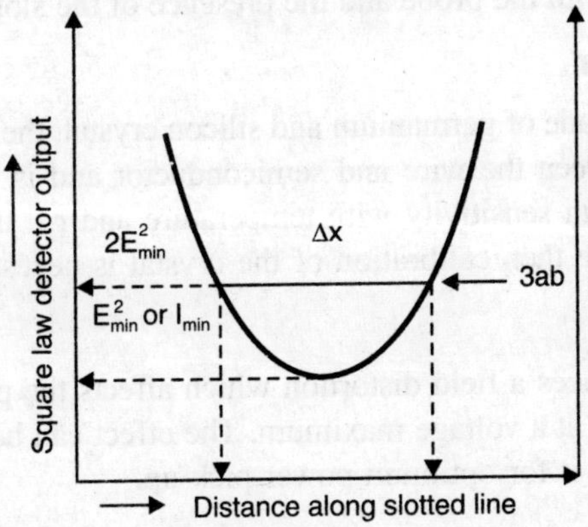

Fig. 11.11 (*b*) Double minima method for high VSWR measurements (illustrated)

Thus, by determining the value of field at any point x rather than at maximum, one can compute VSWR. However, if we consider Δx to correspond to twice the distance between minimum power points, calculations are simplified. Thus if $\left(E_x / E_{min}^2 \right) = 2$

Then
$$\text{VSWR} = \frac{E_{max}}{E_{min}} = \frac{\sqrt{2 - \cos^2 \theta}}{\sin \theta}$$

$$= \frac{\sqrt{1 + \sin^2 \theta}}{\sin \theta} \qquad ...(11.8)$$

For high VSWR,
$$\theta = \frac{\pi \Delta x}{\lambda} \text{ is very small}$$

So
$$\sin \theta \to \theta$$

$$\cos \theta \to 1$$

$$\text{VSWR} = \frac{E_{max}}{E_{min}} = \frac{1}{\theta} = \frac{\lambda}{\pi \Delta x} \qquad ...(11.9)$$

In this way very high and moderate VSWR can be found out with the help of a slotted section. The slotted section method of measuring standing wave is susceptible to many errors. These are listed below.

11.3.6 Errors in Standing Wave Measurement

Slotted section system errors

1. **Error due to frequency instability**—The frequency instability in the source will cause standing wave to shift while measurements are performed. They may arise mainly due to mechanical vibrations, thermal changes, too close coupling between the oscillator tube and the measuring bench or inadequate regulations of the source tube.
2. **Coupling error**—Coupling between the components of the measuring bench should be tight and properly aligned to avoid leakage and reflections.
3. **Standing wave detector error**—These are mainly due to deviation from square law behaviour, the admittance associated with the probe and the presence of the slot in the slotted section.

(a) Square law detector error

Microwave crystals are usually made of germanium and silicon crystal, their characteristics are strongly dependent upon the contact between the wire and semiconductor and is subjected to defects such as uncertain response law, changes in sensitivity with temperature and r.f. impedance. When the crystal current is above one microampere then calibration of the crystal is necessary.

(b) Probe error

The presence of the probe introduces a field distortion which affects the pattern of the standing wave. The distortion is more pronouned at a voltage maximum. The effect can be minimized by reducing the depth of the probe and by tuning it for optimum power pick up.

(c) Slot error

The slot in the slotted section loads the waveguide and produces slight shift in its guide wavelength, and the end of the slot introduces a discontinuity effect. The slot should be narrow and cut symmetrically with proper tapering so that the error due to radiation is eliminated and the other two errors are also minimized.

11.4 IMPEDANCE MEASUREMENT

An unknown load impedance can be found just by measuring the VSWR using the techniques described earlier, and finding the position of voltage minimum. The basic set up is as shown in Fig. 11.12. We know from transmission line theory that the waves incident from the generator on the load get reflected (if the load is not the characteristic impedance) and standing waves are formed between the load and generator. The magnitude and phase of the VSWR are the characteristic property of the load. The input impedance of a transmission line is given by the transmission line equation.

$$Z_{in} = \frac{V_s}{I_s} = \frac{V_L \cosh \gamma l + Z_0 I_L \sinh \gamma l}{I_L \cosh \gamma l + V_L / Z_0 \sinh \gamma l} \qquad \text{...(11.10)}$$

For a lossless line,

$$\gamma = \alpha + j\beta, (\alpha = 0)$$

$$= j\beta$$

$$Z_{in} = Z_0 \frac{Z_L + jZ_0 \tan \beta \Delta l}{Z_0 + jZ_L \tan \beta \Delta l} \qquad \text{...(11.11)}$$

where Z_L is the load impedance (unknown impedance), Z_0 is the characteristic impedance and Z_{in} is the impedance at the input of the transmission line. $\beta\Delta l$ is the electric distance, being measured in wavelengths between position of termination and standing wave minimum. In terms of Z_L, Eqn. 11.11 is written as

$$Z_L = \left(\frac{Z_{in} - jZ_0 \tan\beta\Delta l}{Z_0 - jZ_{in} \tan\beta\Delta l}\right) Z_0 \quad \text{...(11.12)}$$

or

$$Z_R = \frac{(1 - j(\text{VSWR}) \tan\beta\Delta l) Z_0}{\text{VSWR} - j \tan\beta\Delta l} \quad \text{...(11.13)}$$

where Z_{in} corresponds to Z_{min} and so Z_0/Z_{in} is VSWR. Steps to be followed for measurement of an unknown load are:

1. Arrange the set-up as shown in Fig. 11.12.

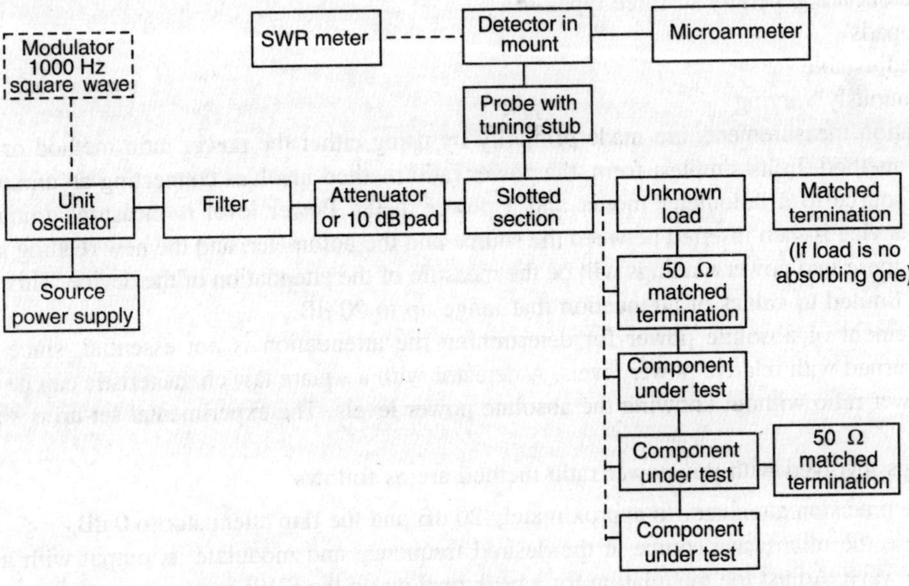

Fig. 11.12 Set-up for the measurement of VSWR at the input of the component under test

2. Terminate the load end with the unknown impedance, measure λ the wavelength and VSWR.
3. Record the position of a minimum with the help of the probe, at around the center of slotted section. Call this reading say x_1.
4. Replace the load by short circuited termination and move the probe carriage to new standing wave minimum and record it as the value x_2.
5. Find the shift in minimum $(x_2 - x_1)$. It will be positive if it is shifted towards load and negative if it is shifted towards generator. Convert the shift Δl in wavelength units, *i.e.* $\frac{x_2 - x_1}{\lambda}$ wavelengths.

   Compute impedance using Eqn. 11.14. Substituting for $\beta\Delta l$,

$$\beta\Delta l = \frac{(x_2 - x_1)}{\lambda} 2\pi \quad \text{...(11.14)}$$

6. Unknown impedance can easily be found by using Smith chart. The only parameters to be measured are VSWR and shift in minimum position as explained in step 4.

11.5 ATTENUATION MEASUREMENTS

The term attenuation is defined as the ratio in decibels, of either the power level or voltage level between two points in a circuit. Hence, attenuation of a component under test will be the ratio expressed in decibels, of the power absorbed by the load without the component in the line to the power absorbed by the load with the component in the line, when the signal sources under load are perfectly matched.

$$\text{Attenuation } A = 10\log_{10}\left(\frac{P_1}{P_2}\right)\text{dB} \quad \text{...(11.15)}$$

P_1 = Power absorbed by the load without the given component in the line.

P_2 = Power absorbed by the load with the given component in the line.

Attenuators are generally of three types—

(*i*) Fixed pads

(*ii*) Step-adjustable

(*iii*) Continuously varying

Attenuation measurements are made primarily by using either the power ratio method or the RF substitution method. In its simplest form, the power ratio method involves connecting an unmodulated microwave source to a bolometer mount and a power meter. Power level is measured initially, the attenuating device is then inserted between the source and the bolometer, and the new reading is taken. The ratio of these two power readings will be the measure of the attenuation of the device. This method is normally limited to values of attenuation that range up to 20 dB.

Measurement of absolute power for determining the attenuation is not essential, since we are mostly concerned with relative power levels. A detector with a square law characteristic can be used to measure power ratio without knowing the absolute power levels. The experimental set-up is shown in Fig. 11.13.

The steps involved with this power ratio method are as follows:

(*i*) Set the precision attenuator to approximately 20 dB and the flap attenuator to 0 dB.

(*ii*) Energise the microwave source at the desired frequency and modulate its output with a 1 kHz square wave. Adjust the modulation for a peak reading on the SWR meter.

(*iii*) For the crystal detector to operate within its square law characteristics, the microwave power level at the crystal should be around 0.1 mW. This is achieved by switching the meter to its 30 dB range. The meter's gain control or, if necessary the precision attenuator's control can be used to bring the needle to the 0-dB mark.

(*iv*) Rotate the control of the flap attenuator to provide the unknown loss. The amount of this loss can then be read directly off the decibel scale of the SWR meter. If necessary the experiment can be repeated for a number of required frequencies.

In this method, we have assumed that both power levels with and without the unknown attenuation are capable of operating the crystal detector within its square-law characteristics. If this is not so there will be an error in the measurement. Greater accuracy can be provided by the RF substitution method, the steps for which are as follows:

(*i*) Assemble the set-up as shown in Fig. 11.13.

(*ii*) Set the precision attenuator to approximately 40 dB higher than the expected attenuation of the component under test.

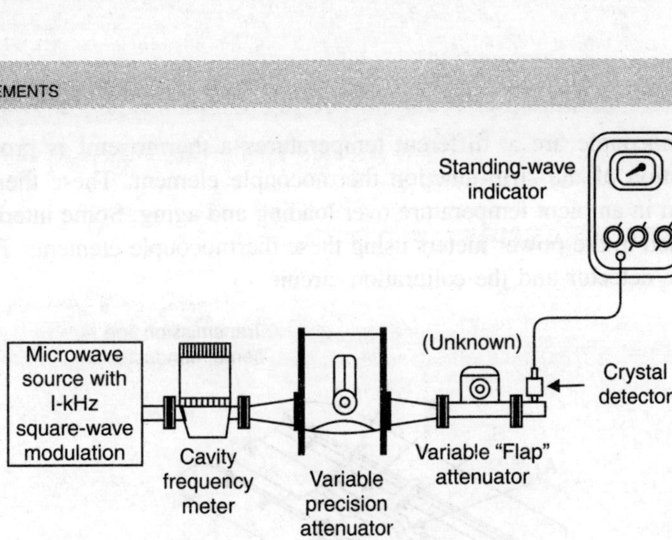

Fig. 11.13 Attenuation measurement set-up

(*iii*) Energise the microwave source and repeat the second and third steps of the power ratio method described earlier and record the reading of the precision attenuator.

(*iv*) Rotate the control of the flap attenuator to provide the 'unknown' loss. The needle of the SWR meter will fall back but should be restored to the 0 dB mark by reducing the value of the precision attenuator. Then the difference between the new reading of the precision attenuator and its reading in step (*i*) is equal to the unknown attenuation.

The advantage of the substitution method is that the power level at the crystal detector is the same when taking both readings of the precision attenuator.

11.6 POWER MEASUREMENT

Microwave power is one of the few fundamental quantities that can be measured directly and conveniently. It is helpful in ascertaining the proper functioning of all microwave devices and circuits.

Other electric properties can be derived by determining relative or absolute power levels of the equipments. Although microwave power is defined in the same way as power at low frequencies, that is, it is the product of rms voltage, rms current and power factor ($P_{av} = I_{rms} E_{rms} \cos \phi$). However at microwave frequencies measurement of current and voltage is difficult as the propagation of microwave power is in terms of electric and magnetic field [E and H]. Most microwave power measurements are accomplished by using suitable calibrated power detectors. Such detectors convert the r.f. electric power into heat and by measuring the temperature rise by temperature sensitive elements, power is read from the directly or indirectly calibrated power meters.

The power levels encountered in the field depend upon the specific purpose, the microwave source is used. For example, radars use many kilowatts of microwave power, while a solid state microwave circuit may work at few milliwatt power. Hence, the power measurement is divided into three classes:

(*a*) Very low power < 1 mW
(*b*) Low power < 10 mW
(*c*) High power (10 mW to few hundreds of watts).

11.6.1 Very Low Power Measurement

Specially designed thermocouples are used to measure low power levels below 1 mW. These thermocouples consist of thin films of antimony and bismuth deposited on thin resistive strips. When two junctions of a thermocouple are at different temperatures a thermo emf is produced. Figure 11.14 depicts the construction of the strip-junction thermocouple element. These thermocouple, units are subjected to variation in ambient temperature over loading and aging. Some internal calibration provisions must be included in the power meters using these thermocouple elements. Figure 11.15 shows a typical thermocouple detector and the calibration circuit.

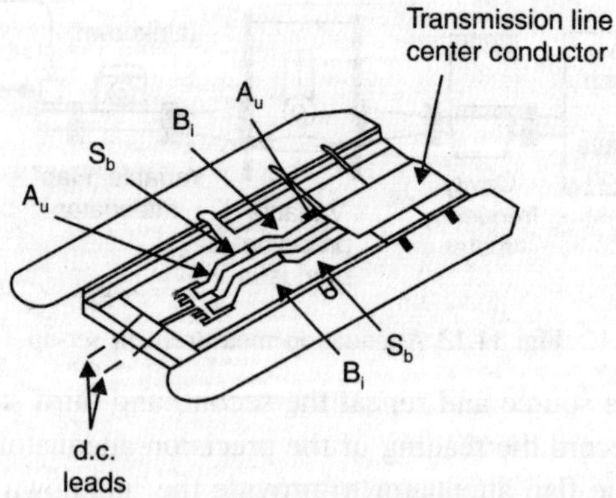

Fig. 11.14

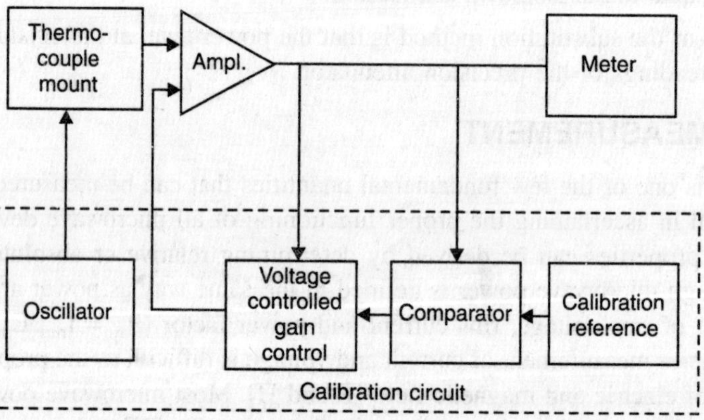

Fig. 11.15

11.6.2 Low Power Measurement (Bolometer Method) [1 to 10 mW]

The low power measurements make use of the temperature sensitive elements known as bolometers. Bolometers are microwave power detectors and operate on the property of change in its internal resistance with rise in temperature caused by heat dissipating microwave power. They are of two types:

(1) **Thermisters**—Thermisters are negative temperature coefficient semiconductors. It consists of a small bead of semiconducting material such as complex metallic oxide of Ni, Mn, Co, etc., supported between two wires as shown in Fig. 11.16.
The bead is coated with a thin film of glass to make the assembly strong, heat resistant and stable. Thermisters are much more sensitive than baretters but need higher bias current.

(2) **Baretters**—These are positive temperature coefficient elements. The baretter element consists

of a very fine platinum wire whose diameter is less than 0.0001 inch, mounted as shown in Fig. 11.17.

The baretters are biased to have optimum resistance approximately around 200 Ω.

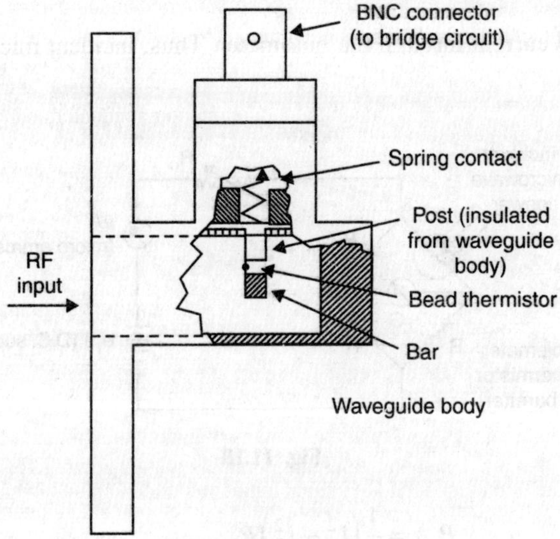

Fig. 11.16

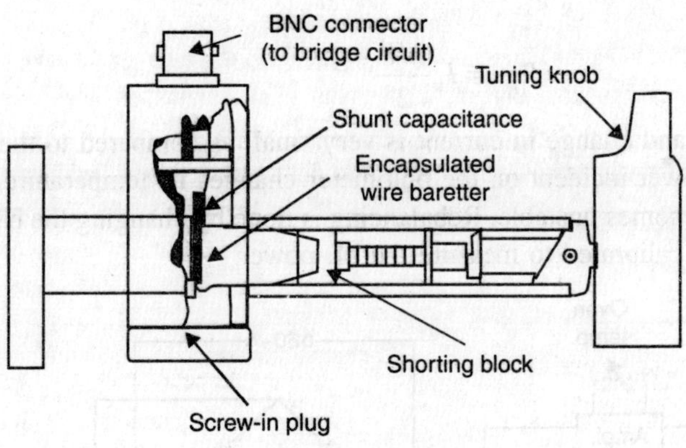

Fig. 11.17

The basic measuring unit is a simple wheatstone bridge arrangement as shown in Fig. 11.18. The bolometer forms one of its arms. The value of the resistance in the bolometer arm is determined by bolometer mount characteristics. To start with, biasing resistance R_c is varied to obtain null point in the galvanometer. In this case, $R_b = R$ and the total power dissipated in the bolometer is

$$P_1 = \left(\frac{I_1}{2}\right)^2 R_b \qquad \text{...(11.16)}$$

Now, bolometer is exposed to the microwave power to be measures and R_c is adjusted to balance bridge. The power dissipated in the bolometer in this case is

$$P_2 = \left(\frac{I_2}{2}\right)^2 R_b \qquad \text{...(11.17)}$$

where $\frac{I_2}{2}$ is the changed current through the bolometer. Thus, incident microwave power is $(P_1 - P_2)$, *i.e.*,

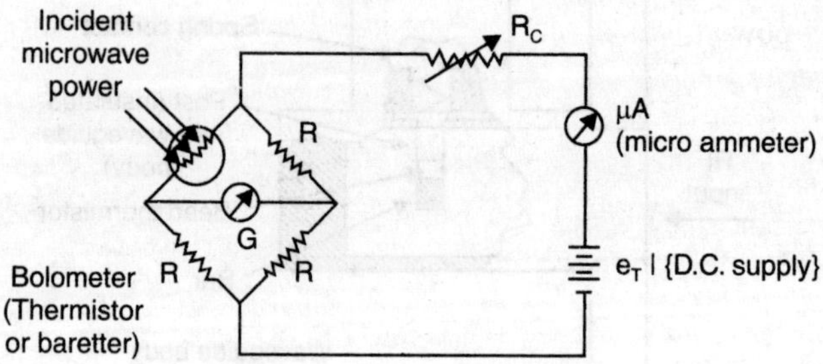

Fig. 11.18

$$P_{(rf)} = \frac{1}{4}\left(I_1^2 - I_2^2\right)R_b \qquad \text{...(11.18)}$$

I_1 and I_2 are read in connected ammeter and microwave/power is thus measured. Equation 11.18 can be written as

$$P_{(rf)} = I_1 \frac{\Delta IR}{2} \qquad \text{...(11.19)}$$

because $(I_1 - I_2) = \Delta I$ and change in current is very small as compared to the biasing current.

The microwave power incident on the bolometer changes its temperature, hence in turn its resistance and the bridge becomes unstable. Rebalancing is done by changing the battery power, the amount of decrease is directly calibrated to measure the r.f. power.

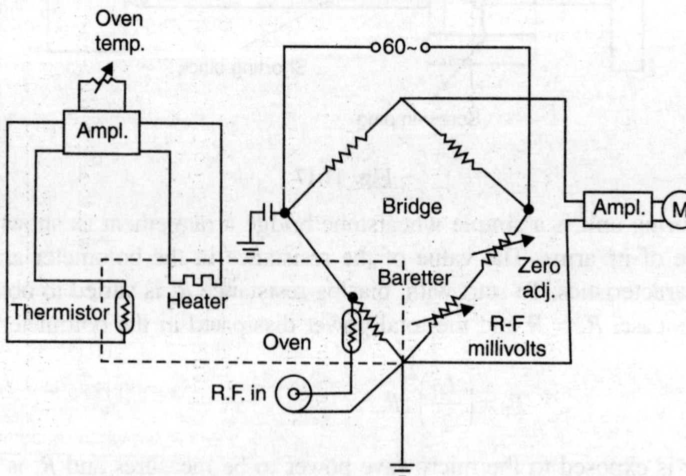

Fig. 11.19 Substitution of audio frequency power method set-up

Since the change in resistance is not accurately known function of temperature and in turn of the power, a substitution of audio frequency (*af*) power method is used to improve the accuracy in the measurements. In this method, the bridge is initially balanced with low frequency (*af*) power. When r.f. power is applied to the bolometer, low frequency power is withdrawn until the bridge is again balanced. The amount of power removed is amplified and displayed on the indicating meter and is equal to the true power substitution.

In such simple bridge arrangements, the bridge can by imbalanced due to ambient temperature change from applied power, which can produce imbalance variations in highly sensitive bridge measurement. The answer to this is, use two thermister mounts placed in close thermal proximity so that both are affected equally by changes in ambient temperature and can compensate for the characteristics of each other. Hence a dual bridge arrangement, one for each bolometer is used. The second bridge is called metering bridge and performs two functions. Firstly, it controls the audio frequency bias which balances the detection bridge (first bridge) under ambient temperature conditions. Secondly, when microwave power is applied to the detection bridge it causes a change in dc bias of the metering (second) bridge. The change in the bias is applied to the current squaring circuit which measures the output power. Thus, the meter becomes direct reading and automatic. A dual bridge circuit is shown in Fig. 11.20.

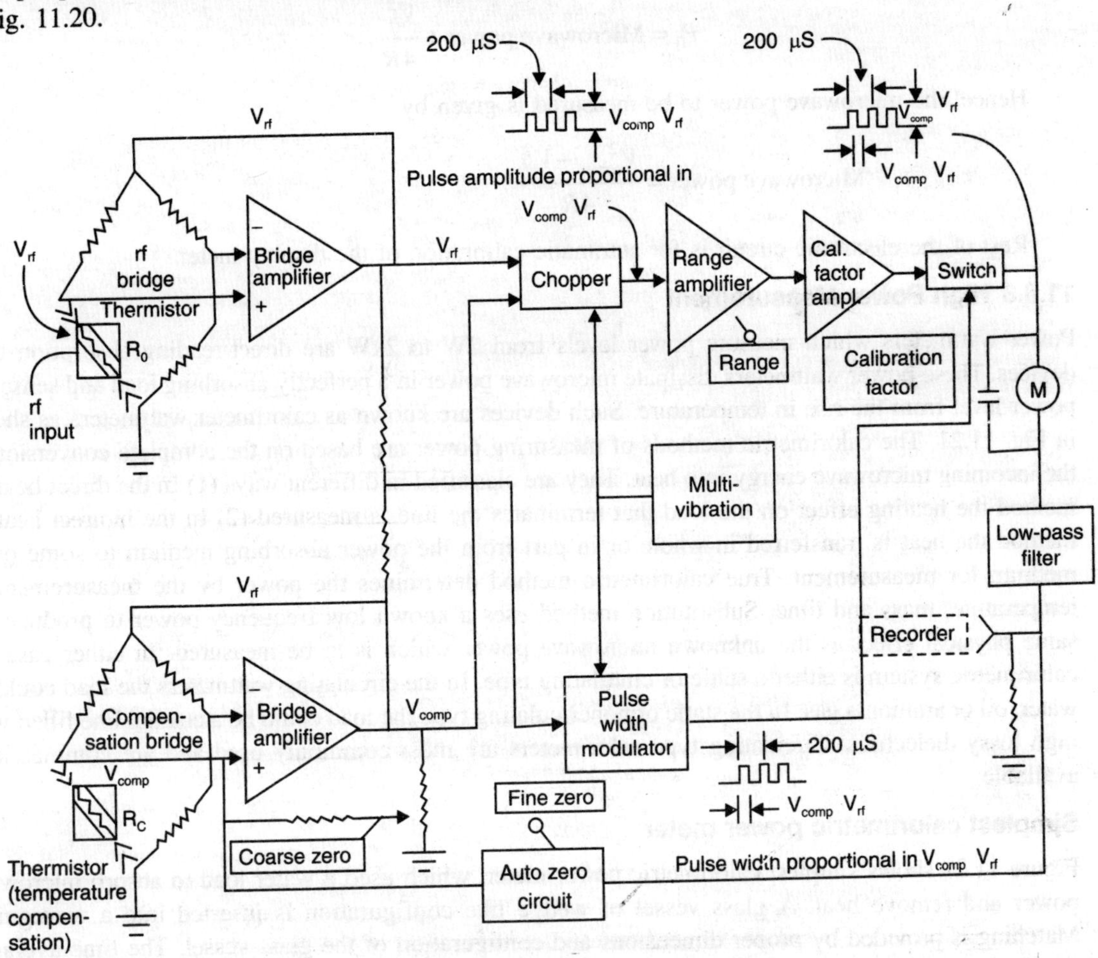

Fig. 11.20 Dual bolometer bridge circuit

Bridge amplifiers provide proper bias to the bolometers used in the bridges so that they have the proper value of bolometer resistance. When microwave power falls on the thermister of the detecting bridge, its resistance varies and the bridge becomes imbalanced. The error voltage thus produced is amplified by the amplifier and then applied at top of the bridge to change the power dissipation in the thermister and its resistance to rebalance the bridge. The two thermisters are placed in identical physical conditions, so a change in the ambient temperature will cause an error signal in both the bridges and thereby get cancelled. Thus the *rf* bridge voltage V_{rf} and compensating bridge voltage contains complete information about the incident microwave power.

Initially V_{comp} is adjusted equal to V_{rf}, when no microwave power is incident on the thermister. In this case total power dissipating in the thermister is

$$P_0 = \frac{V_{comp}^2}{4R}$$

When microwave power is applied and the bridge is rebalanced, the thermister power becomes equal to

$$P_0 = \text{Microwave power} + \frac{V_{rf}^2}{4R}$$

Hence, the microwave power to be measured is given by

$$\text{Microwave power} = \frac{V_{comp}^2 - V_{rf}^2}{4R}$$

Rest of the electronic circuit is for automatic calibration of the display meter.

11.6.3 High Power Measurement

Power wattmeters which measure power levels from 2W to 2kW are direct-reading absorption type devices. These power wattmeters dissipate microwave power in a perfectly absorbing load and sense the power level from the rise in temperature. Such devices are known as calorimeter wattmeters as shown in Fig. 11.21. The calorimetric methods of measuring power are based on the complete conversion of the incoming microwave energy into heat. They are classified in different ways (1) In the direct heating method the heating effect on the load that terminates the line is measured (2) In the indirect heating method the heat is transferred in whole or in part from the power absorbing medium to some other medium for measurement. True calorimetric method determines the power by the measurement of temperature, mass and time. Substitution method uses a known low frequency power to produce the same physical effect as the unknown microwave power which is to be measured. In either case the calorimetric system is either a static or circulating type. In the circulating wattmeters the load could be water, oil or ammonia gas. In the static or noncirculating type the load could be a coaxial line filled with high lossy dielectrics. Circulating type calorimeters are most commonly used and are commercially available.

Simplest calorimetric power meter

Figure 11.21 shows simplest calorimetric power meter, which used a water load to absorb microwave power and remove heat. A glass vessel of wedge like configuration is inserted into a waveguide. Matching is provided by proper dimensions and configuration of the glass vessel. The time averaged power absorbed by the load can be determined from its effect on the temperature of the water, *i.e.*, from the difference between the water temperature at the load input (t_1) and output (t_2), provided the rate of water flow (v) is known.

The following equation can be written

$$P_{av} = \frac{1}{0.24} cv(t_2 - t_1) \text{W}$$

where c is the specific heat capacity of water, which is taken as approximately independent of temperature and equal to 1 cal per gm. per degree centigrade; the rate of water flows should be measured in gm. per second. The rate of water flow is measured by means of flowmeter.

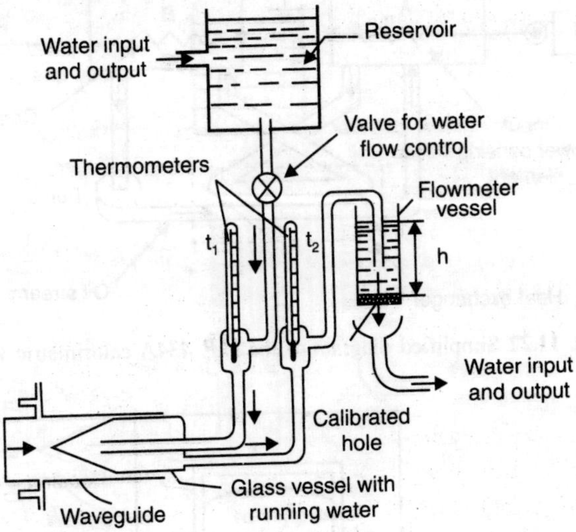

Fig. 11.21

Figure 11.22 shows a commercially available self balancing type wattmeter. It has two identical temperature-sensitive gauges (one in each arm), a high gain amplifier, an indicating meter and two identical load resistors—one for known input and other for comparison of power. The microwave power heats up the input sensing resistor via the heat carried by the oil which in turn imbalances the bridge. The imbalanced signal created is amplified and applied to the comparison load resistor, the power delivered to it to rebalance the bridge is measured and recorded by the indicating meter. Thus, the incident microwave power is measured.

The disadvantages of this method are (1) Inherent inertia of the system, (2) It cannot be adopted for low microwave power measurement, (3) It has low accuracy 5 per cent. Still it is the most prevalent method for measuring power as this measures absolute power and is thus used for standard calibration purpose.

11.7 REFLECTOMETER TECHNIQUES

The value of the SWR may be measured without a slotted line by using two directional couplers. Reflectometer is a special system which allows rapid measurement of SWR of an unknown termination over wide frequency ranges. Figure 11.23 shows the reflectometer set-up, in which the directional couplers may/may not have identical characteristics. The forward 20 dB coupler samples incident power from detector 1, will be 20 dB down or one hundred of the power output from the microwave source. The power coupled to the auxiliary arm is detected and applied to the incident channel of the

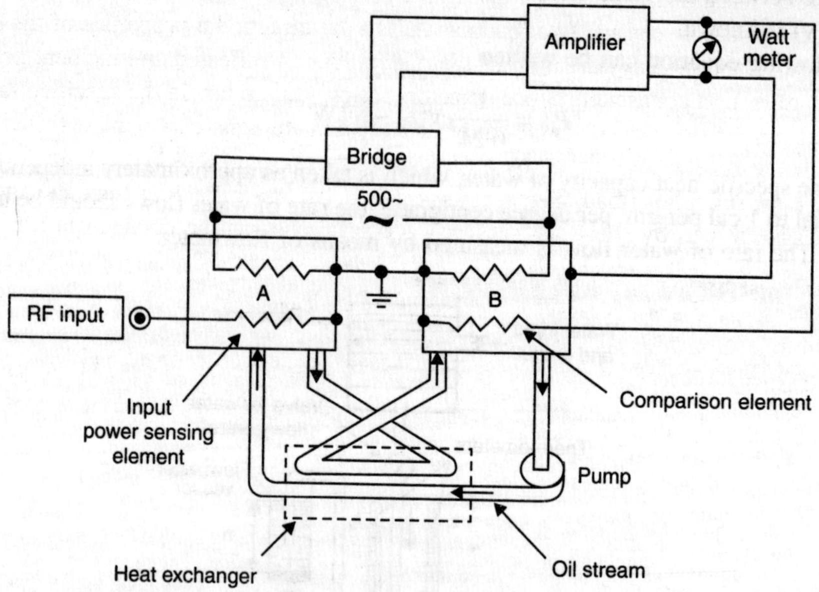

Fig. 11.22 Simplified diagram of the H.P. 434A calorimetric wattmeter

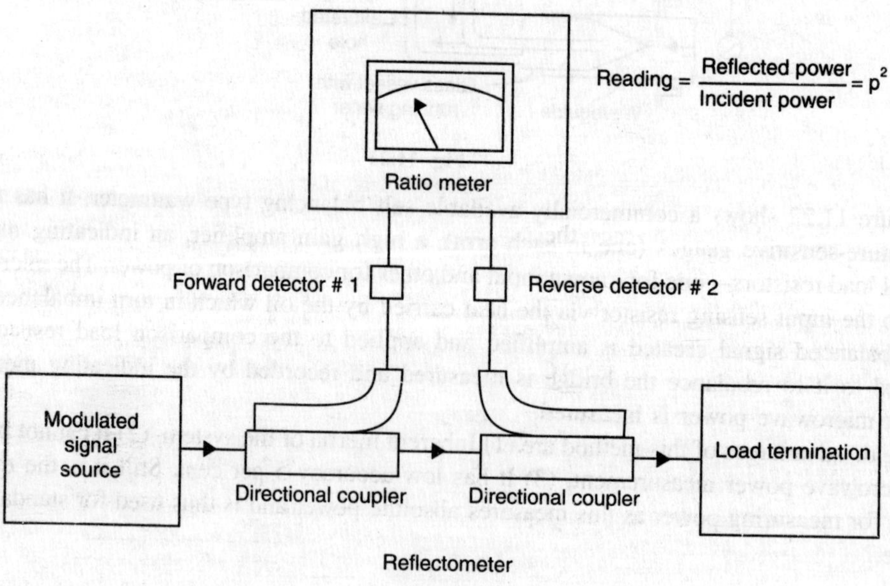

Fig. 11.23

ratiometer. The power coupled to the 10 dB reverse coupler (2) is much small and is absorbed in the internal termination of the coupler and negligible power flows to the connected detector. The remaining power impinges on the load. If the load is not a matched one, a portion of the incident power proportional to the degree of mismatch, is reflected and travels back towards the microwave source. The reverse 10 dB coupler (2), samples the reflected power, which is applied to the reflected channel of the ratiometer via tuned detector. The remaining reflected power passes through the 20 dB coupler (1) connected in reverse direction and hence, the power coupled is absorbed by its internal termination. The variable attenuator is generally provided to avoid overloading of the source due to reflected power. There is no interaction between the two readings due to the directional properties of the couplers. The ratiometer measures and displays the ratio of the magnitudes of reflected and incident power, which is nothing but the reflection coefficient. Since detectors follow square law in the specified range of power we are working in, so the ratiometer in calibrated to indicate square root of detected powers. The calibration of the ratiometer is achieved by a waveguide short circuit termination which corresponds to a hundred per cent reflecting load. Since any load will be less reflecting than the short, so taking reference level determined by short other reflection coefficients can be obtained in dB.

11.8 TRANSMISSION MEASUREMENT

In transmission measurements, we can measure insertion loss, gain, conversion loss, attenuation and S parameters. The measurement set up for S parameters is shown below in Fig. 11.24.

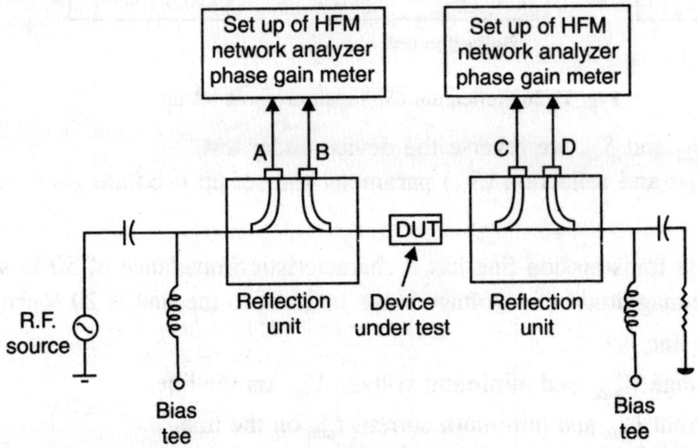

Fig. 11.24 Set up for measurement of S-parameter

The parameters are derived using the expression

$$S_{11} = \frac{B}{A} \qquad \phi_B - \phi_A$$

$$S_{21} = \frac{C}{A} \qquad \phi_c - \phi_A$$

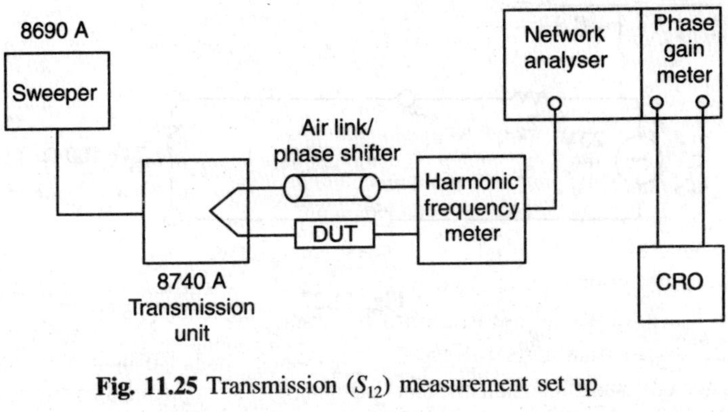

Fig. 11.25 Transmission (S_{12}) measurement set up

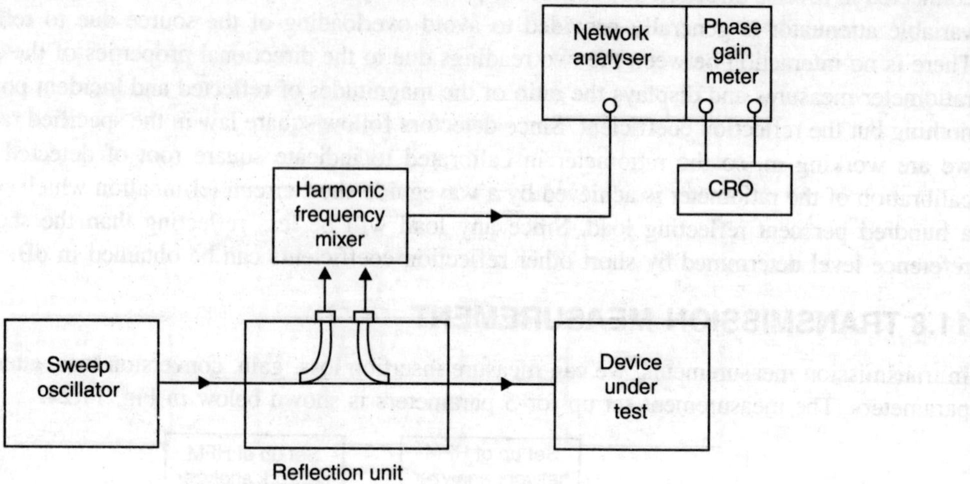

Fig. 11.26 Reflection (S_{11}) measurement set up

For the parameter S_{22} and S_{12}, we reverse the device under test.

For transmission (S_{12}) and reflection (S_{11}) parameter the set up used are given in Fig. 11.25 and 11.26 respectively.

Example 11.1. A lossless transmission line has a characteristic impedance of 50 Ω and is terminated in a load of 100 Ω. The magnitude of a voltage wave incident to the line is 20 V (rms). Determine:

(*a*) The VSWR on the line

(*b*) The maximum voltage V_{max} and minimum voltage V_{min} on the line.

(*c*) The maximum current I_{max} and minimum current I_{min} on the line.

(*d*) The power transmitted by the line.

Solution. Given $Z_0 = 50\ \Omega$, $Z_L = 100\ \Omega$, $V_+e^{-\alpha z} = 20$ V (rms).

$$\text{We know } \Gamma_l = \frac{Z_L - Z_0}{Z_L + Z_0} = \frac{100-50}{100+50}$$

$$= 0.33.$$

$$(a)\ \text{VSWR}(\rho) = \frac{1+\Gamma_l}{1-\Gamma_l} = \frac{1+0.33}{1-0.33} = 2$$

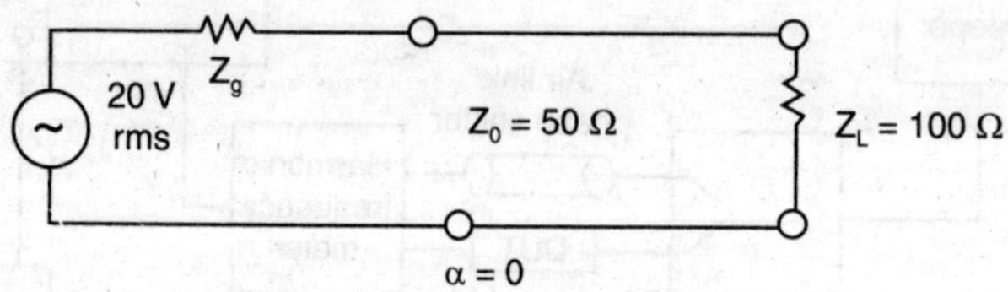

Fig. 11.27

(*b*)
$$V_{max} = V_+ e^{-\alpha z}\left(1+|\Gamma_l|\right)$$
$$= 20(1+0.33)$$
$$V_{max} = 26.6\,\text{V (r.m.s)}$$

or
$$V_{min} = V_+ e^{-\alpha z}\left(1-|\Gamma_l|\right)$$
$$= 20(1-0.33) = 13.4\,\text{V (r.m.s)}$$

(*c*)
$$I_{max} = I_+ e^{-\alpha z}\left(1+|\Gamma_l|\right) = \frac{V_+}{Z_0} e^{-\alpha z}\left(1+|\Gamma_l|\right)$$
$$= \frac{26.6\,V}{50} = 0.53\,\text{A (r.m.s.)}$$
$$I_{min} = \frac{V_{min}}{Z_0} = \frac{13.4}{50} = 0.27\ \text{A (r.m.s.)}$$

(*d*) Power transmitted by the line
$$T = \frac{2Z_l}{Z_l + Z_0} = \frac{2\times 100}{100+50} = 1.33$$
$$V_{th} = V_+ e^{-\alpha z}\cdot T$$
$$= 20\times 1.33 = 26.67\ \text{V (r.m.s.)}$$
$$P_{th} = \frac{V_{th}^2}{Z_L} = \frac{(26.67)^2}{100} = 7.11\ \text{watts.}$$

Example 11.2. A lossless line has a characteristic impedance of 75 Ω and is terminated in a load of 300 Ω. The line is energized by a generator which has an open-circuit output voltage of 20 V (rms) and output impedance of 75 Ω. The line is assumed to be $2\frac{1}{4}$ wavelengths long.

(*a*) Find the sending-end impedance.
(*b*) Determine the magnitude of the receiving-end voltage.
(*c*) Calculate the receiving-end power at the load.

Solution. Given

$Z_g = 75\ \Omega$, $Z_0 = 75\ \Omega$, $Z_L = 300\ \Omega$ and $l = 2.25\ \lambda$

We know $Z_e = \dfrac{Z_l}{Z_0} = \dfrac{300}{75} = 4$

(*a*) For a length of 0.25 λ
$$Z_s = \frac{1}{Z_e} = \frac{1}{4} = 0.25$$

The sending-end impedance is

$$Z_s = Z_s Z_0 = 0.25 \times 75$$
$$= 18.75\ \Omega.$$

(*b*) We know the reflection coefficient is

$$\Gamma_e = \frac{Z_L - Z_0}{Z_L + Z_0} = \frac{300-75}{300+75} = 0.6$$

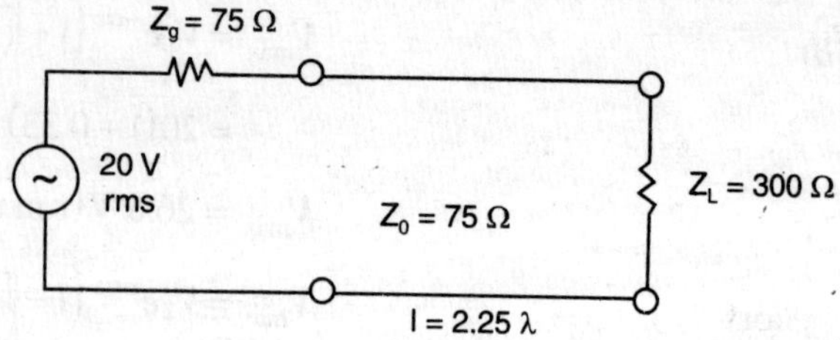

Fig. 11.28

So
$$V_r = V_+ e^{-rl} + V_- e^{-rl}$$
$$= V_+ e^{-rl} + V_+ e^{-rl} \Gamma_l$$
$$= V_+ e^{-rl} (1 + \Gamma_l)$$
$$= 20(1 + 0.6) = 32\ \text{V(r.m.s.)}$$

(*c*) Power at the load is

$$P_L = \frac{(V_r)^2}{Z_L} = \frac{(32)^2}{300}$$
$$= 3.41 \text{ watts.}$$

Example 11.3. A coaxial line with a solid polyethylene dielectric is to be used at a frequency of 3 GHz. Its characteristics impedance Z_0 is 50 Ω and its attenuation constant α is 0.0156 NP/m. The velocity factor which is defined as the ratio of phase velocity over the velocity of light in the free space is 0.660. The line is 100 m long and is terminated in its characteristic impedance. A generator, which has an open-circuit voltage of 50 V (rms) and an internal impedance of 50 Ω, is connected to the sending end of the line. The frequency is tuned at 3 GHz. Compute:

(*a*) The magnitude of the sending-end voltage and of the receiving-end voltage.
(*b*) The sending-end power and receiving-end power.
(*c*) The wavelengths of the line.

Solution. Given, $f = 3$ GHz, $\lambda = 10$ cm, $\alpha = 0.0156$ Np/m

$$v_p = 0.66 \times 3 \times 10^8,\ tr = 2.25 \text{ and } Z_0 = Z_g = 50\ \Omega$$

(*a*)
$$V_s = \frac{V_g}{Z_g + Z_0} Z_0$$

$$V_s = \frac{50 \times 50}{50 + 50} = 25\ \text{V(r.m.s.)}$$

We know
$$V_r = V_s e^{-\alpha l} e^{-j\beta l}$$
$$= 25 e^{-0.0156 \times 100}\ e^{-j\beta l}$$
$$= 5.25\ e^{-j\beta l}$$

So receiving voltage is 5.25 V (rms)

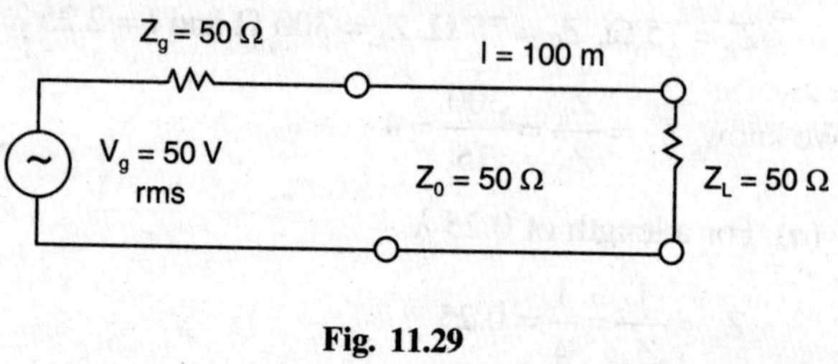

Fig. 11.29

(*b*) We know,

$$P_s = \frac{V_s^2}{Z_s} = \frac{(25)^2}{50} = 12.5 \text{ watts}$$

and $$P_r = \frac{V_r^2}{Z_l} = \frac{(5.25)^2}{50} = 0.55 \text{ watt}$$

(*c*) The wavelength of the line

$$\lambda = \frac{V_p}{f} = \frac{0.66 \times 3 \times 10^8}{3 \times 10^9}$$

$$\lambda = 6.6 \text{ cm.}$$

Example 11.4. An open-wire transmission line has $R = 5\ \Omega/\text{m}$, $L = 5.2 \times 10^{-8}$ H/m, $G = 6.2 \times 10^{-3}$ mho/m and $C = 2.13 \times 10^{-10}$ F/m. The signal frequency is 4 GHz. Calculate:

(*a*) The characteristic impedance of the line in both rectangular form and polar form.
(*b*) The propagation constant of the wave along the line.
(*c*) The normalized impedance of a load 100 + *j* 100
(*d*) The reflection coefficient at the end.
(*e*) The sending-end impedance if the line is assumed a quarter-wavelength long.

Solution. Given, $R = 5\ \Omega$, $L = 5.2 \times 10^{-8}$ H/m, $G = 6.2 \times 10^{-3}$ ℧/m

$C = 2.13 \times 10^{-10}$ F/m and $f = 4$ GHz.

(*a*) We know, the characteristic impedance

$$Z_0 = \sqrt{\frac{R + j\omega L}{G + j\omega c}} = \sqrt{\frac{5 + j2\pi \times 4 \times 10^9 \times 5.2 \times 10^{-8}}{6.2 \times 10^{-3} + j \times 2\pi \times 4 \times 10^9 \times 2.13 \times 10^{-10}}}$$

$$= \sqrt{\frac{5 + j1307}{6.2 \times 10^{-3} + j5.353}} = 244\ \Omega.$$

(*b*) We know, the propagation constant

$$r = \sqrt{(R + j\omega L)(G + j\omega c)}$$

$$= \sqrt{(5 + j1307)(6.2 \times 10^{-3} + j5.353)}$$

$$r = 83.64 \underline{|90^\circ} = j83.64$$

(*c*) The normalized impedance $z_e = \dfrac{Z_l}{Z_0}$

$$= \frac{100 + j100}{244}$$

$$= 0.58 \underline{|45^\circ}$$

(*d*) The reflection coefficient $\Gamma_l = \dfrac{Z_l - Z_0}{Z_l + Z_0} = \dfrac{100 + j100 - 244}{100 + j100 + 244}$

$$= 0.49\underline{|126^\circ}$$

(*e*) The reflection coefficient of a distance of λ/4 from the load is

$$\Gamma_{\lambda/4} = |\Gamma_l| e^{-2\alpha d} = 0.49\, e^{-2\alpha d}\, e^{j(\theta_e - 2\beta d)}$$

$$= 0.49\, e^{j\left(126^\circ - 2\frac{2\pi}{\lambda}\cdot\frac{\lambda}{4}\right)}$$

$$= 0.49\underline{|-64^\circ}$$

From the smith chart, the normalized impedance is $Z_s = 1 - j\,1$

Then $$Z_s = 244 - j\,244$$

Example 11.5. A quarter-wave lossless line has a characteristic impedance of 50 Ω and is terminated in a load of 100 Ω. The line is energized by a generator of 20 V (rms) with an internal resistance of 50 Ω. Calculate:

(*a*) The sending-end impedance
(*b*) The magnitude of the receiving-end voltage
(*c*) The power delivered to the load.

Solution. Given $Z_g = 50\ \Omega$, $\alpha = 0$

$$Z_0 = 50\ \Omega,\ Z_L = 100\ \Omega$$

and the length is λ/4 line

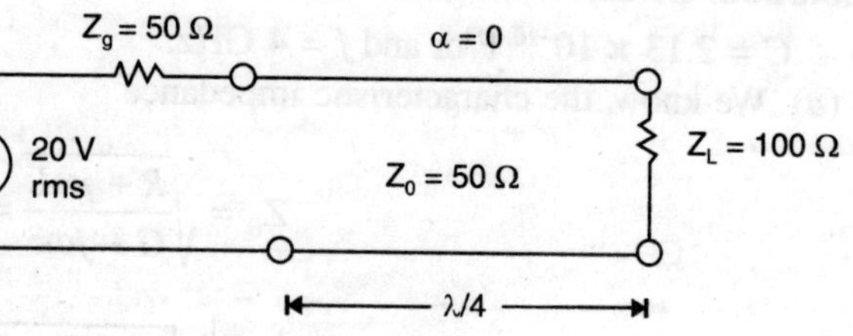

Fig. 11.30

(*a*) $z_e = \dfrac{Z_l}{Z_0} = \dfrac{100}{50} = 2$

and $z_s = \dfrac{1}{z_e} = \dfrac{1}{2} = 0.5$

So $Z_s = z_s z_0 = 0.5 \times 50 = 25\ \Omega$

(*b*) The reflection coefficient $\Gamma_e = \dfrac{z_l - z_0}{z_l + z_0} = \dfrac{100 - 50}{100 + 50}$

$$= 0.33$$

The receiving-end voltage (magnitude) is

$$V_r = V_+ e^{-\alpha l} + V_- e^{\alpha l} = V_+ e^{-\alpha l}(1 + \Gamma_e)$$

$$= 20(1 + 0.33) = 26.6\ \text{V(r.m.s.)}$$

(*c*) The power at the load is

$$P_e = \frac{V_r^2}{Z_e} = \frac{(26.6)^2}{100}$$

$$= 7.08\ \text{watts}$$

REVIEW QUESTIONS AND PROBLEMS

1. Name five types of test equipment used to make microwave measurements.
2. Explain precautions to be taken when connecting microwave equipment to connectors and devices.
3. Explain how a series isolator reduces system mismatch to the generator.
4. Define attenuation and insertion loss. How do they differ?
5. Name three devices that detect power.
6. Explain how bolometers are mounted in a line to detect power.
7. How does a thermo couple differ from a bolometer as a means of detecting power?
8. Explain the operation of a calorimetric power meter.
9. How is frequency measured in a microwave system on a continuous basis using a frequency counter?
10. Explain why an isolator and low pass filter are located after an unknown generator.
11. What is a microwave network analyzer?
12. A 25-dB isolator is added in series at the output of a signal generator to reduce the possibility of frequency pulling due to an affected system VSWR mismatch of 1.75/1. If the generator power output is 234 mW, what is the value of the reflected signal received at the generator?
13. The calibrated power from a generator, as read at the power meter is 25 mW. When a 3-dB attenuator with a VSWR of 1.30/1 is inserted between the generator and the detector what value should the power meter read?
14. What is the average power for a periodic wave if the peak power is 1300 W and the pulse width is 0.56 μs? The periodic frequency of the wave is 1500 Hz.
15. How long a slot would be required on a slotted waveguide section to measure the wavelength of a microwave whose frequency is 5.4 GHz?
16. A slotted line is used in association with an X-band microwave source. When the line is terminated by a short circuit, adjacent nulls are found at positions which are shown as 9.27 and 10.05 cm. What is the value of the guide wavelength?
17. An unmodulated microwave source is connected to a bolometer mount and an appropriate power meter. The microwave power level reads as 25 mW. When an attenuating device is inserted between the source and the bolometer, the power reading falls to 5 mW. What is the amount of attenuation (in decibels) provided by the device?
18. A slotted line is used to determine the SWR value of a waveguide. Adjacent null positions are located at 13.31 cm and 15.45 cm. If the separation between the twice-minimum power points is 2 mm, what is the value of the SWR?
19. Two identical directional couplers are placed in a waveguide to sample the incident and the reflected power. The meter readings show that the power level of the reverse coupler is 10 dB down from the level of the forward coupler. What is the value of the SWR on the waveguide?
20. What is the peak power of a reader where average transmitted power is 200 W, pulse width of 1 μs and pulse repetition frequency of 1000 Hz?

$$P_{\text{per}} = 220 \times \frac{\frac{1}{1000}}{1\mu s} = 200 \times \frac{10^{-3}\,\text{sec}}{1\mu s}$$

Index

N

O

P

Q

R

S

T